795

D1126639

Steinmetz

Johns Hopkins Studies in the History of Technology

Merritt Roe Smith, Series Editor

Steinmetz

Engineer and Socialist

Ronald R. Kline

The Johns Hopkins University Press
Baltimore and London

© 1992 The Johns Hopkins University Press
All rights reserved
Printed in the United States of America

The Johns Hopkins University Press
701 West 40th Street, Baltimore, Maryland 21211-2190
The Johns Hopkins Press Ltd., London

The paper used in this book meets the minimum requirements of American National
Standard for Information Sciences—Permanence of Paper for Printed Library
Materials, ANSI Z39.48-1984.

Library of Congress Cataloging-in-Publication Data

Kline, Ronald R.
     Steinmetz : engineer and socialist / Ronald R. Kline.
          p.   cm. — (Johns Hopkins studies in the history of technology ; new
     ser., no. 13)
     Includes bibliographical references and index.
     ISBN 0-8018-4298-0
     1. Steinmetz, Charles Proteus, 1865–1923. 2. Electric engineers—United
States—Biography. 3. Socialists—United States—Biography. I. Title.
II. Series.
TK140.S77K58   1992
621.3′092—dc20
[B]   91-31050

To my parents and Marge

Contents

Preface

I first heard of Charles Proteus Steinmetz while working as an engineer for the General Electric Company, which I joined fresh out of college in 1969. I remember seeing a photograph on the cover of a company magazine showing this queer, gnomelike man doing calculations while sitting in a canoe floating on a peaceful river. My first thought was that my employer must not be a soulless corporation after all if it had given a genius such freedom from office routine. A little reflection suggested that the public relations department knew the enduring value of Steinmetz in humanizing the company nearly fifty years after his death.

After leaving GE to attend graduate school in the history of science and technology, I steered clear of electrical history as a research topic until I read David F. Noble's *America by Design*. GE managers and engineers played pivotal roles in his account of science, technology, and the rise of corporate capitalism during a second Industrial Revolution marked by the advent of the automobile, mass production, and the electrification of home and factory. But the motives attributed to these men—the stabilization and growth of corporate capitalism—were hardly the altruistic ones associated with Steinmetz working in a canoe.

Noble observed that Steinmetz deserved more study "because of his unique career and perspective, reflecting the convergence of the socialist movement, corporate development, and the growth of modern technology."* Instead of focusing on the convergence, I became intrigued by the incongruities in Steinmetz's life: a prominent Socialist serving as chief engineer of a large capitalist corporation; the head of a home laboratory who institutionalized innovation inside the firm; a part-time college professor who protected the autonomy of a department funded by GE; an

*Noble, *America by Design: Science, Technology, and the Rise of Corporate Capitalism* (New York: Knopf, 1977), p. 42n.

internationally known writer of abstruse mathematical treatises on electrical machinery who was revered as a cultural hero in the 1920s alongside Thomas Edison and Henry Ford.

I began by studying his socialism and found that it supported corporate values, rather than contradicted them. I then completed a doctoral dissertation on his technical contributions, many of which are still used by today's engineers. I came to realize that his need to resolve inner contradictions—between his political views and his position at GE, for example—was a central theme in his life. Steinmetz discovered that he could be both a mathematician and an engineer, a corporate manager and a college professor, a Socialist and an American hero. It was a unique career, one that embraced several themes that recent historians have used to characterize the period under study: technology and the corporation, the advent of a political liberalism based on the corporation, and the growth of new professions. His physiognomy and publicity value were also part of this synthesis (the convergence noted by Noble). The image of a dwarfed, hunchbacked German Socialist hurling thunderbolts about the GE laboratory (from a high-voltage lightning generator) and the Horatio Alger tale of a penniless, misshapen immigrant who rose to fame and fortune as GE's chief engineer fascinated the public and helped define both the man and the myth.

Because of the scarcity of archives about his private life, I have written an intellectual biography set in the economic, political, and social contexts of the Progressive Era, as well as a cultural history of the Steinmetz myth. Part 1 examines the relationship between science and technology in his early career and in this crucial period when electrical engineering was transformed from a craft discipline into one based on science and laboratory research. Part 2 investigates the "incorporation" of science and engineering in the electrical manufacturing industry as seen through the eyes of Steinmetz, who left an inventor-entrepreneur to establish science-based innovation in a gigantic corporation. Part 3 describes his attempts to "engineer society" by working for reforms in education and his professional society, developing a theory of corporate socialism, and trying to put it into practice at General Electric, in Schenectady, and in Soviet Russia. When GE and the national media created Steinmetz, the Wizard of Science, who astonished even Edison, the Wizard of Menlo Park, with his artificial lightning, they ensured that his voice would be heard on the pressing political and social issues of the day. But there are indications throughout a life full of contrasts—not all of which were resolved—that even Steinmetz may not always have been able to separate the myth from the man.

During the preparation of this book I was fortunate enough to receive good advice, support, and encouragement from numerous colleagues, friends, and institutions.

As a graduate student at the University of Wisconsin, I benefited enormously from living and working in a vibrant scholarly community of students and faculty in the history of science and American history. My dissertation committee, Terry Reynolds (chair), Ronald Numbers, Theodore Bernstein, and Daniel Siegel, guided me through the many issues of a technical thesis, while Reynolds and Numbers encouraged me to widen the scope of that work into a full biography. They, along with Victor Hilts, Morton Rothstein, and Daniel Rodgers, provided good examples with their teaching and scholarship of how to place the history of technology in its political, social, and economic contexts.

Receiving the Fellowship in Electrical History from the Institute of Electrical and Electronics Engineers (IEEE) for 1979–80 persuaded me to choose Steinmetz as a topic for my dissertation and gave me a much-needed research year at the beginning of this project. I have also been fortunate to work in four scholarly communities since leaving Wisconsin: the IEEE Center for History of Electrical Engineering, the Humanities and Social Sciences Department at Cooper Union, and the School of Electrical Engineering and the Program in the History and Philosophy of Science and Technology (now part of a new Science and Technology Studies Department) at Cornell University.

I would like to thank several people for their helpful comments on earlier drafts of this book. John Anderson, Ross Bassett, James Brittain, Herbert Carlin, Jonathan Coopersmith, W. Bernard Carlson, Robert Cuff, Terrence Fine, Margaret Kline, Bruce Lewenstein, Terry Reynolds, James Thorp, and George Wise read all or parts of earlier versions. A. Michal McMahon, Brittain, Coopersmith, Robert Frost, Edwin T. Layton, Jr., Arthur Norberg, Donald Novotny, Ruth Oldenziehl, Edward Owen, Leonard Reich, Reynolds, Wise, and audiences at colloquia at the University of Delaware and Rutgers University commented on papers that were incorporated into the book. Brittain, Layton, and Reynolds were more than generous with their time and seasoned advice. I would especially like to thank Merritt Roe Smith, series editor for Johns Hopkins University Press, for encouraging me at an early date to broaden the

xii **Preface**

scope of this work, and Smith and an anonymous referee from the Press for carefully reading earlier versions of the manuscript. Robert Brugger and Mary V. Yates were exemplary editors.

Several archivists went out of their way to provide full and informed access to their collections: Elsa Church (Schenectady County Historical Society), Ellen Fladger and Ruth Anne Evans (Union College), Pauline Wood (Schenectady City Hall History Center), George Wise (GE R&D facility), John Anderson and Ruth Shoemaker (Hall of History Foundation), Joyce Bedi (IEEE Archives), Paul Israel (Thomas Edison Papers), Helen Samuels (MIT Archives), and Christine Bain (New York State Library). Anne Millbrooke (United Technology Archives), Judith Endelman (Henry Ford Museum), and Lynn Ekfelt (St. Lawrence University) thoughtfully sent me copies of material from their archives. The staff of the Tamiment Library at New York University, the New York City Public Library, and the libraries and archives at the University of Wisconsin, Columbia University, Cornell, and Cooper Union were very helpful. Sharon Calhoun at Cornell's School of Electrical Engineering expertly turned my rough diagrams and columns of data into printable figures and tables. Two of Steinmetz's grandchildren, Marjorie Hayden and the late Joseph Hayden, were kind enough to answer my questions about their grandfather.

Finally, I would like to pay a small tribute to those informal conversations at the annual meetings of the Society for the History of Technology and the History of Science Society that help graduate students and more experienced historians mould good-sized research projects into publishable shape. Thankfully, "cooperation," the byword of Steinmetz and his colleagues, is more than an ideal in one corner of today's society.

Part 1

From Mathematician to Engineer

1 German Origins

Charles Proteus Steinmetz's youth is shrouded in myth and legend, a romance created in part by Proteus himself. John Winthrop Hammond, a General Electric public relations writer, wrote the first full-length biography and based it on interviews he conducted with Steinmetz in the early 1920s. Fascinated by the Wizard of Schenectady, Hammond painted an idyllic picture of a budding genius, spoiled by a doting grandmother in a loving German home of limited means, who led an adventurous life at the local university. He paid little attention to the boy's pronounced physical deformity (hunchbacked dwarfism inherited from his father) and other factors that shaped his personality. Contemporary documents tell a less romantic but more realistic story of the young Steinmetz.

Breslau

He was born on April 9, 1865, to lower-middle-class German-Polish parents in Breslau, Germany, now Wrocław, Poland, and baptized Carl August Rudolph Steinmetz in the Evangelical Lutheran Church. The rather ponderous name came from his father, Carl Heinrich Steinmetz, and his two uncles, August and Rudolph. In the early 1800s his grandfather emigrated from the Prussian province of Silesia across the border to Ostrowo, Poland, where he married a Polish woman and fathered three sons. While Rudolph stayed in Ostrowo and became a carpenter, his brothers August and Carl Heinrich returned to their father's homeland of Silesia to seek work in Breslau, its capital city.

Built on the wide and fertile plain of the Oder River in the Middle Ages, Breslau became a center for government, education, commerce, and industry during the industrialization of Germany in the mid-nineteenth century. When Carl Heinrich arrived in 1860 Breslau was a thriving provincial city that boasted a population of about 140,000, many schools and churches (about half Protestant and half Roman Catholic), several light industries, and a bustling transportation network. Carl Heinrich got a job as a lithographer with August's employer, the Upper Silesian Railway, and moved in with August and his family—his wife, Caroline Neubert of Breslau, and their two girls, Marie and Clara. When August died of tuberculosis in 1864 at the age of thirty-seven, Carl Heinrich—whether from the biblical injunction to marry one's brother's widow, or for more romantic reasons—married Caroline. She gave birth to Carl in April 1865, then died of cholera during an epidemic that took over four thousand lives in the summer of 1866. Carl Heinrich summoned his widowed mother and sister Julia from Ostrowo to help him raise the children—baby Carl and his two half-sisters, Marie and Clara, who were then thirteen and eight. By 1871 the grandmother had returned to Ostrowo, and Aunt Julia had married, leaving eighteen-year-old Marie with the task of raising Clara and young Carl, who had started kindergarten.[1]

We hear little more from Hammond about the family until Carl graduates from the local gymnasium in 1883. Yet Hammond left out an important part of Carl's formative years. In his last will and testament Steinmetz left money to a third half-sister, Margarethe Mache, or, if she was not living, to her half-sisters Clara and Hedwig Mache.[2] Why did Hammond leave the Maches out of his story? The answer comes from another will, that of Steinmetz's father, and family letters. Carl Heinrich left his household goods to Margarethe Renner, born in 1874 as the illegitimate daughter (*aussereheliche Tochter*) of him and Bertha Mache, nee Renner. The will, dated less than a year before his death in 1890, states that Bertha, widow of a government official, nursed (*gepflegt*) Carl Heinrich for many years up to his last days. Letters from Carl Heinrich to his son in the late 1880s reveal that Margarethe, Bertha, and two daughters from a previous marriage had lived with the Steinmetzes for some time.[3]

It is clear, then, that Carl grew up in a nontraditional household, a fact that would have a bearing later on. He was raised by a grandmother and aunt until he was six, then by a half-sister. To complicate matters further, Bertha Mache and her daughters arrived when he was about nine. We do not know when he discovered that the youngest girl, Margarethe, was his third half-sister, not on his mother's side—as were Marie and Clara—but the daughter of an illicit liaison of his father's. Compounding this confusion of family ties was the deformity he inherited from his father.

No records survive to indicate how he felt about having such a visible handicap, but it undoubtedly shaped his outlook on life and reminded him of his hunchbacked father and mixed-up home life.

George Moser, a gymnasium student whom Steinmetz tutored in mathematics while attending the local university, vividly recalled how these conditions affected young Carl. Coming from a well-to-do Breslau family, Moser remembered Carl's shabby dress and referred to the Steinmetzes as the "deserving poor in the grimmest sense of the word." He described the twenty-one-year-old Steinmetz as a "small, misshapen young man, whose torso was fully shriveled up and bent, of the size of a twelve-year old boy." Carl had ungainly manners, and a hard, nasal voice, but Moser's mother "was not intimidated by the poor, embittered student's repulsive manner" and invited him to eat with the family.

As they became friends Moser realized that Carl's "outer uncouthness and brusque manner were rather the armor which the thin-skinned and sensitive youth had put on his misshapen form" to protect himself from the condescension and taunts of others. Underneath the defensive shell Moser found a "kind, magnificent" young man who liked to debate, perform card tricks, and play practical jokes. He also admired Carl's "finely cut mouth, small, slightly curved nose, magnificent gray-blue eyes and powerful high forehead," which made him forget "the rest of the man."[4] (Student photographs indeed show the powerful brow, but they also portray Steinmetz as either overweight or unkempt—hardly the boy with the head of a Greek athlete remembered forty years later by Moser. Steinmetz's family was also able to provide a better material life for him than Moser recalled.)

Moser's family made a good decision in choosing Steinmetz, a student of advanced mathematics at the university, as a tutor for their son. More important, he had done exceptionally well in the same exams young Moser was about to take at the Johannes Gymnasium. Steinmetz scored the highest marks in his graduating class on the written tests in mathematics and science and was exempted from taking the oral exams in these fields. The testing commission noted that his knowledge of mathematics far surpassed that taught at the school and that he was able to solve the most complicated problems. They also reported that his knowledge of physics was equally extensive and certain.[5]

He did not do as well in other subjects. While receiving *Sehr Gut* in mathematics, physics, and history, he got a *Genügend* (sufficient) in French, and *Gut* for everything else (religion, German, Latin, Greek, and geography). His teachers noted his logical but terse use of the German language—a characteristic that carried over into his English prose style many years later. They also observed a certain carelessness (*Flüchtigkeit*) and superficiality (*Oberflächlichkeit*) in his upper-level courses, except for those in mathematics and science—a foreshadowing of his later forays into fields outside his expertise. Although he had little money by Moser's standards, he had received an excellent preparatory education at the gymnasium, and by tutoring mathematics he could afford to enroll at the University of Breslau in the fall of 1883.[6]

A relatively young institution, the university had been founded in 1811 by the merger of a new branch of Frankfurt University at Breslau with a Jesuit college established in the early eighteenth century. Like most European universities, it offered courses in the faculties of law, medicine, theology, and philosophy. The philosophy group at Breslau in the 1880s was nearly as large as the other three combined and included such well-known scholars as the chemist Robert Bunsen, inventor of the laboratory burner bearing his name, and the astronomer Johann Galle, who discovered the planet Neptune. Bunsen was there for only two years in the 1850s, but Galle spent over forty years at Breslau.[7] In view of how well Steinmetz had done in mathematics and physics at the gymnasium, it is not surprising that he specialized in these fields at the university. Of the fifty lecture courses he completed during five years at Breslau, none was outside of mathematics and science.[8] History, his third best subject at the gymnasium, succumbed to stronger Muses.

Steinmetz studied astronomy at first, because of Galle's reputation, then switched to pure mathematics to work under Heinrich Schröter. A pupil of the famed geometer Jakob Steiner at the University of Berlin, Schröter came to Breslau in 1861 and started a new program in his (and Steiner's) specialty—synthetic geometry. Schröter became an authority in the field, writing textbooks and conducting important research on the theory of surfaces well into the 1880s. In 1876 he hired Professor Jakob Rosanes, a specialist in analytical geometry, to create a strong and balanced program in geometry. The department cultivated the other areas of mathematics as well, with Schröter, Rosanes, and their *Privat-dozenten* of the 1880s, Otto Staude and Adolf Kneser, teaching a full complement of courses in higher mathematics.[9] Steinmetz took full advantage of these offerings and selected mathematics for nearly half of his university courses. After completing a course in synthetic geometry in his first semester, he specialized in the subject and worked toward a doctorate under Schröter. His remaining mathematics courses—in analytical geometry, algebra, differential equations, calculus, and number theory—covered the entire range of topics in Schröter's department.

Steinmetz broadened this education by studying physics in a department with a rich tradition in electrophysics. Following Georg Pohl's experimental research in the 1830s, Gustav Kirchhoff taught at the university from 1850 to 1854, shortly after he had derived the circuit equations known today as Kirchhoff's laws. In 1864 Schröter encouraged the university to hire mathematical physicist Oskar Meyer to balance the experimental leanings of the department. The brother of chemist Julius Meyer, a Breslau graduate who independently discovered the principle of the periodic table, Oskar specialized in the kinetic theory of gases. He directed the Physical Institute and headed the department from 1865 until he retired at the turn of the century. Although he wrote several papers on geomagnetism, Meyer did not carry on the electrophysics tradition of his predecessors. During Steinmetz's stay in the 1880s Meyer left this to Associate Professor Leonhard Weber and *Privatdozent* Felix Auerbach. Weber performed some research in electrophysics but was much better known for his invention of an improved photometer. Auerbach wrote many papers on magnetism and a book on the theory of dynamos. Meyer, Weber, and Auerbach also taught a wide variety of courses in theoretical and experimental physics.[10] Steinmetz chose to specialize in the theoretical side and took nearly half of his scientific courses in mathematical physics: optics, thermodynamics, mechanics, photometry, and electricity and magnetism. The remainder of his scientific subjects were almost equally divided among experimental physics, astronomy, chemistry, and the life sciences (zoology and botany).

We will see that this education gave Steinmetz an enormous advantage over most electrical engineers in the United States. Mathematics was of primary importance, but we should not overlook his training in science, especially electrophysics. He attended lectures on electricity and galvanism from Meyer, on the theory of potentials from Weber, and on the mathematical theory of electricity and magnetism from Auerbach. His notebooks for the latter course show that it covered the standard topics of electrostatics, electrodynamics, and magnetism, including the laws of Green, Ohm, and Kirchhoff.[11]

The university offered few courses in electrical engineering, mainly because German schools relegated engineering to the younger *Technische Hochschulen* (technical colleges). The *Hochschulen* added electrical engineering to their curricula after European cities began to light their streets with arc lamps in the 1870s and Edison displayed a practical system of incandescent lighting at the Paris International Electrical Exhibition in

1881. German professors taught both *Starkstromtechnik* (strong-current technology, i.e., electric light and power) and *Schwachstromtechnik* (weak-current technology, i.e., telegraphy and telephony). The polytechnic at Darmstadt established the first electrical engineering course in Germany in 1882, two years before the German Edison company and the firm of Siemens & Halske introduced Edison's system in Germany. Erasmus Kittler, who had a doctorate in mathematics and physics from the University of Munich, held the chair at Darmstadt and wrote a two-volume handbook that set the standard in *Elektrotechnik* for many years.[12]

At the University of Breslau Professor Weber gave lectures on *Elektrotechnik* in the physics department in 1887. Steinmetz registered for this course but did not obtain a sign-off signature for it, nor have notebooks for it survived. The only lecture notes on electrical engineering for this period are four pages on the mathematical analysis of direct-current motors and dynamos by Oskar Frölich, a Ph.D. physicist who worked at Siemens & Halske. Steinmetz probably took these notes in a course by Auerbach, because Auerbach published a critique of Frölich's theory and wrote a widely used textbook on the "laws" of dynamos while Steinmetz was at Breslau.[13]

Auerbach wrote in the time-honored tradition of physicists who developed theories of electric machines, which began with the work of Emil Lenz, James Prescott Joule, and others in the 1840s. After C. W. Siemens in Germany and Charles Wheatstone in England independently announced the invention of the dynamo principle in 1867, physicists, including the renowned James Clerk Maxwell, devised theories to explain how the dynamo could produce energy to power its own electromagnets. Inventors and engineers paid little attention to these theories, however, until they were improved in the early 1880s by Rudolph Clausius—better known for his work in thermodynamics—and Frölich, who had designed dynamos at Siemens & Halske.[14] Auerbach thus exposed Steinmetz to a German tradition in mathematical physics that was beginning to have some bearing on the theory and practice of electrical engineering. But his main interests and training at the university were in higher mathematics and the physical sciences.

It was not all study for Steinmetz, who enjoyed a busy social life, despite his deformity, in the German student organizations that were famous for promoting scholarship and beer drinking. Steinmetz savored both pursuits—to the point of producing a handwritten satirical publication, *Bier Zeitung*, as a freshman. He joined the mathematics club, which gave its members nicknames descriptive of their talents. The group chose

Proteus for Steinmetz, after the sea god in Greek mythology who changed into a lion, serpent, falling water, and other terrible shapes when his captors tried to hold him still. If they held on long enough he would divulge all things, past, present, and future, once he returned to his proper shape—a wrinkled Old Man of the Sea. Although the name brought to mind Steinmetz's deformity, it was meant to symbolize the encyclopedic knowledge attributed to him. The name must have held some fond memories for him, because he later adopted *Proteus* as his middle name.[15]

Steinmetz did well in his studies. He won the mark of summa cum laude in seven dean's tests (*Dekanatsprüfungen*): four in mathematics, two in physics, and one in chemistry. In May 1887, after four years of study, the university awarded him a scholarship (*Stipendium*) to continue working toward a doctorate. Reading notes, bibliographic lists, and mathematical papers from this period indicate that he had nearly completed a dissertation on synthetic geometry under Schröter.[16]

Although Steinmetz later published three papers based on this research,[17] he did not finish his education. The troubles began in the summer semester of 1887, when he failed to obtain a sign-off signature from Schröter for two courses in synthetic geometry. He completed three other courses that semester (two in chemistry and one in physics) and also one in mineralogy during the next semester. But those were his last. In late May 1888, two weeks after registering for another chemistry course, he fled his homeland during the dead of night without having earned his doctorate.[18]

The usual explanation for this remarkable turn of events is straightforward: Steinmetz left Breslau to avoid being arrested for advancing the Socialist cause against Bismarck and the German state. But the flight from Breslau has become such an integral part of the Steinmetz legend—a story told over and over by uncritical admirers—that it has become difficult to separate myth from reality. Part of the problem is the scarcity of original records. Part is also due to Steinmetz himself. In 1893, only five years after leaving home, he confided to his Socialist mentor in Breslau that he had told the story "so many times with appropriate embellishments, that I can no longer distinguish fact from fiction."[19] By comparing his account—as told to Hammond and others—with sources closer in time to the events, we can, however, reconstruct a more accurate history of this turning point in his life.

It all began when Steinmetz joined a Socialist group at the University of Breslau in 1884. The 1880s were an exciting, if not dangerous, time to be a Socialist in Germany, because of the Anti-Socialist Law passed in 1878 at the urging of Chancellor Bismarck. It was legal to campaign for Socialist deputies and for them to sit in the Reichstag, but all other Socialist activities, including establishing a party organization and a press, were illegal. While enforcement of the law was more lenient in the second half of the 1880s, it was fairly strict in Breslau—a situation that enlivened Steinmetz's university life considerably![20]

He became a Socialist through his friendship with Heinrich Lux, an older student of mathematics who came to Breslau in the spring of 1883 and became a leader of the Socialist circle at the university. Impatient with the evolutionary aspects of Marxist theory, Lux, Alfred Plötz—a political science student who had recently embraced socialism—and several other students turned to the utopian writings of the French republican leader Etienne Cabet, whose *Voyage en Icarie* (1840) had created a large, communistic, working-class movement in France in the 1840s. Cabet outlined a centrally planned, democratic, egalitarian nation-state with no money, private property, or police. Workers in immense state-owned mechanized factories produced abundant goods, which were distributed from well-stocked warehouses to the Icarians. Although central committees, chosen from the national assembly, directed industry, the Icarians avoided a strict technocracy by electing members to the assembly. Above all, they valued education. Each person was trained in his or her chosen occupation and then received a general education. Peace, harmony, happy workers, and equality of the sexes prevailed in a culture based on such technological marvels as glass-covered sidewalks and well-lit streets and factories.[21]

Enthralled by Cabet's ideas, Plötz and his friends formed an association, the Gesellschaft Pacific, to establish an Icarian colony in the United States. As president, Plötz laid out a secular commune consisting of political leaders, a forester, a botanist, and so forth. Slated for the ministries of education and fine arts were Gerhart Hauptmann, who was just beginning his career as a playwright, and his brother Carl. (No longer at Breslau, the Hauptmanns had belonged to Plötz's circle at the university.) Confident that they were not violating the Anti-Socialist Law, Plötz and Lux even registered the association with the Breslau police in the fall of 1883. The police took no action against the group's members, none of whom belonged to the Socialist party (the Sozialdemokratische Arbeiterpartei, which was renamed the Sozialdemokratische Partei Deutschlands in 1890). In preparation for settling their commune the young idealists raised over 3,000 marks, with the help of Gerhart Hauptmann, and sent

Plötz in the spring of 1884 to investigate the Icarian colonies in America, which Cabet had founded in the Midwest in the early 1850s. Plötz returned to Breslau in late 1884 with bad news: the few remaining colonists in Iowa lived in wretched conditions and would probably not make it through another year. That ended the dream of a Gesellschaft Pacific. Plötz moved to Zurich, and Lux turned away from utopian ideals to the more "scientific" tenets of Marxism and joined the Socialist party.[22]

It was at this time, in the winter of 1884–1885, that Lux invited Steinmetz to join the student Socialists—beginning the "most exciting" time of his life. Heated debates about the failure of the Icarian dream dominated his first Socialist meetings (and influenced his later writings). There were also close brushes with the police, especially after the group affiliated with the local branch of the Socialist party in the summer of 1885, probably at Lux's urging. Steinmetz recalled that when a Socialist member of the Reichstag wanted to address their meetings, he and the group had to take evasive actions. They would enter a tavern with the police tailing the deputy, drink up before the police could order, then head off in different directions. With the police safely out of the way, they reassembled in another hall to hear the deputy's speech.[23]

All of this seemed like harmless fun. But associating with the outlawed Socialists thrust Steinmetz and the student group out of the ivory tower of the university into the harsh reality of the city where the Anti-Socialist Law was enforced. In Breslau, as in many German cities, strict enforcement of the law created a strong party. Provincial Breslau regularly sent two Socialist deputies to the Reichstag in Berlin, one from Breslau-East and one from Breslau-West.[24] The city's Socialists also enjoyed a vigorous press. Breslau served as the center for smuggling the Zurich *Sozialdemokrat*, the party's official organ, into that part of Germany. After the party's newspaper closed in 1878, Breslau leaders published several others, all of which were shut down under the Anti-Socialist Law passed that year. Two less strident papers filled the vacuum, the *Breslauer Gerichtszeitung* (1879) and the *Neue Breslauer Gerichtszeitung* (1883). Representing the two Socialist factions in the city, the papers were often embroiled in bitter lawsuits against each other, but they served the Socialists in Breslau and the rest of Silesia as substitutes for a local party paper.[25]

In 1885, shortly after the student group affiliated with the Socialists, the Breslau branch of the party changed markedly. That August the government shut down the party's printer, which was supported by Julius

Kräcker, the Socialist delegate from Breslau-West. Local Socialists could still read the two *Gerichtszeitungen*, but losing their printer for all other publications came as a severe blow, especially since the print shop was the party's last major property in the city. Then Wilhelm Hasenclever, the Socialist delegate from Breslau-East, and Wilhelm Liebknecht, a national Socialist leader, authorized the publication of a new Breslau paper by Robert Conrad if he could raise the money. A Berlin bricklayer and active member of the trade union movement, Conrad raised over 3,000 marks and issued the first number of the *Breslauer Volkstimme* (People's Voice) in January 1886. By that fall the *Volkstimme* won its spurs as the official Socialist paper in Breslau, with its incisive criticism of the public denunciations accompanying the growing strife between the two *Gerichtszeitungen*.[26]

Steinmetz met Conrad in the spring of 1886 and began to work with Lux on party matters. Things went well until the following March (1887), when the Breslau police arrested Lux at home on the charge of forming a secret organization (*Geheimbund*) to overthrow the government. During the raid police confiscated a photograph of Lux and eight colleagues gathered around a bust of Ferdinand Lassalle (1825–1864), the famous German Socialist whose supporters merged with a Marxist group to form the national party in 1875. Lassalle, who advocated universal and equal suffrage and state-supported producers' cooperatives, held a special place in the hearts of Breslau comrades because he was a native of the city and had graduated from the university. Taken in 1886 to honor a student leaving the university, the photograph included students, two professors, and Conrad. All were former members of a "Free Scientific Society" at the university except Conrad, whom the police suspected of being a member of the Socialist party. Standing next to Lux in the photograph was the unmistakable figure of Steinmetz. Although he recalled being present when the police apprehended Lux, Steinmetz was not arrested. It was not a crime to be photographed with suspected Socialists, but it did bring him under investigation. The police also found many letters between Lux and Plötz, which led them to suspect that the Gesellschaft Pacific was a recruiting organization for the illegal *Geheimbund*. The letters were rather incriminating, contained some youthful bravado about using force to overthrow the government, and stated that forming utopian colonies was the best way to usher in a Socialist state.[27]

Lux's arrest led to a lengthy investigation. The police threw their net around other groups in Breslau, arresting Kräcker (at the end of the Reichstag session) and fifteen suspected Socialists by mid-July. By the

The Lassallean photograph, 1886. Steinmetz is standing at the far right, next to Lux. Source: Hall of History Foundation, Schenectady, N.Y.

time the case came to trial in November, thirty-six of the thirty-eight charged had spent from three to nine months in jail awaiting trial. Lux was one of those imprisoned for the entire nine months. The only other student charged was Julian Marcuse, in whose honor the infamous Lassallean photograph had been taken. Three of those charged belonged to the professional class: publisher Conrad, author Johann Kasprowicz, and editor Bruno Geiser, a leader of the moderate wing of the party as Reichstag deputy for Chemnitz. The remaining thirty-one imprisoned defendants were craftworkers—the backbone of the Breslau party.

It is not clear what Steinmetz did during the investigation and trial. All those in the Lassallean photograph, a major piece of evidence for the prosecution, were either called as witnesses (Dr. Kayer, Dr. Stein, and Walter Samuelsohn) or charged in the indictment (Lux, Marcuse, Conrad, and Kasprowicz), except for Ferdinand Simon and Steinmetz. A student at the university, Simon moved to Zurich to continue his studies in 1886, shortly after the photograph was taken. Theodor Müller, the historian of the Breslau Socialist party, states that Steinmetz was in Zurich during the trial and returned to Breslau after it was over. If Müller is correct, Steinmetz probably joined Simon in Zurich to avoid being arrested or called as a witness. Although this part of the story does not appear in Hammond's biography, Müller confirms Steinmetz's recollection of helping out the defendants before the trial by communicating with the imprisoned Lux via messages written in invisible ink, thus outwitting the prosecutor. Apparently he never figured out why the testimony of those inside and outside of jail meshed so perfectly, even though he had barred communication between them.[28]

Hammond says that the secret correspondence led to the release of all those charged, but Steinmetz was not that successful; the ten-day trial ended on November 17 with more convictions than acquittals. The thirty-eight Socialists were charged with eight counts. The most serious was against Kräcker and the entire group, except Marcuse and Kasprowicz, for forming a secret society that violated the Anti-Socialist Law. All but one of the remaining counts cited Lux and one or two comrades for distributing the Zurich *Sozialdemokrat* and other contraband, collecting money for the party, participating in the Gesellschaft Pacific (with Marcuse and Kasprowicz), and harboring a nihilist (with Kasprowicz). The prosecutor produced an abundance of physical evidence: almost thirty documents and newspapers, a red flag, and the Lassallean photograph. Nearly sixty witnesses were called, including Gerhart Hauptmann, for his role in the Gesellschaft Pacific; Professor Oskar Meyer, who taught both Lux and Steinmetz; and Samuelsohn, a family friend of the Steinmetzes. Despite the large body of evidence, the *Sozialdemokrat* thought the case was extremely weak. According to its account, the prosecutor argued that a secret Socialist organization must exist in Breslau because the *Sozialdemokrat* was distributed from there, Breslau elected two Reichstag deputies and sent a delegate to the party conference in Copenhagen, and many notable Socialists were to be seen in the city. In regard to the students the evidence consisted mainly of letters from Plötz to Lux about the Gesellschaft Pacific and Plötz's trip to the Icarian colonies in America.

The prosecutor could not prove all the charges, but he obtained several convictions. As the target of the investigation, Lux was convicted on four counts and sentenced to a year in jail (he was also expelled from the university). The other sentences varied from one to seven months (Kasprowicz got six months, Marcuse four). Eight defendants were acquitted, including such prominent party members as Geiser and Conrad. All those convicted were released from jail except for Lux, Kasprowicz, and Otto Matschoke, a distributor of the *Sozialdemokrat*. Not a day of the time Lux spent in jail awaiting trial counted toward his sentence.

The *Sozialdemokrat* was livid at this "most scandalous" of all anti-Socialist trials. The paper decried the long terms of imprisonment during the investigation, compared their comrades' treatment with that found in Russia, and was incensed at the punishment meted out to Lux. Lux's only crime was to have been a student *and* a Socialist. The government wanted to stop the growing Socialist movement in the universities and gave Lux a stiff sentence to discourage students from participating. The harsh penalties given to Marcuse and Kasprowicz underscored the monstrous character of the trial.[29]

Although Steinmetz avoided Lux's fate, he got into trouble as a result of the trial. During the latter part of the investigation in mid-August, Conrad, the publisher of the *Volkstimme*, was imprisoned for two months for libel. Steinmetz then secretly edited the paper until the police banned it on November 1 for publishing an inflammatory article about the rights of Socialists. Steinmetz says he avoided arrest at this time because the police had only an illiterate "sitting editor" to question. (He probably also went to Zurich before the trial started on November 7.) He returned to Breslau after the trial and helped establish the *Schlesische Nachrichten* (Silesia News) with the assistance of Johann Maxara, a master tailor, and Fritz Trappe, a cabinetmaker who had represented the Breslau party at a Socialist congress. Geiser, who edited the paper for its first two years, was acquitted at the trial, then expelled by the national party in 1887 for refusing to sign a document announcing a party congress. Despite these troubles, the *Nachrichten* served as the local party paper in place of the *Volkstimme* and formed the basis for the party's official organ well into the 1920s. Steinmetz also helped edit the *Volksbibliothek* series of books for Geiser and a popular scientific magazine while at Breslau.

In the spring of 1888 Steinmetz and Trappe learned that the police were preparing legal proceedings against them for their work on the *Nachrichten*. According to Steinmetz, the prosecutor made two reports about him to the rector of the university, who took no action. He then heard that more drastic measures were in store and decided to flee.[30] Coming home late from saying goodbye to his friends, he woke his father early on the morning of May 29, 1888, and said he was taking a short trip across the border to Austria to visit his clergy friend Wilhelm Lehmann.

Instead, he fled with Trappe to Zurich with little intention of returning home this time. He thus forfeited his chances of completing his doctorate and obtaining a post teaching mathematics in Germany.[31]

Before following Steinmetz to Zurich, let us examine his reasons for leaving more closely. Hammond and other biographers have depicted him as being "hounded" out of Germany by Bismarck, as if it had been a personal struggle between a hated tyrant and a peaceful genius looking for political freedom. Steinmetz wanted this type of freedom, but there were other factors behind the decision to leave his homeland. Though not as overriding as the fear of arrest (which did not prove to be groundless), these considerations help make the Steinmetz legend more understandable.

The first concerns his status at the university. Usually in financial straits, Steinmetz tutored to help earn his way through college, as we have seen from Moser's account. But this was not sufficient. When Steinmetz left town under the cover of darkness he owed the university 548 marks (about $140), or nearly three-fourths of his total bill for private and public lectures. He was able to hold off the university for so long by obtaining a half-tuition extension from the Philosophy Faculty for his first year and full extensions for each of the next three years. In the fifth year he managed to work out similar arrangements with his professors.[32] Professor Poleck, for example, deferred payment until Steinmetz finished college, which apparently was also the de facto policy of the university. In any case, since Steinmetz had run up a bill far beyond his or his family's means to pay, he would have had difficulty getting his doctorate in 1888 even if he had turned in his dissertation.[33]

The second factor concerns his home life, which was not the most tranquil. Moser remembered Steinmetz living on the verge of poverty, with little opportunity to escape from it. Steinmetz was better off than Moser recalled, but there was an underlying tension in the house caused by the presence of Bertha Mache and her three daughters, one of whom was illegitimate. Carl's older stepsister Clara detested this arrangement and would not enter the house as long as the Maches lived with her father. Steinmetz apparently got along better with the Maches than Clara did, but the home was far from peaceful. In fact, a year after Steinmetz left, Carl Heinrich blamed himself for losing his son. He apologized for not having been closer to young Carl and asked him to forgive the type of life he led. He hoped Carl would do well and find elsewhere what was missing at home.[34]

A third factor concerns the flight to Zurich, which biographers have portrayed as a lonely exile from his beloved homeland. Actually, Steinmetz traveled with comrade Trappe and probably had supporters waiting for him in Zurich. With little hope of raising enough money to complete his education, disliking his home life, and threatened by the police, Steinmetz might well have thought that he had little to lose by seeking his fortune among Socialist friends in a more liberal country.

Zurich

Like other German political refugees, Steinmetz found a safe haven in the flourishing Socialist life of Zurich. He first presented a letter of introduction to Gerhart Hauptmann from his friend Simon, who was still in Zurich. Steinmetz had not met Hauptmann when he visited Breslau in 1885, and their paths probably crossed when he went to Zurich during the Breslau trial, at which Hauptmann testified about the Gesellschaft Pacific. At Zurich Steinmetz became a regular with Simon and Plötz at the weekly salon of philosophy professor Richard Avenarius, a positivist who advocated "empiriocriticism" as an alternative to Marxism. He was also exposed to the theory of eugenics, which held more than academic interest for the hunchbacked Steinmetz. Plötz, who studied with physiology professor Justus Gaule, became a leader in the German eugenics movement. Gerhart Hauptmann's brother Carl, a student of the social Darwinist Ernst Haeckel, also came to Zurich to study under Gaule and Avenarius. Gerhart recalled that his brother formed a close friendship with Steinmetz and was quite taken with the encyclopedic gifts of the young Proteus.[35]

Making a living with these talents was more difficult. First, Steinmetz tutored mathematics, as he had in Breslau. Then in July he received a commission through a politician he had met at the Hauptmanns' to write a monthly series of popular science articles for the *Zuricher Post* Sunday supplement. Calling on his experience editing a popular science magazine in Breslau, he wrote at least eleven articles on a variety of subjects, ranging from airships to comets and cyclones to electrical technology. The latter covered Edison's newly invented wax-cylinder phonograph, submarine telegraph cables, and the history of electric lighting. He also got a small advance of 37 Swiss francs per month (about $7.30) toward the publication, in Geiser's *Volksbibliothek,* of a popular book on astronomy he had written at Breslau. Every franc was needed, because Steinmetz had spent most of his money, which he had saved for publishing his thesis, traveling to Zurich and paying his first month's rent.[36] A loan from his sister Clara eased the burden considerably. His father ordered books and chemicals and sent him copies of an astronomy magazine.[37]

Steinmetz recalled that before he left Breslau he "intended to become a chemist. Because I was a Socialist, I could not bring myself to accept a government position, and I could not have earned a living at mathematics except in a government position."[38] He did well in the two chemistry courses he completed at Breslau and had been interested in the subject since he was a boy playing with a small chemical laboratory at home. But in Zurich he opted for a career in engineering and decided to study at the prestigious Swiss Federal Polytechnic Institute in that city. Required for admission was a certificate of domicile, which his father applied for from the Breslau authorities in June. Having little sympathy for the plight of the exiled student, they waited until August to deny the certificate. Nevertheless, Steinmetz entered the polytechnic for the winter semester and managed to obtain a residence permit from the Zurich authorities that October.[39]

He enrolled in a three-year program that emphasized the theoretical principles of civil, electrical, and mechanical engineering. Although he attended only one semester and completed only the mechanical engineering courses, they were an important part of his education because they stressed engineering theories of machines, rather than scientific theories of natural phenomena. He learned mechanical design principles applicable to motors and generators, which complemented what Auerbach had taught him about electric machines at Breslau. Professor Fliegner's course on the slide valve of the steam engine introduced him to engineering graphical analysis, one of the foundations of his work on electric circuits.[40]

It is surprising in light of his subsequent career that Steinmetz received almost no formal education in electrical engineering, outside of the dynamo theory from Auerbach. His Zurich notebooks show that he attended only seven electrical engineering lectures. He probably dropped Professor H. F. Weber's introductory course because it began with elementary electrical theory, which he already knew. Professor Denzler's first lectures on arc lighting covered the practical aspects of regulators and arc-light generators. Steinmetz received most of his electrical engineering education from technical journals and textbooks read outside the classroom. During the semester, he read over one hundred articles from the back issues of the *Centralblatt für Elektrotechnik*—the major German journal in this field—from its founding in 1879 to 1888, when it merged with the *Elektrotechnische Zeitschrift*. The articles were on devices and materials, rather than theory, and covered a wide range of topics: incandescent lighting, the telegraph and telephone, electrical instruments, batteries, magnetic materials, protective devices (fuses), and dynamos. He studied more practice and some theory from four handbooks, including Friedrich Uppenborn's *Kalender für Elektrotechniker* and volume 1 of Kittler's *Handbuch der Elektrotechnik*, which his father had sent him from Breslau.[41]

Steinmetz's college record and reading notebook suggest that he had decided to become a mechanical engineer specializing in electrical technology. This was a common practice at the time, because electrical machines like dynamos and motors were largely mechanical. (His father addressed a letter to him in 1889 as "Chas. Steinmetz, Mechanical and Electrical Engineer.")[42] He took courses in mechanical engineering, in which he had no background, and relied on outside reading to supplement his education in electrophysics. Auerbach's writings on dynamo theory at Breslau probably influenced his career choice. The introduction of alternating-current systems in the late 1880s also had a bearing, since they made electrical engineering the most mathematical of all engineering disciplines, thus giving mathematicians like Steinmetz an advantage in this field.

Electrical manufacturers and utility companies turned to alternating current because of the limitations of Edison's direct-current system. (AC varies in direction and intensity at a prescribed frequency, 60 times per second for household current in the United States. DC flows in one direction and is steady, like the current from a battery.) The DC system was practical only in densely populated urban centers, because the greater the distance from the dynamo to the customer, the larger in diameter the distribution wires had to be to make the system economical (energy losses are inversely proportional to the diameter of the wire). Engineers first attacked the problem by raising the voltage of the system with a three-wire 220-volt feeder network, which increased the service area to only two miles or so. AC proved to be a better solution because of its flexibility. It could be generated at a low voltage, stepped up to a high voltage (in the thousands of volts) for distribution, then stepped down to a low voltage for home use—the system in general use today.

When Steinmetz attended the Zurich polytechnic European and American companies had engaged the Edison interests in the so-called Battle of the Systems between AC and DC. By the end of 1887 the Hungarian Ganz company had installed two dozen AC central stations in Europe, including large plants at Rome and Vienna; Lucien Gaulard and

John Gibbs had erected a small plant serving Grosvenor Gallery and vicinity in London; and the Westinghouse company had sold AC equipment to supply nearly seventy central stations in the United States, about one-half of the number of Edison plants. AC was not as popular in Germany at the time. One historian lists only three German AC plants before 1890 (in Marienbad, Reichenhall, and Köln), compared with twenty-four DC stations, including a three-wire installation at Breslau. The key AC technology was the transformer, a device with two coils of insulated wire wrapped around an iron core that stepped the voltage up and down. Based on experiments conducted by the English chemist and physicist Michael Faraday in 1831, the transformer was not developed into a commercial device until the 1880s, with the work of Gaulard and Gibbs in Britain, Messrs. Zipernowsky, Bláthy, and Déri in Hungary, and William Stanley and others in the United States.[43]

While AC solved one problem, it created another; its variation in direction and intensity made it much more difficult to analyze mathematically than DC. If engineers simply multiplied the magnitude of voltage and current together to calculate the power expended in a circuit, as they did with DC, for example, they would get the wrong value because of the phase difference between them (i.e., the voltage and current did not rise and fall at the same time). Ohm's law for electric circuits (current equals voltage divided by the resistance of the circuit) also seemed to fail with AC. George Prescott, an electrician with an American instrument firm who had learned his trade as a telegrapher, remarked in 1888 that "it is a well known fact that alternating currents do not follow Ohm's law, and nobody knows what law they do follow." Maxwell had shown over twenty years before, that AC obeyed Ohm's law, but most electrical men of the 1880s and early 1890s did not have the training to understand Maxwell's equations (expressed for electric circuits as integral and differential equations).[44]

Steinmetz, of course, understood higher mathematics and made a career out of applying it to electrical systems. While at Zurich he contributed his first two papers in electrical engineering to the *Centralblatt für Elektrotechnik*. The first was on the apparent resistance of a current-carrying conductor; the second—written at the request of editor Friedrich Uppenborn—was a mathematical theory of the transformer.[45] Both were in the tradition, learned from Auerbach, of mathematical physicists writing theories of electrical equipment. These theories (composed of equations relating the input of a device to its output) formed one part of the body of knowledge that made up the new discipline of electrical engineering. There were at least six other parts, all of which were taught in college and learned on the job: (1) elementary principles of physics, chemistry, and mechanical engineering; (2) mathematical techniques; (3) empirical data on materials and machines; (4) design rules of thumb; (5) design equations; and (6) technical skill. Examples of each for a dynamo would be Ohm's law and dynamics, graphical analysis, resistance of conductors and the efficiency of various dynamos, simple proportions between the machine's dimensions and how much power it produced, formulae that related these dimensions to variables in the equations of the dynamo, and how to wind its coils. We will see that Steinmetz and other electrical engineers applied all seven components of this knowledge to design and build electrical apparatus, rather than simply applying physics directly to these tasks, as the "applied science" model of the relationship between science and technology would suggest.[46] Since theories of devices were expressed in mathematics and often drew heavily on science, it was quite appropriate that Steinmetz's first contributions to electrical engineering knowledge were theories of the electric circuit and the transformer.

Though too theoretical for most practicing engineers, these articles mark an intellectual shift from the fields of mathematics and physics to the "science-based" discipline of electrical engineering. The change is evident in the relationship between Steinmetz's theory of the transformer and the approach taken by physicists, who typically used Maxwell's theory of the induction coil (an early form of the transformer). Published in his classic paper on electromagnetism in 1865, Maxwell's theory contained equations with coefficients indicating the amount of electricity induced in each winding or between windings. Maxwell assumed that these coefficients were constant for the ironless induction coil. But with the advent of the iron-core transformer as the heart of the AC system in the mid-1880s, many engineering theorists, including John Hopkinson in Britain, realized that Maxwell's equations were not accurate enough for practical work. The iron core, necessary for the higher power of a commercial system, caused the coefficients to vary with the magnetization of the core. In 1887 Hopkinson modified Maxwell's equations by using the magnetic-circuit theory he had adapted from American physicist Henry Rowland. (Rowland's equation did for the magnetic circuit, i.e., the path followed by magnetic lines of force, what Ohm had done for the electric circuit in the 1820s.) Hopkinson replaced the coefficients with one term, which he obtained from experimentally derived curves or from magnetic-circuit theory. Although his theory still contained differential equations, he took into account the existence of iron-core transformers—a vast improvement over Maxwell's method.[47]

John Hopkinson (1849–1898) was the type of inventor-engineer-scientist that Steinmetz would become. Receiving the highest mathematical prize from Cambridge University (senior wrangler in the mathematical tripos), he was awarded Britain's top scientific honor (Fellow of the Royal Society) in 1877 while working as an engineer for an optical company that supplied lighthouses. Becoming interested in dynamos to generate power for electric lighthouses, he began publishing papers on the experimental data, design rules, and theory of the dynamo in 1879. While serving as a consultant for the English Edison Company, he independently invented the three-wire distribution system and redesigned the Edison dynamo by drastically altering the dimensions of its magnets. The resulting "Edison-Hopkinson" dynamo produced twice the power for the same weight and cost. Earlier, in 1879–1880, Edison and his assistant Francis Upton had tried to derive experimental equations for the magnetic circuit to assist them in designing the Edison dynamo. Hopkinson succeeded in an 1886 paper, where he used his empirical data on dynamos and Rowland's magnetic-circuit theory to derive a set of equations that came to revolutionize the design and building of dynamos.[48] This theory was much simpler than his later one for the transformer, because it did not contain differential equations.

Steinmetz, who had probably learned about Hopkinson's dynamo theory from Auerbach, adopted the magnetic-circuit approach in his paper on the transformer. Although Hopkinson had solved the transformer equations in terms of engineering design parameters, Steinmetz went beyond him to consider efficiency, regulation, open and closed magnetic circuits, and energy losses in the iron core. He thus addressed the problems of transformer design more thoroughly than Hopkinson did—a remarkable feat, since Steinmetz later recalled that he had not even seen a transformer when he wrote this paper. But his theory, as well as Hopkinson's, was still of little practical value because of its higher mathematics, which in the 1880s was a foreign language to most practicing engineers.[49]

Steinmetz probably decided to specialize in electrical engineering because of his expertise in mathematics, but there were other considerations as well. One was the influence of Lux, who encouraged him to take electrical courses at Breslau and showed him, by becoming a prominent electrical engineer and Socialist writer, that electricity and socialism were not incompatible.[50] In fact, the ideology of electricity as a progressive social force, an idea that dated to the late eighteenth century, was a common theme in German Socialist discourse in the 1880s. Even though the high price of electric lighting limited it to the affluent, many writers predicted an electrical utopia. Universal electrification, made possible by inexpensive waterpower and other improvements, would decentralize industry, clean up the cities, and relieve drudgery in the home and factory. Lux wrote about electricity and evolving social conditions in the late 1890s.[51] Twenty years later Steinmetz argued in a similar manner that electrical engineers had a vital role to play in bringing about a Socialist commonwealth.

Steinmetz began his new career sooner than he expected. In early March 1889 his father wrote that the Breslau police had issued a warrant for his arrest (*Steckbrief*), which appeared in a local paper at the end of February. Steinmetz replied that he planned to leave Zurich when the semester was over. Since the warrant had no authority in Zurich, he probably decided to leave the city for other reasons. His father asked if it bothered him to be photographed again—a reference to the Lassallean picture and possible police surveillance in Zurich. Steinmetz also needed a certificate from the Breslau authorities to renew his residence permit, which expired at the end of March. The difficulty of obtaining another permit probably prompted him to leave. In any event, instead of moving to another European city, he took the gigantic step of emigrating to the United States on the invitation of his roommate Oskar Asmussen. A Danish engineering student at the polytechnic and a fellow Socialist, Asmussen was ordered to the United States by his wealthy American uncle and guardian to prevent him from marrying his Swiss sweetheart. Before leaving Zurich Steinmetz obtained a letter of reference from Uppenborn and accepted a loan of 2,000 Swiss francs from Asmussen. In early May they traveled to Le Havre, France, and took passage on the steamer *La Champagne* to the "Promised Land."[52]

2 Eickemeyer's

When Steinmetz boarded *La Champagne* at Le Havre, he embarked on a new life in a strange land, as did thousands of his compatriots in the 1880s. For Steinmetz it was the second uprooting in little more than a year. The first was sudden and traumatic—fleeing home under the cover of darkness to avoid arrest. The second held traumas of a different sort—moving to a foreign land to pursue a new career. Having left without finishing his education at Breslau or Zurich, he must have felt anxious about finding a job in engineering. Although he probably learned at Zurich that he was better educated in mathematics and physics than most engineering students, he lacked their detailed knowledge of engineering theory and practice. His first American employer gave him an opportunity to overcome this shortcoming.

Steinmetz and Oskar Asmussen arrived in New York City on Monday, May 20, 1889, after a weeklong voyage. Steinmetz told Heinrich Lux, "We traveled steerage where we had no decent food and sour wine to drink, and filth!!!! . . . On board ship I had begun to learn English. Either I could not understand the people or when I understood them I could not reply, and so I usually took a chance and said either 'yes' or 'no,' which, as you can imagine, caused the greatest confusion."[1] The manner in which he arrived—as a penniless, deformed political refugee—formed a crucial episode in the rags-to-riches Steinmetz legend. When an immigration officer at Castle Garden asked if he could speak English, he supposedly replied, "A few." The officer tried to stop him from entering the country because he had no money and one side of his face was swollen, apparently as a result of a severe head cold. Asmussen then came to the rescue and offered to take responsibility for the sad-looking creature.[2] The real reason for the swelling was that nearly all of Steinmetz's teeth, neglected from childhood, were decayed—the actual condition that almost kept him out of the country. And, in fact, he probably had some money in his pocket as well, since the fare for traveling steerage was only a small fraction of the loan he had received from Asmussen.[3]

Having made it past Castle Garden, they stayed at the Grand Union Hotel in Manhattan and then with Asmussen's relatives in Brooklyn while they looked for work. The prospects were fairly good. Neither had any practical experience, but they had attended one of the best technical schools in Europe. Trained in mechanical engineering, Asmussen quickly found a job with De La Vergne Refrigerating Machine Company in nearby Yonkers, New York.[4] Several of the company's engineers belonged to the Deutsch-Amerikanischen Techniker-Verband (National Association of German-American Technologists), which provided a social and technical forum for German engineers and helped them find work in the United States. Asmussen and Steinmetz joined it in the fall and may have obtained their jobs through the Association, because engineers were asked to join it after finding work through its free labor exchange.[5] Steinmetz had less of an engineering education than Asmussen, but he had a letter of introduction from Friedrich Uppenborn, author of a well-known electrical engineering handbook and editor of the *Centralblatt für Elektrotechnik*, the journal that had accepted two of his articles for publication. When Uppenborn learned that Steinmetz planned to emigrate, he commissioned him to "write a series of letters from America" for the journal.[6]

It was an opportune time to take up electrical engineering in the United States. While telegraphy was a mature industry, telephony and electric light and power were barely a decade old. The latter field was booming. Arc lamps lit the streets of hundreds of cities, and incandescent lights burned in many stores, offices, hotels, and homes after Edison opened a central station in New York City in 1882. By the fall of 1888 Edison had installed 185 central stations in urban centers and about 1,300 "isolated plants" for individual customers. When Steinmetz arrived in 1889 the Battle of the Systems between the proponents of direct and alternating current was at a fever pitch, with the Edison forces conducting a publicity campaign against the dangers of Westinghouse's high-voltage equipment. By the fall of 1888—only three years after installing the country's first AC system in Buffalo, New York—Westinghouse Electric had 116 AC plants in operation or under construction. It was also developing a self-starting AC motor invented in 1887 by another immigrant engineer, Nikola Tesla from Serbia. That same year Frank Sprague, a former Edison engineer, began installing a DC streetcar system in Richmond, Virginia. Electric trolleys operated in more than a dozen cities by then—in Pennsylvania, New York, New Jersey, Alabama, Ohio, Michigan, Wisconsin, California, and Canada—yet no system had more than twelve cars or ten miles of track. Sprague ran forty cars on a twelve-and-one-half-mile route with more than thirty sharp curves and grades up to 10 percent. Completed in early 1888, the Richmond line proved the feasibility of electric traction on a large scale and over difficult terrain. It ushered in the age of electric power in the United States two decades before the widespread electrification of industry and started an electrical business that was larger than electric lighting for many years.[7]

Most of the early inventors in the American electrical industry had little college training. They were practical "electricians"—a term applied to physicists as well as to inventors at the time—many of whom had worked in telegraphy or the mechanical arts. Edison was the best-known telegrapher, an itinerant operator who made his mark as a telegraph inventor. Arthur Kennelly, Edison's chief electrician in the late 1880s and early 1890s, learned his trade as a telegrapher in Britain. The "mechanicians" included George Westinghouse, who built his company on his invention of a railroad air brake, and the instrument maker Edward Weston, a self-taught chemist who got his start in electroplating. Hiram S. Maxim, a professional inventor in electric lighting, machine guns, and many other fields, served his apprenticeship to a scientific instrument maker; the streetcar inventor Charles Van Depoele had owned a furniture factory. Elihu Thomson, a prominent inventor of arc lighting and AC systems, was a science teacher at the Philadelphia High School, his alma mater; and William Stanley, who designed the transformer for Westinghouse's AC system, studied liberal arts at Yale for a year before embarking on apprenticeships with Maxim and Weston.

Some electricians had a college education in engineering or science. Tesla attended the polytechnic in Graz, Austria; Albert Schmid (Westinghouse) received a degree in mechanical engineering from the Zurich polytechnic in 1879. Frank Sprague, Oliver Shallenberger (Westinghouse), and William Le Roy Emmet (Edison General Electric) graduated from the U.S. Naval Academy in mechanical engineering. Cummings Chesney, hired by Stanley to work on the AC system, had a bachelor's degree in chemistry from Pennsylvania State College; and Charles Brush, a leading inventor-entrepreneur in arc lighting, held the same degree from the University of Michigan.[8]

The number of college-trained electricians was rising rapidly when Steinmetz emigrated, because universities in the United States, Britain, and Europe had established electrical engineering programs in the early

1880s. The first American programs grew out of physics departments at the Massachusetts Institute of Technology in 1882 and at Cornell University the next year; many later ones split off from mechanical engineering. By 1889 most major engineering colleges were offering degrees in the new discipline, whose graduates found good jobs as "operating engineers" for street railways and electric light companies and as design, installation, testing, and sales engineers for electrical manufacturers. As Monte Calvert has observed, compared with civil, mechanical, and mining engineering, electrical engineering was "born yesterday and had no long-standing tradition, no professional culture." Lacking a field, shop, or mine culture and shedding fairly quickly the ethos of the telegraph operator-inventor, electrical engineering adopted the values of the "school culture" being formed in American technical colleges in the late nineteenth century. The introduction of the AC system accelerated this process, because a far better knowledge of mathematics and physics was required for the design and maintenance of AC equipment than for DC equipment, as we have seen. The adoption of magnetic-circuit theory in the late 1880s had a similar effect. Shop culture, the older route of training mechanical engineers, was at a disadvantage in the related field of electrical engineering because it did not teach higher mathematics. The career of Benjamin Lamme provides a good example of how graduates of the school culture rose rapidly in the profession. After completing a degree in mechanical engineering with a specialization in electrical engineering from Ohio State in 1888, Lamme, who had an aptitude for mathematics, joined Westinghouse Electric and rose through the ranks to become the company's chief engineer after the turn of the century.[9]

A small number of electricians in the United States had graduate training in physics, and many of them taught the new electrical engineering courses. When Steinmetz arrived at Castle Garden, this group included Francis Upton, one of Edison's chief assistants in developing the DC system, who studied physics under Hermann von Helmholtz at the University of Berlin; Michael Pupin, professor of electrical engineering at Columbia College, who earned a doctorate under Helmholtz in 1889; Louis Duncan, head of the applied electricity program at Johns Hopkins University, where he had studied under Henry Rowland; and Louis Bell, who also earned a doctorate at Johns Hopkins. Bell taught electrical engineering at Purdue University in the late 1880s, then worked as editor of a technical journal, a corporate engineer, and a private consultant. In contrast, in the 1880s Francis Crocker and Harris Ryan became renowned professors of electrical engineering at Columbia College and

Cornell University with only bachelor's degrees in engineering from their respective institutions (Crocker later finished a Ph.D.). Some engineers did not study electricity formally until graduate school. Carl Hering completed a bachelor's degree in mechanical engineering at the University of Pennsylvania in 1880, then studied with Kittler in Germany. He headed the electrical engineering program at the University of Pennsylvania in 1886–1887 before turning to private consulting. After receiving a B.A. from Ohio State in 1885, Charles Scott attended Duncan's program at Johns Hopkins for two years, specialized in AC equipment at Westinghouse for many years, then joined the faculty at Yale in 1911. Dugald Jackson, a civil engineering graduate from Pennsylvania State College, studied electrical engineering at Cornell for two years as preparation for a distinguished career in industry and academia.[10]

These inventors, engineers, and educators were Steinmetz's professional colleagues and competitors in the United States. He knew their work through several New York journals—including the *Electrical Review, Electrical World,* and *Electrical Engineer*—and met them at meetings of trade and professional societies like the National Electric Light Association and the American Institute of Electrical Engineers. Although Steinmetz never received his doctorate, he was better educated in higher mathematics than most American engineers *and* physicists, since American physicists were not as well educated in mathematics as their European brethren.[11] He also had a solid education in mathematical physics and knew the fundamentals of mechanical and electrical engineering.

But this training did not guarantee him a job when he began looking for work in the summer of 1889. He applied first at the Edison Machine Works in Manhattan, which manufactured dynamos for the Edison system. Turned down there, he presented a letter of introduction from Uppenborn to the firm of Eickemeyer & Osterheld in Yonkers, where Asmussen had landed a job with another firm. Steinmetz was received warmly by Rudolf Eickemeyer, the principal owner of the small company and a German Socialist émigré who had fled the revolution of 1848. The two had much in common, and Eickemeyer hired the inexperienced young man as an assistant draftsman at $12 per week. Steinmetz reported for work on June 10, 1889, and later moved with Asmussen from Brooklyn to East 119th Street in Manhattan to be closer to work. He entered Eickemeyer's as a student of mathematics, physics, and engineering and left, four years later, as an electrical engineer accomplished in theory and practice.[12]

Eickemeyer as Mentor

Eickemeyer played an important role in this process. He gave Steinmetz the opportunity to learn the practical side of electrical engineering, provided him with a laboratory, and encouraged him to publish his research. Steinmetz told Lux in 1893, "Old Eickemeyer is a very intelligent man, really the most intelligent person I have ever met. You can talk with him about anything, any mechanical problem, or Darwinism, chemistry, Socialism, or anything." Steinmetz later recalled that Eickemeyer "took me in hand, made me a part of his intimate business relationship in the line of his inventions, and showed me as long as I was with him how I could apply my knowledge and make myself useful to myself and the world."[13]

The qualities that so endeared the "Old Man" to Steinmetz grew out of a successful career in business and invention. Educated at the polytechnic institute in Darmstadt, Eickemeyer emigrated to the United States with his friend George Osterheld at the age of nineteen in 1850. Four years later they settled in Yonkers, the center of the wool hat industry, where they opened a machine repair shop serving the hat-making and other trades in the city. Eickemeyer discovered his inventive talent, took out several key patents on hat-making machinery, and established a small factory with Osterheld to make this equipment. The business grew into a virtual monopoly because of the dominance of Eickemeyer's patents. One earned royalties of between $30,000 and $40,000 in 1867. Between 1856 and 1880 Eickemeyer received sixty-seven patents, fifty of which were on hat-making machinery and ancillary equipment. When Alexander Graham Bell and Elisha Gray invented the telephone in 1876, Eickemeyer took up electrical experiments on the side and patented improvements to the telephone.[14]

By the time Steinmetz arrived in 1889 Eickemeyer had turned this sideline into a profitable business. It had begun five years before, when the fifty-four-year-old Eickemeyer converted part of the hat-making factory into an electrical laboratory and workshop to develop an ironclad DC dynamo he had invented two years before. The name came from its unique method of construction. Its field coils (windings producing magnetic lines of force) were wound in large loops around the armature (a wire-covered drum rotating in the magnetic field), instead of being wound on iron poles surrounding the armature, the case for most dynamos. The core for the field coils consisted of the iron of the armature drum and the iron plating enclosing the field coils, thus giving rise to the "ironclad" nickname. This design had several advantages. It was more compact, because less iron was required to produce a given amount of electricity; the iron cladding protected the inner workings from oil and dirt; and no magnetic fields extended outside the machine, a feature that prevented the magnetization of nearby watches and other delicate instruments. Another novel feature was its armature winding, which consisted of standardized coils wound by machine and then placed on the iron core, instead of layers of wire wound by hand directly on the core. The "Eickemeyer winding" came into general use for other DC dynamos and motors. The Edison General Electric Company, for example, adopted the winding for its streetcar motors in 1890. But the ironclad design made it difficult to cool the dynamo, which, like all iron-core machines, heated up while running because of currents induced in the core and the core's magnetic characteristics.[15]

Eickemeyer began manufacturing the dynamo after receiving patents on it in 1887 and on its armature in 1888. Although he sold a large machine to a storage-battery factory to charge battery plates, the dynamo was mostly used for lighting railway trains, where its small size, ruggedness, and high efficiency gave it an edge. The J. H. Bunnell Company of New York City, sole agent for the dynamo, packaged it with a gas engine and sold the combination to the Pullman Palace Car Company as its standard train-lighting plant in 1889. By December 1890 twenty-five units were in operation and had not burned out a single armature—a common complaint with other dynamos. Bunnell sold the machine for other applications, such as lighting yachts, and put nearly thirty sizes on the market in 1889.[16]

In 1888 Eickemeyer was also designing a new type of streetcar motor based on this dynamo. Electricians had known since the 1870s that a motor was essentially a dynamo running in reverse (electricity in and mechanical power out, instead of vice versa) and had built many types of machines on this principle. But the task of designing a good streetcar motor was proving difficult because of the adverse conditions under which trolleys operated: frequent starting under a heavy load, climbing steep grades, vibration of the trolley car on the tracks, and exposure of the motor to harsh weather conditions. By 1888 the Thomson-Houston Company and Frank Sprague independently developed the double-reduction-gear motor to meet these conditions. This rugged motor, which had two gears to reduce its high speed to the slower speed of the streetcar wheels, worked well except for two problems with the gears:

Rudolf Eickemeyer, about 1890. Photograph by Rudolf Eickemeyer, Jr. Source: Hall of History Foundation, Schenectady, N.Y.

they wore down rather quickly, and their loud screeching could be heard up to two miles away on a clear night.[17]

Eickemeyer worked to overcome these problems with a gearless motor. The idea probably came from Stephen D. Field, a versatile electrician in the fields of telegraphy and electric power. When it looked like Field would win an interference suit against Edison on a basic electric traction patent, Edison formed a short-lived company with him in 1883 to develop an electric railway. Field experimented with gearless electric locomotives at a Chicago exhibition in 1883 and on an elevated line in New York City in 1887. In the latter trial Field coupled the motor shaft to the drive wheel by a connecting-rod arrangement similar to that employed on steam locomotives. He used an Eickemeyer dynamo to generate electricity for the New York trial and realized it would make an excellent streetcar motor. Eickemeyer and Field then joined forces and put an experimental car on the Steinway road on Long Island in 1888. The armature shaft drove thick discs on either side of the car, from which connecting rods turned the drive wheels, which were somewhat smaller than the front wheels. Through a careful design of the motor Eickemeyer and Field were able to run it at a speed slow enough to eliminate the need for reduction gears. Other manufacturers, such as the Short and Westinghouse companies, also built gearless motors, but such slow-speed machines were usually very heavy. Eickemeyer's ironclad design reduced the size of the motor, and thus its weight, and enclosed the unit to protect it from the debris of the roadway. When Steinmetz joined the firm Eickemeyer and Field were testing this system and working on an improved version.[18]

At the same time the Eickemeyer motor was beginning to prove itself in another application—electric elevators. From Eickemeyer's point of view an elevator was simply a vertical streetcar, starting and stopping at floors rather than at trolley stops. Their motors thus had similar electrical requirements, and Eickemeyer achieved an early success in this field with his well-designed motor. In 1889 the Otis Brothers Company, also of

Yonkers, installed an elevator powered by two Eickemeyer motors in the Demarest Building in New York City—their first electric elevator and the second direct-connected electric elevator in the country (the first had been installed in Baltimore in 1887). Eickemeyer designed the motors and controllers, while Otis engineered the elevator equipment—a combination that soon beat out the steam and hydraulic elevators that had made the skyscraper possible in the 1880s. The Otis Electric Company was established in Yonkers in 1892 to make the Eickemeyer motors, which were the mainstay of Otis elevators for over a decade.[19]

Eickemeyer did well in the electrical business. His firm, which made both hat-making and electrical machinery, grew to just over one hundred employees in 1892 and occupied several buildings in Yonkers. An inventor-entrepreneur, Eickemeyer conducted what would later be called research and development. He set up an electrical laboratory next to the factory, licensed his patents to other manufacturers, and developed his inventions with the help of mechanics, draftsmen, and a few other electricians besides Field. The R&D part of this operation was similar to that in other electrical firms, such as Thomson-Houston and the Brush Electric Company of Cleveland. But it was much smaller than Edison's celebrated laboratory complex in West Orange, New Jersey, built in 1887, where about one hundred employees supported electrical, chemical, and metallurgical research. While Eickemeyer shared the machine-shop culture of West Orange, the polytechnic graduate also cultivated a school culture at the Yonkers factory, particularly in his research on electromagnetism. He invented an instrument to measure the magnetic characteristics of iron and read a paper in 1887 before the New York Electric Club describing his experiments on the magnetic circuits of iron-core dynamos. Although aware of Henry Rowland's work in this area, he apparently did not know about John Hopkinson's 1886 paper, which utilized Rowland's approach to derive a workable theory of the dynamo.[20]

Because of the relatively small size of Eickemeyer & Osterheld, Steinmetz gained a wide experience in electrical machinery. During his four years at Yonkers he performed most of the duties of an electrical engineer in manufacturing (drafting, installation, troubleshooting, design, and product improvement), as well as in R&D (invention, materials testing, and experimental and theoretical research). He was fortunate to serve his engineering apprenticeship in an environment similar to that of the early inventor-entrepreneurs, rather than in the specialized departments of the giant manufacturing firms. At Eickemeyer's he saw electrical engineering as a broad endeavor and learned to combine theory and practice in a highly successful manner.

Like many electrical engineers of the period (and up to World War II), Steinmetz began his career at the drafting table. His first assignment was to make drawings of "Street car motor No. 3," which took about a month. He then drew different parts of the Eickemeyer-Field system, such as the trolley (overhead connector) and trolley switch, in addition to numerous other devices ranging from electric water pumps to magnetic-ore separators (in competition with Edison's ill-fated venture in this area). Many of these were patent drawings, indicative of the innovative nature of Eickemeyer's business. In January 1890, after six months at the drafting table, Steinmetz began to take on other engineering duties, first on a new streetcar system.[21]

By 1890 Eickemeyer and Field had redesigned their first unit into a commercial system. They made the drive wheel as large as the front wheel, raised the output power, which also increased the weight by one-half, lowered the speed of the armature shaft, and increased the terminal voltage to the industry standard of 500 volts DC. These changes made the system competitive, and Eickemeyer and Field sold their first units to the Lynchburg (Virginia) Street Railway Company in 1890. Eight cars with the new gearless motors ran on five and one-half miles of track and performed well on grades slightly steeper than those on Sprague's famous Richmond line. This was a good start for Eickemeyer, but he was competing against two giants. By January 1891 the Thomson-Houston Company and Edison General Electric, which had purchased the Sprague company two years before, had made over three thousand of the fifty-five hundred electric streetcars then in service. But since the trolley business was booming in the United States, more than doubling in capacity between the fall of 1889 and late 1890, Eickemeyer and Field thought there was plenty of room for their inventions. Steinmetz played a rather limited role in developing this system. He designed a switch and copper brushes for the motor and did the calculations for the size of the power cables.[22]

This experience brought him more responsibilities for the next Eickemeyer-Field system, installed in 1892. A slightly larger and faster motor was suspended in the middle of the truck (undercarriage), where it drove all four wheels by means of connecting rods. Eickemeyer and Field claimed that their four-ton system damaged the track less than lighter ones, because the motor was suspended by springs and did not bear directly on the wheels. In keeping with their business strategy, Eicke-meyer and Field had the truck and motor manufactured under their patents by the Western Electric Company in Chicago. Several units were sold—twenty-four in Toledo, Ohio, six in Yonkers, and twenty in Long Island—during another big year for electric streetcars. One survey re-ported that eighty-four new roads were equipped in 1892, increasing the number of electric cars to about twelve thousand. Although Eickemeyer and Field did only a fraction of the business of Thomson-Houston and Edison General Electric, technical journals pointed to the Eickemeyer installations in Toledo and Yonkers as good examples of gearless motors, at a time of strong competition from the new single-reduction-gear motors.[23]

Steinmetz again designed the motor switch and other auxiliary appa-ratus. He must have understood the new system quite well, though, because he spent much time on streetcar motor calculations, made two trips to Toledo to debug the system, and ran laboratory tests on the Toledo motor—all in 1891. In 1892 he participated in a debate before the American Institute of Electrical Engineers on streetcar motors. A rising star in the institute, he argued persuasively for the benefits of the gearless motor in general and for the Eickemeyer-Field system in partic-ular. After the meeting he sent the institute a set of efficiency curves for the Eickemeyer-Field motor taken during a test on the West End Street Railway in Boston. According to figures in his notebook, the Eickemeyer-Field motor outperformed a single-reduction-gear motor made by Thomson-Houston.[24]

Steinmetz also worked on the Eickemeyer-Otis elevator motor. He made design calculations, ran tests on the motor's insulation and arma-ture, and even repaired an elevator motor at the Women's Medical Col-lege in New York City. His most innovative work was on an improved regulator to reduce starting current in elevator motors. The assignment grew out of his experience with the streetcar version of the Eickemeyer motor.[25]

Alternating-Current Machines

Although Eickemeyer focused on direct-current equipment, his ex-periments on alternating currents were the most exciting work in the shop when Steinmetz arrived. The main drawback of the AC system in the battle against DC was the lack of a practical motor. (The best AC motor at the time, the synchronous motor, would not start by itself.) Like many electricians, Eickemeyer was working on a self-starting AC motor—an invention that promised to push AC ahead of Edison's system. Westinghouse thought it had solved the problem after purchasing Tesla's patents on the induction motor in 1888. In addition to being self-starting, Tesla's motor seemed very reliable, because current was transferred from the field to the armature winding by means of electrical induction through air, rather than by electrical conduction through an often troublesome commutator (a mechanical switching device on the armature shaft). By 1890, Westinghouse had spent a large sum on the motor with little to show for it. In trying to make the induction motor propel streetcars, Tesla and his assistant, Charles Scott, discovered that it could not start a heavy load from a standstill—a characteristic obviously unsuited to the stop-and-go running of crowded urban trolleys.[26]

Eickemeyer took a different tack. He knew something about AC motors, having patented an induction motor that would run from household current in 1888 (the Tesla motor required special out-of-step currents).[27] But rather than trying to modify the induction motor to run streetcars, Eickemeyer turned his attention to a different type of motor, the series AC motor, so called because its field and armature windings were connected in tandem. This motor had a powerful starting torque, but it used the troublesome commutator. Eickemeyer realized that if he could overcome the disadvantages of the motor, it would revolutionize the streetcar industry by greatly extending the area served by trolley lines, since AC systems were more economical than DC ones over long distances.

Steinmetz shared his employer's enthusiasm. He wrote to his father in 1890 that Eickemeyer's motor would probably work better for streetcars than Tesla's motor. Steinmetz thought that the series motor, which he called a "joint invention" (*gemeinsame Erfindung*) between him and the Old Man, would make them wealthy if they could obtain a patent monopoly on it.[28] The device did not make them rich because it never worked satisfactorily, but it brought a more lasting benefit. While developing the motor Steinmetz learned how to design, build, and test AC equipment. He published papers on energy losses in iron and the theory of the transformer that earned him an international reputation as an expert on alternating currents by 1892. He also learned a good deal about AC motors and generators from reading volume 2 of Kittler's *Handbuch*, Uppenborn's *Kalender für Elektrotechniker*, and Gisbert Kapp's *Alternate-*

Current Machinery. The many critical annotations in these books show the fruitful interaction between theory and practice that characterized his career.[29]

As tests on the motor progressed in 1890, Eickemeyer built Steinmetz an AC laboratory in the factory. The system consisted of an old Westinghouse alternator driven by the factory steam engine, electrical and mechanical instruments, and several transformers—normally set up to run tests on the streetcar motor. Steinmetz measured the electrical characteristics of these transformers and built a high-voltage transformer to test the motor's insulation and commutator in 1891.[30]

The series AC motor had enough torque to move streetcars from a standstill, but its vicious sparking at the commutator and low power factor (a measure of how efficiently it worked on an AC circuit) were serious drawbacks. A German expert had stated that such a motor was impractical because it could not achieve a power factor greater than 70 percent. This did not deter Eickemeyer, who recognized that its problems could be overcome by placing a compensating coil around the armature, and he patented this arrangement in late 1890. Steinmetz ran a long series of tests on a compensated motor in early 1891 and obtained a power factor of nearly 80 percent and an efficiency of 75 percent. A larger motor built in 1892 achieved a power factor of almost 90 percent, and Eickemeyer patented several improvements to the motor later that year. The improved motor gave encouraging results, but not good enough. Commutation was "perfect" at a frequency of 33 Hertz (cycles per second), "fair" at 85 Hertz, the highest frequency available at Eickemeyer's. Since typical frequencies at the time were 125 and 133 Hertz, far above the point of good commutation, Steinmetz and Eickemeyer realized that their motor would not work well for streetcars.[31]

Although Eickemeyer patented the motor, Steinmetz had helped invent it, as he told his father. He identified the offending electrical characteristic of the armature that led to the idea of the compensating coil as early as 1889 and suggested energizing this coil by induction two years later. Eickemeyer patented this improvement as well.[32] (This practice was not uncommon at the time. Edison took out numerous patents in his name on inventions made in collaboration with his laboratory staff.) In early 1891 Steinmetz helped develop an induction motor that would start from household current by means of shorting out some of the field coils. Eickemeyer patented the motor over a year later. The only public evidence of Steinmetz's part in the invention was his signature as a witness on the patent application.[33]

Steinmetz, who had kept a patent notebook since early 1891, applied for his first patent that August: a means to convert single-phase AC into two-phase current and to use the same principle as the basis for a single-phase induction motor. ("Two-phase" systems, patented in 1888 by Tesla, are powered by two independent, alternating currents that are 90 degrees out of phase, i.e., the second one begins when the first one has peaked. "Single-phase" denotes regular household AC.) Consisting of four field windings wrapped around an iron core and a rheostat (variable resistance), the device produced two-phase current if the number of turns of each coil was correctly related to the magnetic characteristics of the iron core and the rheostat was adjusted properly. Steinmetz could turn the device into a single-phase induction motor by placing an armature that resembled a squirrel cage inside the field coils. Rather than using empirical reasoning, he made the invention by applying AC mathematics to a practical problem, a style that became his trademark.[34]

Although Eickemeyer did not manufacture the device, Steinmetz designed another single-phase induction motor for General Electric in 1892 while still at Yonkers. (It became GE's standard fan motor after he joined the company in 1893.) Called the Whirligig, probably because of its small size and high speed, the motor had an auxiliary field winding that created the phase difference needed to start it on single-phase current. Like other induction-motor inventors in industry, especially Charles E. L. Brown in Switzerland and Michael von Dolivio-Dobrowolsky in Germany, Steinmetz ran comparative tests to determine design rules for the best arrangement of field and armature coils.[35] This empirical research complemented the deductive approach from "first principles" used in his first patent, bringing together the theoretical and experimental strands common to engineering research.

Alternating-Current Theory

The experience Steinmetz gained at Eickemeyer's vastly improved the quality of engineering research he had begun at Zurich. The AC papers he wrote at Yonkers marked two milestones in his career. They were written from the viewpoint of electrical engineering, rather than mathematical physics, and they prepared the ground for his path-breaking theories of AC circuits and machines. The articles also illustrate an important aspect of engineering knowledge identified by Edwin Layton and other historians: the translation of information from science into technology. In this case Steinmetz and other scientist-engineers translated the mathematics of physicists (differential equations) into that used by engineers (graphical analysis).

Steinmetz's AC theory resembled the graphical methods introduced by Thomas Blakesley (1847–1929) and Gisbert Kapp (1852–1922) in the 1880s. A mathematical "wrangler," Blakesley received an M.A. from Cambridge University in 1872 and served an apprenticeship with a civil engineering firm. While working as a consultant he wrote articles on the theory of AC circuits in the early 1880s, then accepted a lectureship in science and mathematics at the Royal Naval College. Kapp had a better engineering background. A graduate of the Zurich polytechnic in mechanical engineering, he emigrated to England in 1875 and became a British citizen. After managing the electrical firm of R. E. Crompton, he, like Blakesley, turned to private consulting, a post from which he edited *Industries* and the *Elektrotechnische Zeitschrift*. In 1885 he independently derived a theory of the dynamo similar to John Hopkinson's. Kapp wrote numerous articles and several books on AC theory and practice before taking the newly established chair of electrical engineering at the University of Birmingham in 1905.[36]

Prior to the work of Blakesley and Kapp, electricians had used Maxwell's equations to analyze an AC circuit mathematically (e.g., to calculate the current drawn by several lamps connected to an alternator).[37] Rather than solving these equations, Blakesley and Kapp expressed the relationships between sinusoidal voltages and currents geometrically (by vectors rotating at the AC frequency) and used engineering graphical analysis to manipulate these vectors and obtain the desired values. Their method simplified matters considerably, because it gave only the steady-state solution to the equation: the condition when surges of voltage and current had died out, which was of most interest to engineers at the time. It did not give the complete solution provided by a rigorous analysis using Maxwell's equations. However, if they used trigonometry in addition to a ruler and compass, their values were as accurate as those obtained by the Maxwellians for the steady-state condition.[38]

Engineers preferred the graphical method over Maxwell's differential equations for several reasons. They were more familiar with graphical analysis because of its foothold in civil and mechanical engineering, in which Blakesley and Kapp had been trained. They could handle complicated circuits more easily because they simply dealt with more vectors in these cases, instead of solving simultaneous differential equations. And they could visualize the complicated relationships between the magnitude and phase of voltages and currents in a circuit from a vector diagram, which also let them check whether their calculations were in the ballpark. Differential equations, on the other hand, were a bewildering abstraction to most engineers at the time and bore little resemblance to the real world of practical engineering.[39]

At Eickemeyer's Steinmetz recognized the advantages of graphical analysis and developed an improved method. He remarked in 1891 that "at present all mathematical and analytical theories [of AC circuits], especially if they have to start from the solution of differential equations, are still of very little value for the 'practical engineer,' who is not yet generally expected to master the powerful weapons of mathematics."[40] Steinmetz was correct. Even if engineers had graduated from college, they might not have studied differential equations. A survey of eighteen electrical engineering curricula in the United States in 1899, for example, revealed that only four schools required a course in differential equations (MIT, Armour Institute, California, and Ohio State), although all of them required calculus.[41]

Steinmetz based his graphical method, developed in 1890, on the Zeuner diagram rather than on the clock diagram favored by Blakesley and Kapp.[42] He had learned about the Zeuner diagram—drawn in polar rather than rectangular coordinates—just two years earlier at the Zurich polytechnic, where a professor used it to analyze the slide valve of a steam engine.[43] He was also skilled in this type of mathematics because of his graduate training in synthetic geometry. While the Zeuner and clock diagrams produced essentially the same results, Steinmetz preferred the former because it gave certain values more quickly and could be applied to nonsinusoidal functions.

He adapted it to electrical engineering in order to analyze the transformer, which can be modeled as a branched AC circuit with a common magnetic field. Although he used this magnetic-circuit approach rather than Maxwell's equations in his first theory of the transformer, he wrote it in differential equations, the mathematical language of physics. But now he switched to graphical analysis, which he had used for simple AC circuits. Steinmetz followed in Kapp's footsteps once again. Kapp had introduced rotating vectors in his theory of the transformer in 1885, where they represented the mutual magnetic field and the voltages and currents in the transformer's windings. This method allowed him to analyze the unit's operation under different loads easily (e.g., switching lamps on and off the circuit) and to see the effects of changing design parameters. Kapp improved his theory in 1888 by taking into account the energy lost in the iron core.[44]

Steinmetz published a similar theory in 1890, which went beyond Kapp to consider the phenomena of magnetic leakage (lines of force not linking both windings), nonresistive loads, and nonsinusoidal sources. He also included power calculations and the effects of changing several design parameters.[45] Having gained more experience at Eickemeyer's, he published an improved theory in June 1891. He could now handle energy losses caused by induced currents circulating in the iron core and the core's variable permeability (its ability to "conduct" magnetic lines of force). His theory was the best one to date because it considered, in a rigorous manner, all energy losses in the iron core—caused by induced currents, hysteresis (see Chapter 3), magnetic leakage, and variable permeability—which most theories neglected.[46]

The contrast between the transformer theories written at Zurich and Yonkers shows the influence of working at Eickemeyer's. In just two years Steinmetz dropped the mathematical language of the physicist for that of the engineer and developed the most complete engineering theory of the transformer to date. The *Electrical Engineer* praised the 1891 paper because Steinmetz based it "upon the results of actual experience." The *Electrical Review* simply said that he was "well known as a brilliant electrical mathematician."[47] Steinmetz also derived two empirical equations for transformers. One related efficiency to the number of turns of wire in a coil; the other compared output power with frequency.[48]

These equations are good examples of the design-rule component of engineering knowledge discussed in the last chapter and further illustrate Steinmetz's metamorphosis at Eickemeyer's. Steinmetz entered the Yonkers factory as a student of mathematics, physics, and engineering and emerged four years later as an electrical engineer accomplished in design, laboratory research, and theory. He was not oblivious to these changes. At Breslau and Zurich he stamped his books "Karl Steinmetz, Stud. Math." His father addressed early letters to him in America as "Chas. Steinmetz, Mechanical and Electrical Engineer." His book stamp at Eickemeyer's read "Chas. Steinmetz, Electrical Engineer." When he left Eickemeyer's in early 1893 he signed his published articles as being from the "Eickemeyer Laboratory."[49]

His articles show a similar transition. He wrote the two electrical papers at Zurich from the viewpoint of a mathematical physicist who lacked engineering experience. He wrote the Eickemeyer articles for practicing engineers. Although his papers still dealt mostly with theory, they now considered the practical concerns and mathematical abilities of workaday engineers. The vast majority of the fifty articles and technical letters he published at Eickemeyer's—a prodigious body of work—were on electric light and power, rather than telecommunications, reflecting the interests of his employer. All but six were on AC theory and the magnetic circuit, the fields that brought him international renown.[50]

Steinmetz gained much of his reputation at meetings of the American Institute of Electrical Engineers. Founded in New York City in 1884, the AIEE functioned as the professional group for "electricals" alongside the older societies for civil and mechanical engineers, mainly by publishing papers read at its monthly meetings. Steinmetz made his mark quickly in the AIEE. David Lain, a graduate of the first class of electrical engineers at Cornell who had charge of building an experimental streetcar motor at Eickemeyer's, remembered going with him to a meeting in October 1890. After listening to Professor Thornburn Reid of Columbia College read a paper on the electromagnetic characteristics of armatures, Steinmetz—who had a much better command of English by then—asked Reid if he had given a complete solution to this complicated problem. Reid replied that he had and invited Steinmetz to give a better answer. According to Lain, "Then our unknown, young emigre, without apparent embarrassment, went to the [black] board and within a few minutes had written thereon, in figures readable from all parts of the [large] room, a solution" to the problem. Prominent members of the institute then gathered around Steinmetz for introductions, and the program chairman booked him for a lecture.[51] Steinmetz rose rapidly in the institute, becoming president in 1901 and an active member thereafter, thus helping to create the profession of electrical engineering in the United States (see Chapter 7). Although he was elected to the New York Electrical Society in 1891,[52] the papers he read before the AIEE did more than anything else to establish his reputation as an AC expert at home and abroad.

It was not all smooth sailing. For one thing, he did not easily give up his dream of becoming a mathematician. He subscribed to the *Annalen für Mathematik* and the *Zeitschrift für Mathematik und Physik* shortly after joining Eickemeyer's, published a lengthy paper based on his dissertation in the latter journal in 1890, and told his father earlier that year that he planned to finish his doctorate. Since he decided in July 1890 to pay his debt of over 500 marks to the University of Breslau when he had "superfluous" money, he probably intended to finish his Ph.D. in absentia.[53]

Steinmetz also joined the mathematical community in New York City and became friends with Thomas Fiske, an instructor of mathematics at nearby Columbia College and a founder of the New York Mathematical

Steinmetz, about 1890. Source: Hall of History Foundation, Schenectady, N.Y.

Society. During the first part of 1891 Steinmetz went to Columbia to see Fiske, use the library, and attend meetings of the society when he could get away from Eickemeyer's. He corresponded regularly with Fiske about his research in synthetic geometry, and Fiske saw that two of his papers on the topic were published by the *American Journal of Mathematics*. (They were later listed as "important" in a history of American mathematics from 1875 to 1900.) Fiske also nominated him for membership in the American Association for the Advancement of Science. While Steinmetz was becoming known in mathematical circles, his work at Eickemeyer's began to consume more and more of his time. In March 1891 he apologized to Fiske that "my connection with electrical engineering did not allow me in the last weeks to leave the factory, where I was superintending the installation of my testing room." He also complained that his mathematical researches "proceed now only very slowly." By August he and Fiske were corresponding infrequently, and Steinmetz had apparently decided to drop "pure" mathematics and pursue his "connection with electrical engineering."[54] He published a lengthy paper on the theory of the transformer printed in installments from June to December 1891 and a major paper on energy losses in iron in January, which vaulted him into the front ranks of electrical engineers. Attaining a similar position in synthetic geometry might have taken a lifetime. Electrical engineering also paid well. By 1892 Eickemeyer had raised his annual salary to $1,200, which enabled him to pay off a loan from his sister Clara and take a vacation to Niagara Falls.[55]

But his new profession left little time for politics, which, like mathematics, fell by the wayside (until picked up again twenty years later). There were opportunities to keep up with the movement. He roomed with Asmussen, a Socialist, until Asmussen married the woman he had fallen in love with at Zurich and moved to the Bronx in the summer of 1890.[56] Steinmetz then boarded at the Yonkers home of another Socialist, Edward Mueller, a senior draftsman at Eickemeyer's. Mueller recalled that Steinmetz gave a lecture on electricity to a Socialist group—probably Section Yonkers of the Socialist Labor party, because Steinmetz used its library in 1892. He also discussed socialism with Eickemeyer at Sunday afternoon gatherings at the Old Man's home.[57] But as a prominent inventor and factory owner, Eickemeyer had given up the political struggle in order to enjoy the economic fruits of capitalism. Steinmetz came under the same spell. In a letter to his father in mid-1890 he said that the AC streetcar motor would "be worth millions" if it panned out and if they could obtain a "monopoly" on it for the seventeen-year life of the patent.[58]

Steinmetz may have spoken so warmly of profits in order to quell his father's fears that he would get into trouble with socialism again. Or he may have come to share Eickemeyer's economic philosophy while still believing in some form of socialism. Though he still read Socialist literature in 1892, he curtailed his political activities sharply after coming to America. He resigned as co-worker for the Breslau *Volksbibliothek* in late 1889 and rarely wrote to his Breslau friends, even after the Anti-Socialist

Law expired in 1890 and was not reinstated.[59] When he did write to Lux in the fall of 1893, Steinmetz admitted that he had "withdrawn almost completely from politics," mainly because of the Socialists he had met in America. Most were German immigrants who refused to learn English or to become citizens—traits he deplored. He also disliked the personal denunciations in the Socialist Labor party (SLP) press, which was controlled by the vituperative Daniel DeLeon. Coming from Breslau, the home of Lassalle, Steinmetz may also have become disillusioned when the Lassallean faction of the SLP lost control to the Marxists during his first year in America. In any event, Steinmetz told Lux that he "had nothing to do with the Party except to send in my contribution when they send a subscription list, something they never forget to do."[60] He later revealed another reason why he was not politically active in the 1890s: American bias against Socialists would hurt his engineering career. Steinmetz said that America then was unlike Germany in that "it was not illegal to be a Socialist, but no man could be one and be considered respectable or admitted to positions of trust."[61]

American politics may not have suited him, but the country did. In June 1890 he praised the United States lavishly to his father:

I am very satisfied to have left the narrow living conditions of Germany and to have come here where a reasonable man can live reasonably and succeed. It is infinitely better here than at any place where I ever was—even if certain things could still be better. Nobody interferes with your freedom; you can go where you want to and do or not do what you want; there is no war, no soldiery, no possible danger of war, *no taxes* (except for real estate), you can travel where you want without notifying the police or similar nonsense, you can change your name, the income is several times higher, life is not significantly more expensive, and you have much more time for yourself.[62]

Later that month he received a telegram that his father had died on June 26, less than three weeks after this letter had been written. Carl Heinrich had been sick for over a year with tracheal catarrh and heart disease, a condition that worsened and had kept him at home with badly swollen feet since the beginning of April. In February he had advised his son to take good care of his health in order to avoid contracting the same illness. He thought it might be hereditary—a real concern for him, because they were both short and deformed. Although family letters reveal an underlying tension between father and son, mostly over the Maches, young Steinmetz had continued the relationship after moving to America. He sent his father photographs of himself, German-American newspapers, and reprints of his articles, and asked him to order German books through a Breslau bookseller, as he had done in Zurich. He sent money to his half-sisters, wrote about once a month, and, in 1889, promised to return home in two years. His father hoped the relationship between Clara and the Maches would greatly improve by then.[63]

Carl Heinrich's death cut one of the last ties Steinmetz had with the Old World and the unhappy household he had left behind. About the same time he adopted America as his homeland. He changed his name from Carl August Rudolph to Charles Proteus—he told Lux that *Proteus* was "too lovely to be forgotten"—applied for citizenship papers in 1891, and later invited Clara to live with him in America. Charges were still pending against him in Breslau when he told Lux in 1893, "I will not go to Germany in the foreseeable future. Being put on trial does not matter. I will become an American citizen in a few months. What would I do in a stupid and rotten police state like Germany! Why don't you come here for a visit to a sensible country where you can live decently at least without police informers and junior officers loafing around everywhere."[64]

Like other immigrants, Steinmetz exaggerated the differences between the Old and New Worlds, probably to help rationalize his decision to become a citizen of another country. He criticized a homeland that had persecuted him for his political beliefs, but he left no record of how he felt about his father's death. The record shows only an outpouring of research and publications after his father died—an effort that placed him at the head of a new profession in his adopted land.

3   Engineering Research

$\equiv$ **S**teinmetz advanced to the head of his profession on the basis of two research projects: theories of AC circuits and experiments on magnetic hysteresis. The first led to a revolutionary style of engineering mathematics described in Chapter 4; the second brought his first taste of international acclaim. Working in the borderland between the mathematical physics of his university days and the engineering practice he learned at Eickemeyer's, Steinmetz combined the values of both cultures when he studied hysteresis. This case illustrates the contrast between scientific and engineering research in the late nineteenth century and the role of scientist-engineers in bridging the two communities.

The Law of Hysteresis

Magnetic hysteresis plagued electricians from the start of the electrical power industry in the 1870s. The tendency of a material to resist being magnetized or demagnetized, hysteresis caused the iron cores of electrical machines to overheat, which led to ventilation problems and low efficiencies, since the energy given off as heat was not available to do useful work. The phenomenon occurred in direct-current as well as alternating-current machines, because both produced fluctuating magnetic fields in their iron cores. But the problem was usually worse in AC devices, whose magnetic fields rose and fell more rapidly than those in comparable DC machines. To prevent alternators from overheating, for example, engineers designed them to run on magnetic fields weaker than those in DC dynamos. Hysteresis also played havoc with transformers, whose economy of operation depended on high efficiency. Determining the effect of hysteresis was difficult, because another phenomenon caused iron cores to overheat—the circulation of closed-loop electric currents in the core. Engineers learned to eliminate these eddy currents by constructing the core out of thin, insulated sheets of iron. But laminating the core in this manner did not affect hysteresis, which remained a thorn in the side of electrical engineers until the development of low-hysteresis silicon steel in the early twentieth century. Like friction in a mechanical device, hysteresis caused electrical equipment to function quite unlike the ideal machines portrayed in textbooks. How to account for hysteresis was thus an acute problem in electrical engineering in the 1880s.[1]

Engineers first dealt with the problem by estimating how much energy was lost by hysteresis in a known sample of iron. Much of this knowledge came from empirical tables and graphs published by physicists and other researchers, who typically depicted the energy lost in this manner by the area bounded by an S-shaped curve. The curve formed a closed loop, because changes in the strength of a magnetic field lagged behind changes in the alternating current producing the field. The more a material resisted changes in its magnetization, the wider the curve. In 1881 James Alfred Ewing, professor of mechanical engineering at the University of Tokyo, coined the term *hysteresis* from the Greek word ὑστερέω (to lag behind) to describe this phenomenon.[2] Earlier that year Emil Warburg, professor of physics at the University of Freiburg in Germany, proved that the area enclosed by the S-shaped curve equaled the energy dissipated in iron, if no eddy currents were present. Ewing independently discovered the relationship in 1882 and called the curve a hysteresis loop.[3]

Ewing became the leading authority on hysteresis. In 1885 he published empirical curves and tables of the energy lost by hysteresis in numerous kinds of iron and steel in the *Philosophical Transactions of the Royal Society*, a highly respected scientific journal that also published John Hopkinson's theories of the dynamo. Ewing demonstrated that the amount of energy lost depended upon the magnetization reached, the type of iron, and the frequency of the cycle of magnetization.[4] After deriving a molecular theory of hysteresis in 1890, Ewing, who had become professor of engineering at Cambridge University, did research on more applied topics and published his previous scientific work in engineering journals. He wrote papers on hysteresis in transformers, a hysteresis tester he had invented for the factory workshop, and a two-year series of articles on hysteresis for the London *Electrician*. (During these years the *Electrician* published the work of other scientist-engineers, such as Thomas Blakesley, John Ambrose Fleming, and Oliver Heaviside, on electric circuit theory.) The *Electrician* brought out Ewing's articles in book form in 1892 as *Magnetic Induction in Iron and Other Metals,* which quickly became the standard reference for engineers and physicists.[5]

Steinmetz knew something about hysteresis before coming to Eickemeyer's, having studied the subject in Professor Auerbach's course on electricity and magnetism at the University of Breslau. He learned that the area of the S-shaped curved equaled the energy lost by hysteresis, but he apparently did not know about Ewing's 1885 article. In his paper on the theory of the transformer, written at Zurich in 1888–1889, Steinmetz stated that more research was needed on how the energy lost by hysteresis varied with different types of iron and levels of magnetization—a topic Ewing had exhausted three years before.[6]

Steinmetz learned about Ewing's research at Eickemeyer's, where hysteresis was an everyday concern. Although Eickemeyer had engineered a very efficient dynamo and motor based on his research on the magnetic circuit, the resulting ironclad design made it difficult to ventilate the armature and remove heat generated by eddy currents and hysteresis.[7] Hysteresis was thus a topic of more than academic interest when Steinmetz joined the firm in mid-1889. In December, as he began working on the AC streetcar motor, Steinmetz read Gisbert Kapp's recently published *Alternate-Current Machinery.* While working as a consulting engineer Kapp had used Ewing's hysteresis data to design AC equipment. In his book Kapp converted Ewing's data from ergs per centimeter cubed to watts per ton. Kapp made this "translation" from scientific to engineering units because he thought that Ewing's units were "rather inconvenient for application in the workshop or drawing-office."[8] For the same reason Kapp converted Ewing's data to thousands of units of magnetization and rounded off energy lost by hysteresis to the nearest 50 watts per ton.

Steinmetz read Kapp's book with the eye of a mathematical physicist who was becoming an engineer. An equation relating energy lost by hysteresis to magnetization would be useful to scientists trying to understand hysteresis as a natural phenomenon and to engineers designing AC machines. Looking for such a relationship, Steinmetz derived an equation to fit Kapp's data. The formula fit well, giving a total error of less than 2 percent over the observed values.[9] But he did not publish the formula, probably because it had an exponent to two decimal places (1.83), which gave it the appearance of an empirical equation rather than a fundamental relationship that might hold for other materials. (When fundamental equations in physics had exponents—like Newton's law of gravity—they were whole numbers.)

In 1890 Steinmetz's father sent him volume 2 of Kittler's *Handbuch der Elektrotechnik,* which contained both Ewing's original data and Kapp's converted data.[10] Comparing the two, Steinmetz discovered errors in Kapp's figures for energy lost by hysteresis at two values of magnetization. Since his equation no longer described the corrected data very well, giving a total error of nearly 7 percent, he derived a new equation to fit Ewing's data. This equation also had an exponent that was not a whole number (1.6), but it fit Ewing's data very well, giving a total error of a little over 1.5 percent and a maximum error of 8.5 percent at the lowest value of magnetization. In December 1890 Steinmetz published his new equation in *Electrical World* as the "law of hysteresis."[11]

The law entered the corpus of electrical engineering knowledge within a few years, but at this early stage Steinmetz was unsure about the status of his equation. Was it a scientific "law of nature"? Did it hold at all values of magnetization and for all magnetic materials? Steinmetz leaned in this direction when he proposed it as the "law of hysteresis." (He did suggest, however, that his equation might indicate a more general and complicated law.)[12] At the other end of the spectrum, was his equation an engineering formula? That is, was it similar to a design rule but applicable to a natural phenomenon, hysteresis, rather than to the operation of a specific machine? Did it hold for enough values of magnetization and types of iron to make it a useful approximation for engineering purposes? Steinmetz did not know at this point, because he had derived it from only one set of data. His equation would not be a scientific law or a good engineering formula if it did not apply to more measurements. The task of gathering this data, in 1891 and 1892, thrust him into a type of engineering research that came close to being basic science.

Steinmetz had already used a mathematical technique common in science when he derived his equation by applying the method of least squares to Ewing's data. Developed by Adrien Marie Legendre in 1806 for computing the orbits of comets, the method of least squares was a curve-fitting technique for expressing empirical data—which seldom followed a simple curve—by a mathematical formula. Nineteenth-century geophysicists also used the method extensively for magnetic measurements. An 1886 American book on the subject, for example, noted that most magnetic observations could be represented by an exponential equation having the form in which Steinmetz later expressed his law of hysteresis. It is unknown whether Steinmetz had read this or similar books, but we know that he was quite familiar with the method of least squares in mathematical astronomy, because he had taken a course on the subject from Johann Galle at the University of Breslau in 1884.[13]

One difference between Steinmetz's research and that conducted by contemporaries who were considered to be scientists had to do with laboratory technique. Trained in mathematical physics, Steinmetz knew the types of instruments used in the electrical laboratory. Chief among these for hysteresis measurements was the ballistic galvanometer—a very sensitive device for measuring current or voltage. To determine the energy lost by hysteresis in a metal, Ewing and other researchers shaped the metal into a ring and wrapped a coil of wire on either side of the ring, thus forming a transformer of the type Michael Faraday had used in his classic experiments fifty years earlier. Ewing connected the galvanometer to one winding of the ring and energized the other coil with a battery connected to a circuit interrupter. The galvanometer measured the voltage produced when a known value of battery current was applied to the ring. By increasing the current and then decreasing it in a step-by-step manner, Ewing obtained enough galvanometer readings to draw an accurate hysteresis curve. The area enclosed by the curve represented the energy lost by hysteresis. The main advantage of the method was the high degree of accuracy given by the sensitive galvanometer.[14] But Steinmetz disliked the method. It was too slow—a factor of more importance in a factory laboratory, where time meant money, than in Ewing's university laboratory. He also thought that the galvanometer was "generally too delicate for the unstable floor of an electric factory"—a real concern for him because of the vibration caused by the heavy machinery at Eickemeyer's. (Edison had the money to solve this problem by constructing a separate laboratory building for his galvanometer.)[15]

Steinmetz invented two techniques to overcome these difficulties. Devised in 1891 to measure the hysteresis of the Eickemeyer AC streetcar motor, the first method employed standard factory voltmeters and wattmeters for its instruments instead of the ballistic galvanometer. The method also used alternating current for the power source instead of interrupted battery current. Rather than taking step-by-step readings like Ewing, Steinmetz took only one reading from the wattmeter for one maximum value of magnetization. Since this reading was proportional to energy lost by hysteresis, he did not have to plot the galvanometer readings and calculate the area of the hysteresis loop. By taking wattmeter readings at several magnetizations for numerous materials, he obtained all the data he needed to test the generality of his law of hysteresis in a rapid manner.[16] He improved the accuracy of his wattmeter method over that to be expected from using factory instruments by determining the electrical characteristics of the instruments,[17] calculating other energy losses in the circuit, and testing only laminated materials, in order to eliminate losses due to eddy currents.

For samples that were not laminated, Steinmetz used the differential magnetometer—the basis for his second method of measuring hysteresis. Eickemeyer had invented the device in the late 1880s to determine the magnetic characteristics of the iron in his dynamos and motors, but he had not used it to measure hysteresis. The magnetometer was a magnetic bridge that worked on the same principle as the common Wheatstone bridge, except that it measured magnetic rather than electrical conductivity by comparing the unknown conductivity against a known standard. To measure the hysteresis of a material Steinmetz inserted a sample of it into the magnetometer and "balanced" the bridge, indicated by a zero meter reading, by adding calibrated pieces of iron to the standard. The cross-sectional area of the added pieces was proportional to the magnetization of the sample at the current used. Repeating the process for several values of current gave him the data he needed to plot a hysteresis loop.[18]

Although Steinmetz admitted that he used the ballistic galvanometer a "few times for controlling observations,"[19] he preferred his factory methods, mainly for the reasons given above. He also argued that the wattmeter method was more accurate than Ewing's technique because it used alternating rather than intermittent currents, worked at larger values of voltage and current, enabled more data to be taken because of its speed, and better simulated the actual conditions under which hysteresis occurred. The magnetometer was less accurate, but he preferred it over Ewing's approach because it was faster and could test samples with large eddy-current losses.[20]

Equipped with these tools, Steinmetz performed lengthy experiments in Eickemeyer's new AC Testing Room with the help of an assistant and published his results in early 1892. He told Lux that he made over four thousand observations, which "occupied all my free time for two years."[21] To see if the law of hysteresis held beyond Ewing's data on soft iron, Steinmetz tested ten types of magnetic materials available at Eickemeyer's by gathering data with the wattmeter method and plotting hysteresis curves. If certain data points did not fit the curves, he threw them out without repeating the experiment at those points—an indication of his strong a priori belief in the validity of the law of hysteresis.

He then used the method of least squares to calculate the exponent and coefficient of an equation similar in form to his law of hysteresis.

Although he found the exponent to be 1.6 in most samples, it did not equal this value in nonlaminated samples that probably contained eddy currents. He was able to "save" the law of hysteresis by reasoning that energy lost by eddy currents was proportional to the square of the magnetization, because electric power was proportional to voltage squared and magnetization was proportional to voltage. He then derived a new equation for energy loss that combined his law of hysteresis with a term to account for eddy currents. The equation had two coefficients, whose values depended on the type of material and the way it had been metallurgically treated. (The one regarding hysteresis came to be known in engineering and physics circles as the *Steinmetz coefficient*.) Because the part of the equation that represented hysteresis still had an exponent of 1.6 and the equation held from low to high values of magnetization, Steinmetz concluded that the wattmeter tests had "proved" the law of hysteresis. He then confirmed the law with the magnetometer tests by assuming an exponent of 1.6 and comparing values of the hysteresis coefficient at different frequencies.[22]

The engineering community was nearly unanimous in praising this work. The New York *Electrical Engineer* declared that no paper of "more absorbing interest and practical utility has been presented to the [American] Institute [of Electrical Engineers] . . . We are sure that electrical engineers will feel a sense of relief in having finally gotten rid of another factor of uncertainty in the designing" of dynamos and transformers.[23] The London *Electrician* commended his "exhaustive analysis" and "very simple empirical formula." Since his results compared favorably with those obtained by Ewing, the journal thought he had given the "first proof that the static method [using the ballistic galvanometer] yields results applicable to alternate-current work."[24] During the discussion of the paper at the AIEE Arthur Kennelly, then Edison's chief electrician, called it a "classic." Charles Bradley, a prominent inventor of AC equipment, alluded to the European background of both Eickemeyer and Steinmetz as one explanation for the high quality of the research. Bradley observed, "It is very seldom that in America, anything of this kind is taken up. We see it very often in Europe, but our commercial age will hardly permit us to devote our time to such experiments and carry them out as they should be."[25] The perceived contrast between the scientific engineering on the Continent—a legacy of the Ecole Polytechnique seen in French and German journals—and the less rigorous approach of "bustling" American engineers could not be more evident.

In helping to make turn-of-the-century American engineering more scientific, Steinmetz and other college-trained researchers faced criticism from scientist-engineers more firmly committed to the canons of physics. Steinmetz's chief critic on this score was Michael Pupin. A Serbian immigrant who had attained a faculty position at Columbia College by returning to Europe to study physics under von Helmholtz, Pupin knew Steinmetz from the New York Mathematical Society, which met at Columbia. The two rivals, who had similar educations but worked in quite different environments, joined the AIEE at the same meeting in March 1890. Their arguments during discussions of AIEE papers enlivened the rather staid meetings of the institute for a decade. Upon hearing Steinmetz read his hysteresis paper, Pupin did not agree with Bradley that Steinmetz had carried out his experiments "as they should be." Pupin noticed that Steinmetz's law predicted an increase in hysteresis with increasing magnetization, independent of the degree of magnetic saturation, which contradicted Ewing's theory. Professor Pupin told the AIEE that he was "inclined to side with Professor Ewing, until I am convinced by Mr. Steinmetz that his method of measurement and observation could not be objected to in any particular whatever." Pupin asked Steinmetz to describe the theory of his method and the probable percentage of error of observation in a later paper.[26]

Many electrical engineers praised Steinmetz's paper, but Pupin, who saw himself primarily as a physicist, criticized it by the standards of the scientific community. Steinmetz was fair game on this score, because he had pointed out the scientific implications of his research in the paper. Contrary to Ewing's theory of molecular magnetism, Steinmetz found that hysteresis did not vanish at low magnetizations, did not increase rapidly for medium ones, and did not increase slowly at saturation. "Nothing of this is the case, but hysteresis seems to follow the same law over the whole range of magnetization."[27] His scientific aspirations also came out in the discussion of the paper. He compared his law with Rudolf Clausius's theory of entropy, maintaining that the law of hysteresis was the first general formulation of energy loss in a cyclic conversion. And in the German version of the paper, he called hysteresis research a "branch of experimental physics" (*Zweige der Experimentalphysik*).[28]

Steinmetz answered Pupin at the AIEE by retreating to a conservative, Baconian position. He did not "take the time" to compare his results with Ewing's theory. "My aim was to gather facts, being convinced that based upon a large number of facts, a theory will be found in due time to explain them."[29] When he did have time to think the matter over in the next month, he reported to the AIEE that the "law of hysteresis agrees very nicely with Ewing's theory, giving just the phenomenon this theory leads us to expect."[30] He explained that the energy lost by hysteresis increased slowly at saturation because magnetization increased more slowly there too, as shown by hysteresis curves published by him and Ewing. While he agreed with Ewing on this point, he still maintained the validity of his law over the entire magnetic cycle.

The report satisfied one of Pupin's concerns, but Steinmetz did not address the main issue—the validity of his measurements. During the spring and summer of 1892 he conducted further research on this point, ran more tests, and profited from reading Ewing's recently published *Magnetic Induction in Iron*.[31] The result was a hundred-page paper, published in September 1892, that attempted to place the law of hysteresis on a firm scientific foundation. During the discussion of this paper at the AIEE Steinmetz remarked that his investigations "had been undertaken, first, for a strictly practical purpose . . . [but] have since developed into scientific research."[32] He now described his experiments more completely, tested many more types of magnetic materials, wrote the law in scientific terms, and proposed a theory of molecular magnetism. He made all these changes, except for the type of instrumentation, to emulate Ewing's scientific style of research.

In response to Pupin, Steinmetz compared the wattmeter, magnetometer, and ballistic galvanometer methods. He provided an electrical diagram of the first technique, as called for by Pupin, and explained why he preferred it and the magnetometer to Ewing's method.[33] It was not the exhaustive theoretical explanation Pupin had requested, but Steinmetz did improve his measurement techniques. He used pulsating direct current to reduce energy losses in the wattmeter reading, added permanent magnets to the magnetometer to correct for the influence of terrestrial magnetism, and brought the pole pieces of the magnetometer closer together to obtain higher magnetizations. Even with these changes, the magnetometer was not sufficiently sensitive by scientific standards, and Steinmetz could not test cast iron at as high a magnetization as Ewing could—a problem he blamed on the poor quality of American cast iron.[34]

In order to determine whether the law of hysteresis was a scientific relationship or an engineering formula, Steinmetz tested an impressive variety of materials with his improved instruments. His first experiments had been limited to ten engineering materials: the iron in a Westinghouse transformer, well-insulated layers of sheet iron, various steels, cast iron, and so forth.[35] He now tested over twenty more materials, including cast iron and steel of different degrees of hardness, wrought iron, ferrotype, tin plate, magnetite, nickel wire, and a tube of iron filings. He could not obtain a sample of cobalt, the only known magnetic material he did not test, and relied instead on Ewing's data for it. From the results of his experiments Steinmetz concluded that the law of hysteresis held for "all kinds of wrought and cast iron and steel, for nickel, and magnetite, and most likely for amalgam of iron, hence apparently for all magnetizable materials."[36]

Convinced by these tests that the equation was a law of nature and not merely an engineering approximation, he redefined it in scientific terms. Steinmetz now claimed that it applied between any two values, as well as to equal changes of magnetization. He modified his law in this way to counter Ewing's molecular theory of magnetism, which held that energy lost by hysteresis depended on the absolute value of magnetization, not on relative values.[37] Kennelly, who had performed experiments on magnetization at Edison's laboratory, agreed with Steinmetz and thought that this latest insight was "enough to immortalize one paper."[38]

Steinmetz also rewrote his law in the scientific terminology favored by Ewing. Ewing commonly used the term *intensity of magnetization*, which Maxwell had introduced in the 1860s.[39] Steinmetz invented a similar term, *metallic induction*, and, using an analogy derived from Kennelly's recent experiments, defined it as being proportional to intensity of magnetization.[40] In another move toward physics, he adopted the scientific term *magnetic field intensity* in place of the engineering term *ampere-turns*.

The scientific validity of Steinmetz's law depended upon its application. He had tested it from very low magnetizations to saturation (85 to over 19,000 lines per centimeter squared) in his first paper. Now he considered whether the law applied outside this range by means of theory rather than experiment, because his instruments did not extend much beyond these limits. He derived a formula to give the lowest value of flux density where the law held, then used the English physicist Lord Rayleigh's experimentally determined figure for a limiting value of permeability to calculate this density (0.19 lines per centimeter squared) for a sample of sheet iron he had tested earlier. Steinmetz remarked that "such low values, equal to 1/200th the intensity of the magnetic field of the earth, are still rather out of reach for experimental research."[41] The obvious inference was that the law of hysteresis held over a wide range of magnetizations, even at experimentally unobtainable values, thereby "proving" that it was a universal law of nature.

If the law was universal, the next step was to compare it with Ewing's theory of magnetism, then standard in physics. In 1890 Ewing modified Maxwell's magnetic theory by dispensing with the internal restoring force acting on a molecule of iron and explaining the shape of the hysteresis curve by the mutual attractions and repulsions of the molecule alone. By assuming that the length of a molecule was not much greater than the distance between molecules, Ewing calculated that the field intensity required to upset the equilibrium of molecular pairs was inversely proportional to the square of the distance between the molecules.[42]

Steinmetz criticized Ewing's theory, saying that it contradicted "all our present knowledge of molecular physics. All the facts of the kinetic theory of gases, of thermodynamics, etc., carry to the conclusion that the *dimensions* of the molecules are *infinitely* small compared to their *distances*" apart.[43] He then modified Ewing's theory by assuming that the distance between molecules was much greater than the length of the molecules. He calculated values of field intensity for the three stages of the hysteresis curve: low hysteresis, rapid increase of hysteresis, and slower increase of hysteresis. This theory supported the law of hysteresis because hysteresis now began at a field intensity inversely proportional to molecular distance cubed, instead of squared, as in Ewing's theory. Thus, energy lost by hysteresis would appear at lower field intensities and lower magnetizations than in Ewing's theory. The theory supported Steinmetz's law over Ewing, who held that hysteresis would vanish at low magnetizations.[44] In a further attempt to explain the scientific character of his law, Steinmetz said that it "is not a *differential* law, like for instance the quadratic law of gravitation, but is an *integral* law like the law of probability [of molecular distances] with which it seems to be connected in some way."[45]

Reception of the Law of Hysteresis

In comparing his equation with Newton's law of gravity and the law of probability, Steinmetz crossed the boundary between engineering and scientific research. The excursion drew fire from both sides. Physicists questioned whether the law held for all magnetizations and all magnetic materials, as Steinmetz claimed. Engineers were much less interested in these claims and wanted to know if it was valid at magnetizations used in practice (1,000 to 10,000 lines per centimeter squared) and for commercial types of iron and steel.

The severest criticism came from Britain, home of Ewing's theory.

Sydney Evershed, a British instrument maker who had done research on the magnetization of transformers and who was known for his "quaint quips and odd humour," countered Steinmetz's scientific claims with a sharp piece of satire in the *Electrician*:

Could infatuation go further? What can be done with a man who prefers to accuse American founders of casting bad iron rather than abandon a more or less clumsy empirical formula—which it is ridiculous to call a law—and which was only made to fit the facts as far as they were known? Is it possible that Mr. Steinmetz is unable to distinguish between a physical law and that miserable abortion an empirical law? Mr. Steinmetz warns iron founders "to keep manganese out of cast iron, since manganese is known to be the enemy of all the magnetic properties." May we warn him to keep empirical formulae out of his researches since empirics are known to be the enemy of scientific progress![46]

Stripped of its vehemence, Evershed's criticism reveals a major flaw in Steinmetz's research: his inability to distinguish between an empirical equation and a physical law. Although Steinmetz called his law an empirical formula, he compared it with such universal scientific theories as gravitation, entropy, and probability. Evershed conceded that the law of hysteresis might be related to the theory of probability, yet he remarked that Steinmetz "is just a little out of his depth here; at all events we are." Evershed also thought Steinmetz was out of his depth when it came to criticizing Ewing's grasp of molecular physics: "Truly a most perfidious monster this Ewing. It seems sufficient to point out to the author first, that solid iron is *not* a gas."[47]

Ewing criticized Steinmetz in a more civil manner. In a major paper published in the summer of 1893 he downplayed Steinmetz's challenge to his theory of magnetism and concentrated on the validity of the law of hysteresis. Ewing was interested in the law because an "empirical formula of the kind is of use to designers of transformers."[48] He applied the method of least squares to hysteresis data he had taken on transformer sheet iron and found that Steinmetz's law, with a suitable coefficient, held over the entire range of magnetization, from the lowest values to saturation. But he noticed variations of up to 15 percent between the calculated and observed values of energy lost by hysteresis. Investigating these high values, Ewing found that the hysteresis exponent was not constant at 1.6 but varied between 1.475 and 1.9, depending on the range of magnetization. He concluded, "While, therefore, a formula of this type cannot be admitted to have any physical significance, it may still be serviceable in giving rough approximations for the electrical engineer [because] an index of 1.6, or a number approximating that, gives a curve lying generally in the neighbourhood of the true curve throughout the range of [magnetization] which is of most practical importance."[49] While ostensibly evaluating Steinmetz's law for the transformer designer, Ewing dismissed it from the realm of science and relegated it to the domain of engineering. He also maintained that the changes he observed in the hysteresis exponent corresponded to the three stages of magnetization in the hysteresis curve, thus upholding his own theory of magnetism against Steinmetz's revisions of it.

Several experimenters confirmed Ewing's results. Between 1893 and 1901 Reginald Fessenden, professor of electrical engineering at Purdue University and later a radio pioneer; Frank Holden, an engineer at General Electric; J. W. L. Gill, a British physicist; and H. Maurach, a German physics professor, found the exponent to vary between 1.22 and 2.47.[50] Francis Bailey, professor of physics at University College, Liverpool, demonstrated that the exponent depended on the value of magnetization. In 1894 he found that Steinmetz's law did not hold at saturation and a year later discovered that energy lost by hysteresis followed the stages of magnetization and approached a constant value at saturation, just as Ewing had predicted. Bailey suggested that the "function expressing the hysteresis in terms of the magnetic condition of the iron should depend not on induction, but on the intensity of magnetisation, with at most a small correction in terms of the former. The results accord well with Ewing's theory of magnetism."[51]

Bailey's experiments answered Pupin's query of 1892: Was Steinmetz right about hysteresis at saturation, or Ewing? Bailey decided in favor of Ewing, although Steinmetz should be given credit for expressing energy lost by hysteresis in terms of intensity of magnetization in his second paper through the term *metallic induction*. Since metallic induction approached a maximum value at saturation, so did energy lost by hysteresis. Steinmetz, however, did not point this out, and most physicists and engineers interpreted the law of hysteresis in terms of magnetic flux density instead of metallic induction.[52]

British engineers soon recognized the significance of Bailey's work. Silvanus Thompson, one of the leading electrical engineering professors in Britain, thought Bailey had shown Ewing's theory to be true and Steinmetz's false.[53] With the best of hindsight, the London *Electrician*, which had praised Steinmetz's law three years earlier, opposed it in 1895 on scientific grounds: "We have always thought that there existed an element of unjustified empiricism in [Steinmetz's law]. It was difficult to believe that any such connection between the two things really existed; and it was simply impossible to rationalise the formula on the basis of any modern theory of the magnetisation of iron."[54]

Steinmetz did not ignore this criticism. In 1894 he reported that Ewing had found the hysteresis exponent to vary with the three stages of the S-shaped curve. But since he incorporated the law of hysteresis into his mathematical theories of transformers and electrical machinery, he apparently agreed with Ewing that it was a useful engineering approximation.[55] In 1900 Steinmetz acknowledged the empirical nature of the law. In a discussion on a related topic at the AIEE he admitted, "We know that the hysteresis loss follows the 1.6th power, but we know that it is not an inherent, physical law. It must sometimes deviate from the law. It is a somewhat complex phenomenon which can be closely represented by the 1.6th power, but it is not a physical law."[56]

Why hadn't Steinmetz reached this conclusion earlier? Why didn't he discover the limits of his law and the variability of its exponent? Bailey laid the blame on his measurements, which Bailey considered to be less accurate than Ewing's.[57] Although Ewing's ballistic galvanometer was more accurate than Steinmetz's factory instruments, Steinmetz did obtain data of sufficient accuracy to find the lower limit of the law in his tests of laminated horseshoe magnets, reported in the first hysteresis paper. If he had not been so enamored of his law, these tests would have shown him that its exponent was not equal to 1.6 at low magnetizations, as Ewing had found, and he would not have had to make an unwarranted assumption regarding eddy currents.[58] Steinmetz also recorded a very large error (15 percent) on a test of iron filings at a low magnetization (about 300 lines per centimeter squared). Unable to blame it on eddy currents, which could not circulate in the tube of filings, he simply ignored this reading as an isolated anomaly at the lower limit of the experiment.[59]

He did not perform more research to investigate these anomalies, probably because he retained the goals of engineering rather than of scientific research. Although he said that the experiments in his second paper had "developed into scientific research," he tested the law for as many magnetic materials as possible rather than at all possible magnetizations. While physicists were interested in both types of experiments, engineers were more concerned with whether the law applied to most engineering materials than whether it was valid outside the magnetizations used in practice. Thus, when Steinmetz attempted to reach back to his university days and do basic science, he still directed his research toward engineering needs—a further indication of how far he had moved along the path from mathematical physics to engineering at Eicke-meyer's.

His methodology was a respected one in science, however. Despite invoking the Baconian claim of only gathering facts, Steinmetz used a standard hypothetical-deductive method. He proposed the law of hysteresis as a hypothesis based on one set of data, then conducted numerous experiments on other materials to test the hypothesis. Finding the empirical law valid, he modified Ewing's theory of molecular magnetism and proposed it as a higher-order hypothesis to explain the law of hysteresis. Ewing employed a similar methodology. But as a Cambridge professor of engineering with close ties to the Royal Society and other bastions of the British scientific community, Ewing was more attuned to the goals of scientific research. Consequently, he went directly to the heart of the matter and tested Steinmetz's law for one iron sample at all magnetizations, the converse of Steinmetz's experiments. Finding that it held only for magnetizations used in practice, Ewing encouraged engineers to incorporate the law into the body of engineering knowledge.

Steinmetz's law became a well-respected part of this field because several other investigators confirmed Ewing's results. From 1894 to 1897 Alexander Siemens, head of the English branch of the German Siemens & Halske company; Horace F. Parshall, Steinmetz's first boss at General Electric and a prominent consulting engineer in Britain after 1894; John Ambrose Fleming, professor of electrical engineering at University College, London; and C. P. Feldmann, a German engineer, reported that the law of hysteresis held for iron and steel within the magnetizations used in practice.[60] Engineering researchers also discovered that it held for other magnetic materials. From experiments conducted at Edison's laboratory in 1892 Kennelly confirmed the law for nickel wire using a wattmeter method. Fleming verified it for a sample of cast cobalt in 1899.[61]

Confident that the law of hysteresis held within practical limits of magnetization for iron, steel, nickel, and cobalt, engineers added it to their kit of design tools. Such terms as *Steinmetz coefficient, Steinmetz curve, Steinmetz law,* and *Steinmetz exponent* became familiar to succeeding generations of electrical engineers. Several textbooks published in Germany, France, Britain, and the United States between 1892 and 1899 cited the law as Steinmetz had formulated it.[62] Later writers qualified the law, following Ewing's example in the third edition of *Magnetic Induction in Iron* (1900). Ewing stated that Steinmetz's equation was an "empirical formula, which, although it fails to apply when the magnetisation is very weak or very strong, may be used as a good approximation throughout the range of magnetisation generally employed in electrical engineering."[63]

The law of hysteresis illustrates an important characteristic of engineering theory: its status midway between physics and ad hoc design rules. Steinmetz's equation gave values less accurate than those obtained by calculating the area under a hysteresis curve—the method used in physics. But it was more accurate than the rules of thumb engineers had used to design electrical machinery before Steinmetz's work (e.g., that the increase of magnetization in the ratio of 1 to 9 increased energy lost by hysteresis in the ratio of 1 to 24.5).[64] The law was derived empirically like the design rules, yet it had a scientific structure that the rules lacked, expressed in this case by an exponential formula. Because the equation defined the relationship between two design parameters, Steinmetz and other engineering theorists could easily incorporate it into their higher-order mathematical theories of transformers, motors, and generators.

The lengthy study of magnetic hysteresis shaped Steinmetz's later career. Disappointed that his equation was not a physical law yet gratified that engineers used it in practice and named the formula after him in textbooks, he concentrated thenceforth on engineering rather than scientific matters. We will see that he conducted this research—theoretical and experimental studies of machines and engineering materials—scientifically. As in the case of hysteresis, Steinmetz used scientific methods and developed mathematical theories that were as general as possible yet of value to practicing engineers.

Part 2

The Incorporation of Science and Engineering

4 General Electric

Life changed rapidly for Steinmetz after he had finished the extensive experiments on hysteresis. Engineers saw his name often in the electrical journals and began to notice the queer figure of a four-foot-tall hunchback reading papers in a heavy German accent at meetings of the American Institute of Electrical Engineers. In this way he came to the attention of the General Electric Company, which hired him after purchasing the Eickemeyer company in late 1892. In the words of one biographer, the rest of his life story "is written on a General Electric letterhead."[1] Yet Steinmetz was hardly the typical corporate engineer working in a faceless bureaucracy. He joined GE as an accomplished engineer and gained considerable freedom in a corporate culture while paradoxically helping to promote the "incorporation" of electrical engineering in the United States.

Steinmetz's career move from the proprietary capitalism of Eickemeyer's to the corporate capitalism of GE reflected a larger pattern in American business in the late nineteenth century. The 1880s and 1890s witnessed the rise of the multifunctional industrial corporation that utilized new production technologies and expanded markets to integrate mass production and mass distribution vertically. The giant enterprises thus replaced Adam Smith's "invisible hand" of market mechanisms with the "visible hand" of salaried management described by Alfred Chandler. While the 1880s were the decade of internal expansion of corporations forward into marketing and backward into purchasing, the 1890s saw two merger movements. The merger wave of 1890–1893 followed the passage of the Sherman Antitrust Act in 1890 and liberal incorporation laws in New Jersey. The "merger mania" of 1898–1902 followed the severe business depression of the mid-1890s. Both movements tended to replace the illegal activities of the horizontally integrated "trust" and holding company with the legal, vertically integrated corporation.[2]

General Electric was formed during the first merger wave as the consolidation of two vertically integrated firms: Thomson-Houston and Edison General Electric. Organized in 1883 by Elihu Thomson and Edwin Houston, Thomson's teacher at Philadelphia's Central High School, Thomson-Houston became a leading electrical manufacturer in the fields of arc lighting, AC equipment, and streetcars. Employing four thousand people in 1891, the company owed its success to the business leadership of Charles Coffin, formerly a shoe salesman in Lynn, Massachusetts, the technical expertise of Thomson and Edwin W. Rice, Jr., Thomson's former student, and the purchase of many electrical firms. The Edison General Electric Company, formed in 1889 as the consolidation of the Edison Electric Light Company, the Edison manufacturing firms, and the Sprague Electric Railway and Motor Company, had a work force of six thousand in 1891 and held a dominant position in DC systems and electric traction. Prominent financial backers—Boston investment bankers for Thomson-Houston and J. P. Morgan's firm for EGE—merged the two companies in April 1892. Westinghouse Electric, the third largest electrical manufacturer and the leader in AC systems, apparently resisted an offer to join the group. The House of Morgan controlled GE's board of directors, while Thomson-Houston's men filled most of the top executive positions. Coffin became president. Rice managed the engineering department at Lynn (Thomson-Houston's main factory) and plants at Schenectady, New York (EGE's main factory), Harrison, New Jersey (site of the Edison Lamp Works), and Lynn. Although a member of the board of directors, Edison concentrated on developing new inventions at his West Orange laboratory, which also did research for the Edison utility companies. Thomson, who had retired as chief inventor for Thomson-Houston in 1890, acted as a consultant and retained a laboratory at Lynn.[3]

Like its parent firms and Westinghouse Electric, GE hired many college-trained engineers. They made a relatively easy transition to the corporate world, because the school culture had prepared them well to perform engineering jobs in the bureaucratic corporate culture of a science-based industry. In order to give college graduates up-to-date practical training, GE and Westinghouse also set up engineering apprenticeship courses in which the apprentices tested electrical equipment before it was shipped to customers. While these two-year courses were open to those trained in the shop, they were, in the words of Monte Calvert, the "final step in the process of superseding shop culture as an educational and professional institution." David Noble has shown that they also played a major role in inculcating engineers with the corporate values of the giant electrical manufacturers.[4]

Joining General Electric

Like other turning points in Steinmetz's life, his move to General Electric became a central part of the Steinmetz legend. The most unlikely version of the story claims that GE bought out Eickemeyer in order to get Steinmetz, who would not leave his close friend for a rival firm. One writer even invents a conversation in which Charles Coffin says, "If this young man is as good as Rice says he is, buy the company outright." Hammond stays closer to the evidence, stating that GE desired Eickemeyer's patents, but he says that "secondarily, it is quite evident that the General Electric Company wanted the services of the youthful mathematical master, Steinmetz."[5]

Particularly valuable was his mathematical expertise in alternating currents. The formation of the company combined the resources of the strongest direct-current company in the country, EGE, with those of a firm well on its way to becoming a major force in AC, Thomson-Houston. The merger allowed GE to compete with Westinghouse in the promising field of AC—a tacit admission that AC would win the Battle of the Systems with Edison. The merger also strengthened GE's hand considerably in arc lighting and the important DC field of electric streetcars.

Steinmetz made a strong impression on the GE officials who came to look over the Eickemeyer company in the winter of 1892–1893. Rice recalled their first encounter in vivid terms:

I shall never forget our first meeting at Eichmeyer's [sic] workshop in Yonkers. I was startled, and somewhat disappointed by the strange sight of a small, frail body surmounted by a large head, with long hair hanging to the shoulders, clothed in an old cardigan jacket, cigar in mouth, sitting cross-legged on a laboratory work table. My disappointment was but momentary and completely disappeared the moment he began to talk. I instantly felt the strange power of his piercing but kindly eyes, and as he continued, his enthusiasm, his earnestness, his clear conceptions and marvelous grasp of engineering problems convinced me that we had made a great find.[6]

Rice may have first met him at Eickemeyer's, but Steinmetz was well known to engineers at Thomson-Houston and Edison General Electric before the merger. He and Elihu Thomson had participated in a discussion of a paper on alternators at a monthly meeting of the AIEE in December 1891. Thomson, who helped make the decision to hire Steinmetz about a year later, referred to him by name in a favorable manner during the discussion. They may have met earlier, when they were elected to the grade of "member" in the AIEE in 1891.[7] Engineers who joined GE from Edison General Electric also knew the Yonkers engineer. At an AIEE meeting in April 1892 Steinmetz commented on a paper on streetcar motors by Horace Parshall, the former chief design engineer for Edison General Electric who transferred to GE. As we have seen, Steinmetz convincingly argued for the superiority of the Eickemeyer gearless motor over the reduction-gear motors of EGE and Thomson-Houston, citing the recent trials on the West End Street Railway in Boston to back up his claims. Parshall was not present at the meeting to read his paper, but he must have heard about Steinmetz's comments from his colleagues and read them later in the AIEE *Transactions*.[8] (By a twist of fate, Parshall was Steinmetz's first boss at GE almost a year later.)

The Eickemeyer motors outperformed Thomson-Houston equipment on the West End system during the spring trials in 1892 and powered long cars on the same railway later that summer.[9] This strong showing in Thomson-Houston's backyard—which was also a center for electric streetcars, since in 1891 Boston had more trolley lines than Europe—is probably what convinced GE to buy out Eickemeyer. Before the merger Thomson-Houston's share of the West End contracts amounted to $3 million. The 212-mile system, formed by consolidating Boston's horsecar lines and building new lines to the "streetcar suburbs," was the "largest single street railway system in the world," according to a recent study, and represented a vital contract for the company. By the time of the merger GE owned all of the major firms that had supplied generators and motors for the West End system. In 1889 Thomson-Houston purchased the Bentley-Knight company, which had electrified the lines in the city, and Edison General Electric absorbed the Sprague company, the first electrical contractor for the West End railway. It thus made good business sense for GE to purchase Eickemeyer, who threatened to make inroads at West End with a new type of streetcar motor and drive unit.[10]

Buying out firms like Eickemeyer was central to the aggressive business strategy of Charles Coffin, president of Thomson-Houston. Since

1888 Coffin had acquired nearly a dozen arc-lighting and street-railway firms, which kept his company alongside Edison General Electric and Westinghouse as one of the leading electrical manufacturers in the United States.[11] Purchasing Eickemeyer was thus routine business, except for the arrangements with a third party, Otis Brothers & Company, which relied on Eickemeyer motors to run its elevators. Although full records of the transactions are not available, it appears that Eickemeyer's business was split into three pieces. Otis got the elevator motor side, GE gained the streetcar portion, and Eickemeyer retained the hat-making machinery business.

The alliances created to carry out this partitioning were rather tangled, but they were not uncommon during the merger movements of the 1890s. In August 1892 GE and Otis agreed to form the Otis Electric Company to manufacture Eickemeyer motors for Otis elevators. Otis received 51 percent and GE 49 percent of the stock in the new company, which was capitalized at $350,000. The two firms agreed to transfer all patents, goodwill, and future inventions relating to electric elevators to the Otis Electric Company. Otis agreed to do all of its electrical business with the new company, which promised not to compete with GE in the manufacture of electric lighting and railway equipment. In return GE agreed not to manufacture or supply any other company with electric elevator equipment, but it could make mining hoists under the new company's patents. The contracts show that the new company was to pay $15,000 to the Eickemeyer Dynamo Company for its electric elevator patents. The Otis Electric Company prospered under the agreement, employing about one hundred workers in 1896.[12]

The few surviving records of the streetcar deal suggest that Otis and GE made a similar arrangement to divide up this part of Eickemeyer's business. In October 1892 the Eickemeyer-Field Manufacturing Company was incorporated with a capitalization of $1 million as a consolidation of the Eickemeyer Dynamo Company, the Yonkers Machine Company, and the Eickemeyer-Field Company. The latter firm had been established in early 1891 with a similar capitalization to manufacture the Eickemeyer-Field streetcar system. It is noteworthy that Norton P. Otis and William D. Baldwin, president and general manager of Otis Brothers, respectively, were directors of the Yonkers Machine Company and the consolidated company.[13] No GE officials were listed in the public announcements of the Eickemeyer-Field Manufacturing Company, but it is likely that Eickemeyer sold his streetcar patents to this company, which was probably jointly owned by GE and Otis. GE probably held the controlling stock in this company, as Otis did in the Otis Electric Company. Evidence to support this view is that Steinmetz told a German journal in early 1893 that Eickemeyer-Field "consolidated" (*consolidirt*) with GE in 1892. GE purchased Eickemeyer-Field outright in 1895.[14]

The Eickemeyer & Osterheld Company continued to manufacture hat-making machinery and some electrical equipment for several years in Yonkers. Sixty-one years old at the time of the consolidation with GE, Eickemeyer went into semiretirement and took a long journey to the Southwest in early 1893. Later that year Steinmetz said in a letter that Eickemeyer sold out "for a very good price" because of his "valuable patents." When Eickemeyer died in January 1895 his son Carl, an engineering graduate from Cornell University, took over the family business, which was then limited to hat-making equipment, and ran the venerable firm of Eickemeyer & Osterheld for two years. (It was purchased by the Turner & Atherton Company in 1905.)[15]

The division of Eickemeyer's elevator and streetcar business between Otis and GE is reflected in the fortunes of the firm's leading engineers. John D. Ihlder, a native of Bremen, Germany, and a former sea captain and wholesale merchant, joined Eickemeyer after graduating in mechanical and electrical engineering from Cornell in 1887. He specialized in elevator motors and joined Otis Electric with the breakup of Eickemeyer's electrical business.[16] The chief assistant to Eickemeyer and Field in the streetcar line was Steinmetz, who went with General Electric. Steinmetz probably learned of the pending merger in late November 1892, when he dropped most of his regular engineering work and began transferring data to his personal notebooks. He completed research on the physical properties of insulators in December, designed a fan motor for GE later that month, and wrote up some work on the effect of frequency on transformers in January. On February 15, 1893, he closed his Eickemeyer notebooks and moved to a rooming house "kitty corner across the street from the [Lynn] factory" to begin his new career as a corporate engineer.[17]

Steinmetz left few records to indicate how he felt about moving from a small, family-owned firm to a corporate giant. He and Eickemeyer had become friends and were probably even closer after Steinmetz's father died in 1891. He visited the Old Man whenever he was in New York City and seems to have sympathized with his plight in the electrical industry. Steinmetz told Lux that when GE was formed it "was so tremendously powerful that competition [by Eickemeyer] was simply hopeless." He was "well satisfied" with his contract at GE, even though he gave up his patent rights: "These patents would be of no value to me as I could make no use of them. That is the way it is here now—only the two giant companies, General Electric and Westinghouse can make use of inventions . . . Naturally, it is only a matter of time as to when G.E. and Westinghouse will combine," forming a monopoly and driving small companies out of business. We will see that Steinmetz spoke much more favorably about large corporations after GE had treated him extremely well for over twenty years. In 1916, for example, he criticized the penny-pinching nature of a small manufacturer as compared with the liberal resources of a giant corporation, in a book that proposed the corporation as the model for a Socialist society.

GE was not Steinmetz's first choice for a new employer in 1893. He later told a company executive, "It was only accidental [that] I came with General Electric, and my own preference would have been very strongly in favor of Westinghouse, as in my work at Eickemeyer['s] I had a small Westinghouse machine [alternator] and Westinghouse transformer— which had been presented to us—a larger Westinghouse machine was loaned to us when we required more power, and I was thus familiar with and favorably impressed with Westinghouse apparatus."[18] Ironically, he spent his first years at GE designing generators and transformers to compete with Westinghouse equipment.

Hiring engineers like Steinmetz was another key part of Coffin's strategy. While head of Thomson-Houston he had transferred several chief engineers of purchased companies to his engineering staff in Lynn. In this manner Charles Van Depoele, the inventor of the overhead trolley system and head of the Van Depoele company, joined Thomson Houston in 1888, and Edward Bentley and Walter Knight of the Bentley-Knight streetcar firm joined the patent and engineering groups, respectively, in 1889. Coffin also hired the chief engineers of former rivals as outside consultants. Elmer Sperry, for example, consulted for Thomson-Houston after it bought his mining equipment and electric streetcar companies in 1892.[19]

Only a few of these inventor-entrepreneurs were working for GE when Steinmetz joined the firm in February 1893. Knight was chief engineer of the Railway Department at Lynn, and Sperry held the title of electrical engineer at the affiliated Brush Electric Company in Cleveland. The engineering staff at Lynn was then organized into the Railway, Stationary Motor, Mining, Transmission of Power, Alternating [Current], Calculating, Testing, and Student Departments. Louis Bell had resigned as editor of *Electrical World* to head the Transmission Department, Parshall was in charge of the Calculating Department, and Axel Ekström presided over the AC group. Ekström was part of a Swedish contingent of engineers at Lynn, which included Ernst Danielson, the primary designer of GE's induction motors, and Ernst Berg, who became Steinmetz's assistant and protégé. Like a growing number of engineers in industry, Berg and Danielson were college-educated, both having graduated in mechanical engineering from the Royal Polytechnicum in Stockholm. Other GE engineers with degrees in engineering or science at this time included Albert Rohrer, Henry Reist, Walter Moody, and William J. Foster at Thomson-Houston and William Le Roy Emmet at EGE.[20]

Steinmetz's qualifications made him a good candidate for at least three of the engineering departments at Lynn: Railway, AC, and Calculating. After a brief stint in the testing room he was placed under Parshall in the recently formed Calculating Department, which did many of the detailed computations for the designs worked out by the other groups. Parshall, who held a bachelor's degree in physics from Lehigh University, was a good choice to head this department. He had used magnetic-circuit theory, the cornerstone of the new "scientific" approach to electrical engineering, for several years while designing DC motors and generators for the Sprague company and then for Edison General Electric, where he was chief designing engineer.[21] Steinmetz's research on hysteresis and the theory of the transformer, in addition to his advanced mathematical training, made him well suited to be Parshall's assistant.

One of his first tasks was to help transfer the technology of Eickemeyer's streetcar motors to GE. Before he came to Lynn GE had agreed to manufacture twenty Eickemeyer motors for the Steinway road on Long Island at its plant in Schenectady to fulfill a contract negotiated by the Eickemeyer-Field Manufacturing Company the previous November. Rice went to Yonkers in mid-January to discuss the contract with Henry Osterheld and probably met Steinmetz for the first time on this trip. After Steinmetz joined GE Rice asked him to gather all the information he could on Eickemeyer's electrical equipment so that Parshall could incorporate its best features into GE's apparatus. Rice, Knight, and Thomson were impressed by the Eickemeyer brush (a current pickup device on the rotating armature of a DC dynamo or motor) and Eickemeyer's compensating-coil AC motor.[22] But GE did not adopt the Eickemeyer streetcar system, probably because of its radical departure from GE's previous designs. The company did use the Eickemeyer machine-wound armature winding in its new streetcar motor introduced in February 1893. The GE-800, a lightweight single-reduction-gear motor, proved very successful as the company's standard streetcar motor: over three thousand were sold the first year.[23] Because Edison General Electric had licensed the Eickemeyer winding to use in its streetcar motors since 1890, the main effects of buying out Eickemeyer were to gain control over the patents on this winding and to remove an up-and-coming competitor from the streetcar field.[24]

GE had more in mind for Steinmetz than gathering data on Eickemeyer's equipment. After eight months at Lynn Steinmetz told Lux, "I have nothing to do with their regular business, but have unlimited power to work at what I wish, to go and come as I like, and only if I discover anything I inform the Company and let them have the patent. In addition I am a consultant and give expert opinions." His salary was $3,000 per year, much less than the $5,000 received by Sperry. But it was more than twice his salary at Eickemeyer's, which in 1900 was about the same as that of a college physics teacher.[25] Although Steinmetz exaggerated the freedom he was given at GE, he did work some as an inside consultant for the company—a position he created formally much later. At Lynn he gave opinions on the same technical matters as two of GE's outside consultants: Arthur Kennelly, who advised the firm while working for Edison from 1892 to 1894, and William Stanley, who had resigned from Westinghouse to start his own company.[26] In some ways the Calculating Department resembled an internal consulting group. Steinmetz, Parshall, Berg, and a half-dozen other engineers calculated electric and magnetic circuit parameters for most of the other groups, much of which amounted to checking their designs.

This work did not always endear them to their colleagues. Ekström, the head of the AC Department, for example, complained to Rice a week before Steinmetz arrived about the Calculating Department's lack of expertise in AC engineering. (Rice probably placed Steinmetz under Parshall to bolster the group's knowledge in AC. Parshall was a master designer of DC machines, such as the large dynamos for the West End railway, but this skill apparently did not transfer well to the AC field.) Ekström also complained about errors made by the Calculating Department and thought that some of them had been made on purpose because he had criticized the group. Rice told Parshall not to do any more work for Ekström unless he or Ekström requested it, but Ekström was to give data on AC equipment to the Calculating Department. A month later Steinmetz was receiving information on transformers. The hostility between the two groups had not ended, but it probably did later as Steinmetz became more useful to the company.[27]

His most valuable work centered on the polyphase AC system. After Westinghouse dropped the expensive polyphase project with little to show for it, European engineers built practical three-phase motors by using a starting resistance and narrowing the air gap between the field and armature windings. In 1891 they displayed induction motors of 20 and 100 horsepower at an international electrical exhibition at Frankfurt, Germany, which was doubly impressive because the large motors operated at the end of the most spectacular transmission line to date: a 15,000-volt 175-kilometer three-phase line from Lauffen am Neckar to Frankfurt am Main.[28]

The Frankfurt-Lauffen system persuaded Westinghouse to renew its polyphase work and prompted Thomson-Houston to enter the field. Rice returned to Lynn from visiting the Frankfurt exhibition in late 1891, prepared to introduce a similar three-phase system into the United States. During the winter of 1891–1892 engineers at Thomson-Houston built their first three-phase motor. Henry Reist did the mechanical design, while Ernst Danielson handled the electrical part. By the time Danielson returned home to Sweden that spring to become chief engineer of Allmäna Svenska Elektriska, he had designed induction motors of up to 15 horsepower in consultation with Rice, Thomson, and Ekström. GE displayed these motors, which Steinmetz later praised for their high starting torque and low starting current, at the Columbian Exposition in the summer of 1893 and sold the first one to a New Hampshire machine shop that September.[29] In the same month GE turned power on a plant at Redlands, California, that is usually considered to be the first commercial three-phase line in the United States. The Stanley Electric Company's two-phase system at Pittsfield, Massachusetts, preceded the Redlands plant by six months, but GE had pulled ahead of Westinghouse, which had yet to install a commercial polyphase central station.[30]

Steinmetz's role in developing these systems is revealed in a notebook he kept as a member of the Calculating Department in 1893. He performed computations on transformers, motors, and generators for Ekström's AC Department and on power lines for Bell's Transmission Department. The notebook illustrates his virtuosity in using magnetic-circuit theory to calculate the design variables of new and redesigned AC machines and in using electrical theory to determine design parameters of proposed transmission lines. He also developed new machines on his own, progressing from coils in April to a fan motor in June to a complete line of polyphase motors in October.[31]

Two projects indicate how he combined theory with practice at Lynn. In May 1893 he investigated problems with the Farmington River Line, a 4,000-volt three-phase transmission line of the Hartford Electric Company that ran eleven miles from a nearby hydroelectric plant to Hartford. The utility allowed Thomson-Houston to experiment with the line in exchange for financing most of its improvements. Asked to analyze its high transmission losses, Steinmetz measured the line's inductance (a measure of the magnetic field produced per unit of current) and capacitance (a measure of the electric charge produced per unit of voltage), compared these with his calculated values, and computed the line's impedance (AC resistance, which was a function of frequency). He probably determined that the line's high frequency (125 Hertz, Thomson-Houston's single-phase standard) caused the abnormal energy losses, because later notebook entries and other evidence show that GE soon switched to frequencies half this value (those in common use today).[32]

This work led to Steinmetz's first patents at GE. In September 1893 he applied for three patents on ways to reduce the inductance of transmission lines: using capacitors, induction motors run above synchronism, and synchronous motors. In the last scheme, which he had considered as early as April, Steinmetz realized that a synchronous motor with an "overexcited field" connected to the end of the line would act as a capacitor and compensate for the inductance electrically. According to Silvanus P. Thompson, who discovered the principle independently, GE tried out the invention on the Farmington River system. Thompson reported that "later, after Mr. Steinmetz had submitted the question to calculation, this feature was introduced into the synchronous motors used in all their recent power transmissions." The synchronous condenser, as the device was called, became a standard component on GE's polyphase systems, even after GE reduced the line frequency, because it compensated for the considerable inductance of the increasing number of polyphase motors connected to the company's lines.[33]

Steinmetz invented the synchronous condenser by using applied mathematics, as he had done at Eickemeyer's. This approach was much different from that of Elihu Thomson, the company's chief inventor before the merger. W. Bernard Carlson has shown that Thomson invented electrical machines in a highly nonverbal manner, changing the form and dimensions of prototypes in a "Model Room" attached to the factory.

Edison made most of his inventions in a similar fashion, as revealed by countless drawings of devices having only slight variations. Thomson thus shared Edison's machine-shop culture, but he also cultivated a school culture at Lynn similar to that at Eickemeyer's. He was known as "Professor" from his days of teaching at Philadelphia's Central High School and published theoretical articles in the technical journals. He became more involved in scientific pursuits after 1890 and struck up a friendship with the scientist-engineer Steinmetz.[34]

Local legend at the Hartford Electric Light Company has it that when Steinmetz came out to the hydroelectric plant to make measurements on the Farmington River line, he "ran an uninsulated iron wire into the water of the tail race from the water turbine. Steinmetz carefully picked up the wire and received such a shock as to nearly . . . 'terminate his valuable career.'" Helped to his feet, he ran to his carriage—the juxtaposition of the horse and buggy with the high-voltage line is striking—and had an inordinate fear of high voltages thereafter. Although the story has probably been embellished over the years, co-workers have reported that Steinmetz, later hailed as Modern Jove for creating artificial lightning, wore rubbers over his shoes to protect himself from electric shocks in the laboratory.[35]

An early GE report called the Hartford line the "first successful long distance multiphase transmission in the United States." GE later reassigned this honor to the plant at Redlands, most likely because of the technical problems at Hartford.[36] The Redlands installation was more successful and provides a second view of Steinmetz's work at Lynn. GE received the Redlands contract in the spring of 1893 and built two 250-kilowatt generators to transmit 50-Hertz three-phase power at 2,500 volts from a turbine at Mill Creek to the city of Redlands, a distance of seven and one-half miles. While one of the generators was being tested at Lynn, engineers found that energy losses in its core amounted to 38 kilowatts, over 15 percent of its rated capacity. Steinmetz investigated the problem and used his law of hysteresis to calculate the energy lost by hysteresis to be less than 6 kilowatts. He concluded that the remainder of the "core loss" was due to eddy currents and stated that the "core has to be pulled apart, the Japan [insulation] replaced by paper." This apparently reduced the energy loss by one-half. He later published this data to show the applicability of his law of hysteresis—a further illustration of the symbiotic relationship between theory and practice in his career.[37]

Electricity from the Square Root of Minus One

Steinmetz proved his value to the company as a designer of polyphase systems in his first months at Lynn. During that time he also developed a mathematical technique that stands as his most lasting contribution to electrical engineering: the method of analyzing AC circuits with complex numbers. (Complex numbers consist of a "real" part plus an "imaginary" component expressed by the square root of minus one.) In an obituary Vladimir Karapetoff, professor of electrical engineering at Cornell University, characterized Steinmetz's thirty-year career at GE with the remark that he "was allowed by his employers to try to generate electricity from the square root of minus one."[38]

The complex-number method falls into the category of engineering knowledge called "mathematical techniques" in Chapter 1. Unlike the empirically based law of hysteresis, it was applicable to the entire field of AC engineering, not just to a limited number of materials and magnetizations. Steinmetz's method had an enormous influence on another portion of electrical engineering knowledge—theories of devices—because it changed the way engineers calculated AC circuits and machines. This influence later extended well beyond the discipline of power engineering to that of telecommunications, electronics, and lasers, because AC circuit analysis became the backbone of these fields as well. Before Steinmetz, engineers had used differential equations or graphics to calculate AC circuits. His method, which is still used extensively, improved and simplified this work by making graphical calculations more precise and by eliminating esoteric differential equations.

Although the engineering community called the technique "Steinmetz's method," he did not invent it. Between 1879 and 1893, the year Steinmetz introduced his method, several engineers and physicists had already used complex numbers to analyze electrical instruments, telephone transmission lines, and power circuits. The method became associated with Steinmetz because he developed it more fully than his predecessors and spent the better part of his career applying it to the entire range of AC circuits and machines. Another factor was that engineers were largely unaware of prior work in this field. The application of complex algebra to AC circuits occurred along two independent lines. The first was based on exponential functions, the second on engineering graphical analysis.

British and European physicists invented the first approach in the late 1870s and early 1880s while searching for a better method to analyze AC circuits in a scientific instrument, the induction bridge. Invented in 1863 by Maxwell and Fleeming Jenkin, a British telegrapher, the induction bridge measured inductance in much the same manner as the DC Wheatstone bridge measured resistance. In 1879 Johann Victor Wietlisbach, a physics graduate student under Helmholtz, improved Maxwell's bridge by substituting the newly invented telephone for the ballistic galvanometer as the null detector and alternating current for the battery and current interrupter as the power source. He placed capacitance as well as inductance in the legs of the bridge. Faced with the task of solving four simultaneous differential equations for the improved bridge, Wietlisbach turned to complex numbers to simplify matters. His equations were nearly identical to those derived by Steinmetz fifteen years later. Wietlisbach did not use the term *impedance,* which was coined in 1886, but he expressed AC resistance as a complex number.[39]

Wietlisbach obtained his results by employing a technique well known in physics for solving differential equations involving the trigonometric functions used to represent AC. Since the 1870s physicists in many fields had simplified the solution of such equations by using Euler's identity to express periodic functions by an imaginary exponential term. The British physicist Lord Rayleigh (J. W. Strutt) employed the technique extensively in acoustics, optics, hydrodynamics, and the theory of mechanical vibrations. One of the earliest physicists to apply it to electric circuits was Helmholtz, who taught the method to his student Wietlisbach. Wietlisbach derived his equations by substituting the imaginary exponential expressions for voltage and current into Maxwell's differential equation for the voltage across an inductor and capacitor in a leg of the bridge. He then solved for voltage in terms of current multiplied by a complex number (impedance).[40]

Wietlisbach's equations greatly simplified the mathematics of AC circuit theory. Instead of solving Maxwell's differential equations, physicists now worked with algebraic formulae expressed in complex numbers. They could use these equations to analyze AC circuits in the same manner as they had previously solved DC circuit problems, since the equations provided an Ohm's law for AC, with impedance substituted for resistance. The equations provided a simple answer to the vexing problem noted by George Prescott in 1888 when he said (erroneously) that AC did not follow Ohm's law. The German physicist A. Oberbeck improved the method in his analysis of Wietlisbach's bridge in 1882 by writing Kirchhoff's laws in complex-number form and pointing out the relationship between the complex-number expressions for voltage and current and their phases.[41] The method had gained some currency in Britain by 1891 through the writings of Lord Rayleigh, who used it to analyze AC circuits, the skin effect (the crowding of current at the surface of a conductor), and the impedance bridge (an improved version of Wietlisbach's bridge invented in 1891 by the German physicist Max Wien).[42]

At the same time Oliver Heaviside was taking a different approach. An eccentric British mathematician and former telegrapher, Heaviside coined the term *impedance* in 1886, as well as many others in electricity and magnetism. He is perhaps best known for developing vector analysis independently of the American scientist J. Willard Gibbs and for rewriting Maxwell's field equations in their present form.[43] In the late 1880s Heaviside developed an "operational calculus" for solving differential equations, in which the mathematical operations of differentiation and integration were represented by algebraic symbols. This approach enabled him, in 1887, to express Ohm's law for AC in much the same way as Wietlisbach had done eight years before. Heaviside, however, expressed impedance as a *resistance operator,* the mathematical "operator that turns the current into the voltage." He also defined a *conductance operator* as the reciprocal of the resistance operator. He defined *impedance* as the ratio of the amplitudes of voltage to current and named the reciprocal ratio *admittance.*[44]

We can easily derive complex-number expressions for Heaviside's resistance and conductance operators that are equivalent to the expressions for impedance and admittance in the work of Wietlisbach and others.[45] Yet Heaviside's early papers reveal only one case in which he explicitly wrote a resistance or conductance operator as a complex number—in January 1887, for the terminal impedance of a telegraph circuit excited by an AC source. At other times he simply noted that the differential operator was a complex number, but he did not write his equations in complex numbers. He followed this pattern as late as November 1893, when he derived the resistance and conductance operators of an AC circuit consisting of a coil and capacitor connected in series or parallel.[46]

Although Heaviside applied this mathematics to long-distance telegraphy and telephony, he was primarily a mathematical physicist. He was more interested in how these systems demonstrated the propagation of electromagnetic waves along wires and left the detailed, practical application of his work to his readers. In the case of resistance operators he remarked, "Resistance-operators combine in the same way as if they represented mere resistances. It is this fact that makes them of so much importance, especially to practical men, by whom they will be much employed in the future. I do not refer to practical men in the very limited sense of anti- or extra-theoretical, but to theoretical men who desire to make theory practically workable."[47] These prophetic remarks describe Steinmetz's work in AC theory after 1893 remarkably well, except that he used complex-number impedance and admittance instead of resistance and conductance operators.

Several investigators published complex-number papers between Heaviside's work in 1887 and Steinmetz's in 1893. One of Heaviside's chief followers was Max Wien, who employed resistance operators to analyze the impedance bridge in 1891. While Wien preferred to use operators, two German physicist-engineers applied the exponential complex-number method to telephony. In 1887 Wietlisbach, now chief engineer of the Swiss Telegraphy Administration in Bern, used the approach he had taken earlier with the induction bridge to derive a complex-number expression for telephone line impedance in terms of the line's four parameters: resistance, inductance, capacitance, and leakage conductance. Four years later Adolf Franke, a Ph.D. physicist who worked at the Telegraph Engineers Department at the German Post Office in Berlin, used Wietlisbach's method in a more extensive analysis of transmitting signals over telephone lines.[48]

Physicists and scientist-engineers from Wietlisbach to Franke had thus developed a complex-number method for calculating AC circuits several years before Steinmetz's 1893 paper. Most of their articles appeared in scientific journals, such as the British *Philosophical Magazine* and the German *Annalen der Physik und Chemie*, but leading technical journals republished many of them. The London *Electrician* reprinted Rayleigh's papers in 1887 and 1891. The *Elektrotechnische Zeitschrift* published an abstract of Wien's paper on the impedance bridge in 1892. *La lumière électrique* carried a lengthy account of Wien's work in the same year. Wietlisbach's and Franke's papers appeared in German technical journals, and the Chicago *Western Electrician* published a translation of one of Wietlisbach's articles in 1887. Much of Heaviside's work, including that using resistance operators, appeared originally in the *Electrician*, which also published his collected papers in 1892.[49]

Engineers reacted to these papers with mixed reviews. The *Electrician* praised Rayleigh's analysis of the induction bridge yet thought that "anyone unaccustomed to live in a world of [mathematical] symbols would carefully avoid such an investigation, but Lord Rayleigh, no doubt, found it easy enough."[50] The *Electrician*'s attitude toward higher mathematics depended largely on who sat in the editor's chair. Charles H. W. Biggs backed Heaviside, but his successor, W. H. Snell, terminated Heaviside's regular series of articles in November 1887, apparently because Snell had received no replies from a questionnaire asking subscribers if they were reading Heaviside's articles. The French engineers, heirs to the mathematical tradition of Laplace and Lagrange at the Ecole Polytechnique, were more appreciative of higher mathematics. Camille Raveau, who wrote the article on Wien for *La lumière électrique* in 1892, recognized the significance of complex numbers for AC engineering. Raveau remarked that the method was similar in many respects to AC graphical analysis and encouraged electrical engineers to adopt complex numbers in their work.[51]

One year later, in August 1893, Steinmetz presented his famous paper on the use of complex numbers in electrical engineering before the International Electrical Congress in Chicago. The paper went beyond all previous work, which had focused on instruments and telecommunications, to show how complex algebra could be applied to AC circuits, transformers, and polyphase transmission lines.[52]

Although the paper followed on the heels of his British and European predecessors, Steinmetz was unaware of this tradition in telecommunications. He was influenced instead by Arthur Kennelly, who had helped design an experimental AC system at Edison's laboratory in the early 1890s. When Edison did not introduce the system because of the pending merger with Thomson-Houston, Kennelly continued to work on AC equipment after the merger. As a consultant for GE in 1893, he calculated the inductance of the Farmington River line, in order to determine why it was not working properly. Steinmetz did an independent calculation that May and also computed the impedance of the line.[53] A month earlier Kennelly had read a lengthy paper on impedance before the AIEE. Embedded in that paper, which was directed toward the practical engineer, was the statement that engineers could treat complex-number impedances in AC circuits like resistances in DC circuits, a result Wietlisbach had obtained fourteen years before.[54]

But Kennelly did not derive this result by expressing voltage and current as an imaginary exponential function, as Wietlisbach and others had done. Instead, he worked along the second line of development noted above—engineering graphical analysis. Kennelly defined impedance graphically, using the impedance triangle that John Ambrose Fleming, a British professor of electrical engineering, had published in

1889.[55] Kennelly knew that a complex number could be expressed graphically by a point in a plane defined by horizontal and vertical axes. How far the point was from the origin (the intersection of the axes) along the horizontal axis represented the "real" part of the complex number, while the distance up or down the vertical axis represented the "imaginary" part of the number, a geometric convention dating to the eighteenth century. He then recognized that a line drawn from the origin to the point could be considered as the hypotenuse of the impedance triangle. He could thus express impedance as a complex number.

He did not, however, publish a complete method. He did not express voltage and current by complex numbers, nor did he give complex-number equations for Ohm's and Kirchhoff's laws, as had his predecessors in Britain and Europe. In fact, Kennelly stated that "Kirchhoff's laws, while true instantaneously, are only applicable to alternating currents, when the currents are all in step,"[56] an error he could have corrected by writing voltage and current in complex numbers. Kennelly's approach suggests that he had developed his method independently of British and European physicists. As a former British telegrapher self-taught in mathematics and physics, he knew about Heaviside's work and cited his paper on the skin effect in his AIEE article on impedance.[57] Kennelly later acknowledged Heaviside's priority in expressing complex-number impedance algebraically, but he "doubted if [Heaviside] did so geometrically or vectorially." He also recalled that he "did not consciously get the idea [of complex-number impedance] from Heaviside." Instead, Kennelly "worked the matter out for [himself] in the winter of 1892–1893."[58] He probably did it by applying the well-known geometric representation of complex quantities to the impedance triangle, as described above.

As in the case of previous complex-number papers, most electrical engineers did not understand the significance of Kennelly's work. The paragraph-long statement explaining the new method lay hidden in a long, practical paper on a subject, impedance, that many of Kennelly's colleagues did not fully comprehend. Frank Sprague said that he was "prevented by an impedance in my own head from taking any part in the [AIEE] discussion." Nevertheless, Sprague thanked Kennelly for writing clearly "about a subject that is so little understood."[59] Perhaps because of that ignorance, and because those AIEE members like Michael Pupin who understood impedance discussed the AC problems Kennelly treated rather than his mathematical methods, neither the participants in the discussion nor the American electrical journals mentioned Kennelly's reference to complex quantities.[60]

Steinmetz, however, quickly recognized the significance of Kennelly's remarks. Although not at the AIEE meeting, he perceived the kernel of a new approach to AC calculations in Kennelly's paper and wrote a five-page discussion of it for the AIEE *Transactions*. He wanted to "emphasize somewhat more strongly" Kennelly's statement on complex-number impedance, because it was "enclosed between so many remarks of the highest practical value as to be liable to escape notice." Steinmetz thought that the "analysis of the complex plane is very well worked out, hence by reducing the electrical problems to the analysis of complex quantities they are brought within the scope of a known and well understood science."[61]

He did not describe a complete complex-number method at this time. He explained how to express impedances as complex numbers and how to use their absolute values in Ohm's law better than Kennelly did. But he did not establish Ohm's law for AC because he, like Kennelly, did not express voltage and current in terms of complex numbers. That awaited the much longer paper he gave at the International Electrical Congress four months later. Then he presented a method similar to that which had been used by physicists in Britain and Europe since the early 1880s.

What was his debt to these predecessors? In his discussion of Kennelly's paper Steinmetz said, "It is, however, the first instance here, so far as I know, that attention is drawn by Mr. Kennelly to the correspondence between the electrical term 'impedance' and the complex numbers."[62] As we have seen, Kennelly was not the first person to do this. Physicists and engineers from Wietlisbach to Raveau had expressed impedance as a complex number in several papers published between 1879 and 1892.

Steinmetz had ample opportunity to read this work. We know that he read an article in *La lumière électrique* in the same year in which Raveau's paper appeared, but he was much more familiar with the German technical press. Franke published his papers in the *Elektrotechnische Zeitschrift* in 1891 and 1892, when Steinmetz contributed frequently to that journal. In those two years the *ETZ* published nineteen papers or abstracts of papers and three letters to the editor by Steinmetz. In one of the letters he responded to an article by C. Grawinkel that was only one page away from Franke's telephony paper in which impedance was expressed in complex numbers. Steinmetz apparently did not read Franke's articles because he specialized in power engineering. In fact, Grawinkel was criticizing a paper by Steinmetz on telecommunications and power line interference, the only one he wrote during this period on *Schwachstromtechnik*, precisely because Steinmetz was unfamiliar with German telephone practice.[63]

Although not well versed in telephony, Steinmetz shared an interest with his predecessors in the problem of measuring inductance. While he invented new ways to measure the inductance of AC instruments in connection with his hysteresis research, physicists analyzed the induction bridge with the help of complex numbers. Their work even appeared in the same journal. The *Electrician,* for example, published Rayleigh's complex-number paper on the induction bridge in May 1891, four months after it printed a letter to the editor by Steinmetz on determining the inductance of AC instruments. Steinmetz may have read Rayleigh's paper, but the previous October he had criticized the ballistic galvanometer, used in the induction bridge, as being too inaccurate for AC measurements, as we have seen. These remarks reveal that he was unfamiliar with the papers on the induction and impedance bridges by Wietlisbach, Oberbeck, and Wien, who replaced the galvanometer with a telephone receiver for AC measurements.[64]

The only engineering work containing complex quantities that Steinmetz unquestionably read before he saw Kennelly's paper was an AIEE article by Frederick Bedell and Albert Crehore. Instructors in physics at Cornell University, Bedell and Crehore did not use the imaginary exponential function to represent voltage and current, as had Wietlisbach and others. Instead, they used it to obtain a real solution for the discharge of an AC circuit when component values made the solution "imaginary."[65] Steinmetz undoubtedly read their paper, published in June 1892, because he later used the symbol j for the square root of minus one, instead of the symbol i, which was common in mathematics and physics. Bedell and Crehore introduced the symbol j in order to avoid confusion with i, which in electrical engineering normally stood for current. (In general, electrical engineers still use j for the square root of minus one, while mathematicians and physicists continue to use i.)

If Steinmetz had read electrical papers containing complex numbers other than Bedell's and Crehore's before April 1893, it was a well-kept secret. He did not refer to his predecessors in his published papers, not even to Kennelly in the paper he read before the International Electrical Congress (IEC) that August—a practice that became a habit with Steinmetz.[66] Since we lack archival material on Steinmetz's derivation of the complex-quantity method, we must turn to the internal evidence of his published work. This indicates that he derived his complex-number method in a way very different from his British and European predecessors—an indication that it was done independently. Unlike physicists from Wietlisbach to Wien, Steinmetz did not substitute the imaginary exponential expression for voltage and current into the appropriate differential equations. Instead, he noted the similarity between AC graphical analysis and the geometric representation of complex-number impedance devised by Kennelly. Steinmetz combined those two approaches to create a complete complex-number method for solving AC circuit problems.

This derivation is apparent in his IEC paper. He introduced his new method by first describing the graphical technique he devised between 1890 and 1891, in which AC voltages and currents, depicted trigonometrically by sine waves, were represented by a rotating vector. He then made a smooth transition to the complex-number method, stating that "to distinguish the horizontal and the vertical components of sine-waves, so as not to mix them up in a calculation of any greater length, we may mark the ones, for instance, the vertical components, by a distinguishing index, as for instance, by the addition of the letter j."[67]

Using the geometric representation of complex numbers, Steinmetz then defined j as the square root of minus one. He concluded that a sine wave could be "represented in intensity [amplitude] as well as phase by one complex quantity: $a + jb$, where a is the horizontal, b the vertical component of the wave." In this manner he reduced the addition or subtraction of sine waves to the "elementary algebra of complex quantities."[68] He then established Ohm's law for AC circuits and derived results equivalent to that of his predecessors, except that he obtained negative rather than positive signs for the imaginary part of the impedances.

The signs were reversed because of the way in which Steinmetz derived complex impedance. He did not substitute the imaginary exponential function for voltage and current into the appropriate differential equations, as his predecessors had. Instead, he expressed voltages and currents as complex numbers in rectangular coordinates and then used the 90-degree relationship between voltage and current of an inductor, shown in the impedance triangle, to derive the complex form of impedance. Since he did give the imaginary exponential function as an alternative expression for the sine wave, he may have privately substituted it into the differential equations to obtain complex-number impedance. But that is unlikely, because he derived a minus sign where everyone before him had obtained a positive sign. The difference indicates that he derived his complex-number method from his rotating vectors, which produced negative signs for the reactances because of the way in which he defined the direction of rotation of the vectors.[69]

Steinmetz had good reasons for not presenting his complex-number method in the language of differential equations. He designed his method, like his graphical technique, for practicing engineers who in the 1890s were not that familiar with differential equations. As we saw earlier, he made that criterion quite clear in his papers on rotating vectors. He held a similar philosophy in his IEC paper, stating that "in introducing j, first as a distinguishing index and then defining it as $\sqrt{-1}$, my object was to introduce the complex quantity in an elementary and graphical manner, without reference to higher mathematics."[70] A reporter covering the IEC for *Electrical World* wrote that Steinmetz said he introduced complex numbers in this manner "so that electrical engineers who had gone to college and forgotten much when they came out could handle the expressions without the stumbling block of complex quantities."[71] Even though Steinmetz presented his method in simple mathematical terms, he applied it to a host of topics. He treated resistance-inductance-capacitance circuits, the transformer, single-phase transmission lines, and polyphase transmission systems with his new method. The paper became the basis for his influential *Theory and Calculation of Alternating Current Phenomena*, published in 1897. The main focus of the paper was polyphase systems rather than theories of AC machines, which dominated the book.[72]

He concentrated on polyphase networks because GE was then bidding for the lucrative contract to harness the power of Niagara Falls. In January 1893 Rice asked Bell and Ekström to put together a complete demonstration unit of generators, motors, transformers, and so forth for the International Niagara Commission. Chaired by William Thomson (later Lord Kelvin), the commission had the task of deciding which power system to install at the falls. Parshall got into the act when he asked Steinmetz to work on the transformer calculations in early March. Later that month GE and Westinghouse submitted proposals to the commission for polyphase systems, which differed mainly in regard to the number of phases. GE proposed three phases, while Westinghouse favored the two-phase arrangement pioneered by Tesla. The competition heated up in May, when Westinghouse charged GE with industrial espionage and had its alleged spies arrested after Westinghouse blueprints were found at the GE factory at Lynn. That summer both companies displayed their systems at the Columbian Exposition in Chicago. On August 10 the Niagara Commission rejected the proposals and invited bids from Westinghouse and GE on a two-phase generator designed by the British engineer George Forbes, chief adviser for the commission. At the IEC meeting, held later that month at the exposition, Forbes criticized GE's three-phase proposal for its complexity. (The commission had not yet decided on two or three phases for the transmission line part of the contract.) Forbes thought that it would be difficult to keep GE's system balanced when unequal lighting loads were applied and to locate faults in the network because the phases were not electrically isolated, as were Westinghouse's.[73]

Steinmetz anticipated these criticisms in his complex-number paper, presented the day before. He concluded that "we can not fail to see the regularity displayed in the variation of potential [between the three phases], which makes it possible to control this phenomenon."[74] Louis Bell, head of GE's transmission line department, referred to Steinmetz's paper in his reply to Forbes at the meeting as evidence that the unbalanced conditions of a three-phase system could be easily calculated and, thus, avoided with an appropriate design. The Niagara Commission awarded Westinghouse the contract to build Forbes's two-phase generators shortly after the exposition. It later selected GE's three-phase system for the long-distance transmission line, probably because the company had so much experience with this type of network. Engineers turned power on the twenty-six-mile line from Niagara Falls to Buffalo, New York, in late 1896.[75]

There is some indication that, after reading his paper (dated July 1893) to the IEC, Steinmetz read the work of Heaviside. In an AIEE article published in May 1894 Steinmetz used the term *admittance* for the first time and gave it the symbol Y, the same symbol Heaviside had used for the conductance operator in December 1887, when he coined the term *admittance*. (Admittance was the magnitude of the conductance operator.) The IEC had adopted the term *impedance* in 1889, but *admittance* was not well known in 1894. Heaviside himself predicted that it would not "meet with so favourable a reception as impedance" because it was less useful in practice.[76] Although Steinmetz possibly coined the term *admittance* as the opposite of *impedance* independently of Heaviside, it seems too coincidental that he also used Heaviside's symbol Y, which has no obvious relationship to admittance. (Heaviside probably chose Y because it immediately preceded Z, the symbol for the resistance operator, in the alphabet.) Thus, it appears likely that Steinmetz read Heaviside or one of his followers before May 1894, despite his claim in 1900 that the "term 'admittance' is nothing new, but was introduced by me years ago." He did "introduce" *admittance* into electrical engineering through his numerous papers and books—as he did the related term *susceptance*, which he coined in the May 1894 paper. But Heaviside, not Steinmetz, coined the term *admittance*.[77]

Steinmetz may have read an 1891 paper by Fleming on the Ferranti effect, a rise of voltage at the end of an AC transmission line where one would expect a lower voltage. First noticed by Sebastian Ziani de Ferranti, a pioneer in high-voltage engineering in Britain, as a rise from 8,500 to 10,000 volts in a five-mile-long cable, the puzzling effect was a hot topic of debate in the technical journals. Several engineers from the old school of DC even doubted its existence. (The effect is due to a resonance condition between the inductance of a line's transformers and the capacitance of the line.) Fleming, who had corresponded with Heaviside, cited the latter's 1887 paper that contained a complex-number form of impedance, even though Fleming did not employ complex numbers in his own analysis. More important, Fleming defined *admittance* and the conductance operator (Y) in a footnote, attributed the terms to Heaviside, and referred readers to Heaviside's article on resistance operators for more information. Steinmetz had good reason to read Fleming's paper. Kennelly mentioned it during the discussion of his AIEE paper on impedance, where he argued for the existence of the Ferranti effect. Steinmetz then analyzed the Ferranti effect in a contemporary article and in his letter in response to Kennelly's paper—Steinmetz's first circuit analysis using complex quantities.[78]

A better case can be made for Steinmetz's having read Heaviside directly. Ernst Berg recalled that Heaviside once asked him if he remembered the *Electrician*'s 1887 survey of Heaviside's readers. Berg did not and said, "Pity! I know of at least two persons who would have replied favorably had the slips [inserted in the journal] found their way to Lynn, Massachusetts."[79] The two persons were undoubtedly Berg and his mentor Steinmetz. If Berg's memory is correct, Steinmetz was reading Heaviside's articles in the *Electrician* in 1893. (The *Electrician* began publishing Heaviside again in 1891.) An article published in November 1893 included resistance and conductance operators, denoted by Z and Y, and expressed the differential operator as a complex number for AC. It did not mention the term *admittance*, but referred readers to an article in Heaviside's *Electrical Papers* (1892), where he coined the term. It is noteworthy that Steinmetz's copy of this two-volume set is stamped "Factory Library, General Electric Co., May 31, 1893," that is, after Kennelly's AIEE paper but before Steinmetz's IEC article. The only annotation in the permanently "borrowed" set is a signed statement in volume 1 that corrects Heaviside's analysis of commutation in DC dynamos.[80] A plausible reconstruction of the events is that Steinmetz read the section on resistance operators in volume 2 after seeing Heaviside's article in November 1893 and then used the term *admittance* in his AIEE paper the following May.

However Steinmetz learned about admittance, he did not adopt Heaviside's method of solving AC problems. We will see that Berg went back to Heaviside's work in the 1920s in order to simplify Steinmetz's analysis of electrical surges. Berg adopted Heaviside's operational calculus and converted the resistance operators into complex-number impedance for AC, just as Wien and others had done in the early 1890s. After his mentor died Berg stated, "It has always been a regret to me that Steinmetz, with his prestige, did not use Heaviside's operational calculus [in his work on surges]. It would have been the greatest help to the art."[81] Steinmetz probably thought that operational calculus was too complicated for steady-state analysis, the focus of AC circuit theory in the 1890s. His graphical method and the complex-number technique that grew out of it were much better suited for this work, given the mathematical training of electrical engineers at the time.

Steinmetz's IEC paper met a somewhat better fate in the electrical engineering community than had the work of his predecessors. While the American and British response was limited to brief accounts in the electrical journals,[82] German and French engineers received the paper more enthusiastically. The *Elektrotechnische Zeitschrift* printed a translation of the entire paper in October 1893, only two months after Steinmetz had read it in Chicago. This was the first publication of the paper; the AIEE did not publish the IEC proceedings until late 1894. *La lumière électrique* printed a lengthy account of Steinmetz's paper in December 1893. The author, C. F. Guilbert, highly recommended the complex-number method and was surprised that no one had suggested it before Steinmetz. Two weeks later Guilbert, having conducted a literature search, published an account of the prior work of Wien, Raveau, and others. The fact that Guilbert was unaware of this work—especially that of Raveau, which appeared in the very journal in which he himself was writing—indicates the obscurity of the complex-number method in engineering circles before Steinmetz's IEC paper.[83]

Not all readers viewed it as favorably as Guilbert. In a coincidence reminiscent of Charles Wheatstone and C. W. Siemens reading papers on the dynamo principle at the same meeting in 1867, Alexander Macfarlane, professor of physics at the University of Texas, presented a paper on complex numbers in electrical engineering at the same IEC session at which Steinmetz read his paper. An internationally known mathematical physicist, Macfarlane had also seen in Kennelly's AIEE paper the key to an improved type of AC analysis. But Macfarlane developed a complicated method based on quaternions, an early form of vector analysis used by Maxwell. During the discussion of Steinmetz's paper Macfarlane said that Steinmetz's graphical definition of complex numbers was "ambiguous" and implied that he should use more rigorous mathematics. Steinmetz replied that his definition avoided the use of higher mathematics—an accurate criticism that Macfarlane's quaternions were too complicated for engineering work.[84]

The case of complex algebra further illustrates the difference between electrical engineering and electrophysics at this time. Engineers and physicists derived similar mathematical techniques to analyze different AC circuits: power lines for the engineers and the impedance bridge for the physicists. They also derived their methods from different intellectual traditions: engineering graphical analysis versus differential equations. The former was common to branches of engineering knowledge like the strength of materials and statics in civil engineering, which Edwin Layton has called "engineering sciences."[85] The AC circuit techniques developed by physicists and engineering scientists looked similar, but practicing engineers at the turn of the century were better able to understand and use Steinmetz's method.

Steinmetz won this victory by popularizing his method extensively. In paper after paper and in several books he showed electrical engineers how to apply complex numbers to all types of AC circuits and machines. The extent of that promotion is revealed in the AIEE *Transactions*, the foremost American research journal in the field. Between 1893, when Kennelly gave his paper, and 1910, when the method was standardized by the IEC, Steinmetz authored thirteen of the thirty complex-quantity papers published by the AIEE. Steinmetz's two assistants, Berg and Maurice Milch, wrote the other six GE papers, but other prominent GE engineers, such as Ernst Alexanderson, William Le Roy Emmet, and Henry Reist, used complex numbers in their work—a measure of the strength of the Steinmetz "school" at GE. His nearest rival in promoting the new method in the AIEE was Kennelly, who wrote three papers.

Kennelly left Edison to become a private consultant in 1894 and joined the electrical engineering faculty at Harvard in 1902.[86] Six other academics wrote complex-quantity papers, but only two papers came from companies other than GE. Westinghouse engineers did not employ the method, probably because chief engineer Benjamin Lamme preferred the graphical approach. Although Steinmetz did not invent the complex-number technique, the extent of his popularization of it in these formative years makes it clear why his contemporaries knew it as "Steinmetz's method."

5 Theory and Practice

$\overline{\underline{\overline{}}}$ **E**ighteen ninety-three was a pivotal year for Steinmetz. He left Eickemeyer's for a lifelong corporate home in February and invented the complex-number method in July, which set the course of his research for over a decade. General Electric engineers learned the new method firsthand, while "electricals" throughout the world picked it up by reading his numerous articles and books on theories of alternating-current machines. His rather abstruse theories were valued highly because they worked. The reason, of course, was that Steinmetz had been initiated into the alchemy of the symbiotic relationship between theory and practice at Eickemeyer's. GE gave him much better resources to carry out this research program and to become one of the leading electrical engineers of his generation.

Schenectady

The corporation did have its drawbacks. Gigantic firms produced goods more efficiently than smaller companies, but they were still vulnerable to the financial gales that periodically struck American business during the boom-and-bust cycles of the nineteenth century—a fact Steinmetz learned during the Panic of 1893 when that crisis almost destroyed GE. The collapse of stock prices in the fall of 1892 and a shrinking money supply forced the company to the brink of receivership the following summer. Sales dropped to one-half of what GE's parent firms had sold in 1891, five thousand of its eight thousand employees were laid off, and many were reduced to part-time work. Observers placed much of the blame on president Charles Coffin's policy of accepting poor-quality stock in utility and street-railway companies—GE's main customers—as payment for their purchases of electrical equipment. Faced with a severe cash shortage in July, $2 million cash on hand to pay $10 million of immediate debts, Coffin asked the investment banker J. P. Morgan, who had not only set up GE a year earlier but had rescued the U.S. government in a similar cash-flow crisis in 1877, to bail him out. Morgan obliged with an infusion of $12 million, saving GE from bankruptcy in exchange for the best stocks at one-third of their market value.[1]

John Broderick, a clerk in Edwin Rice's office, recalled that Rice was asked to draw up a list of his salaried employees, indicating the conditions under which they could be dismissed or how much their salary could be reduced. Broderick said that conditions were listed for every employee except Steinmetz, whose entry read, "Charles P. Steinmetz; electrical engineer; no conditions; services indispensable; increased compensation warranted." Broderick, who became a close friend of Steinmetz, said this was a "vivid" recollection, but Rice no doubt felt the same way about some of his other engineers. Steinmetz's prospects were not dimmed by the fact that he had signed a two-year contract when he joined the company in February 1893, shortly before the Panic began to take its toll.[2]

After surviving the crisis Coffin became a "born-again financial conservative," in the words of historian George Wise. For the next few years Morgan turned the management of the firm over to two of his representatives: Hamilton Twombley, chairman of the board of directors of GE and a member of Morgan's banking firm, and Joseph P. Ord, the company's comptroller who soon became its second vice president. Given the task of streamlining the firm, Twombley and Ord moved much of its management and manufacturing to Schenectady, cut back the independence of district sales offices, whose policies had helped create the crisis, and established more uniform accounting and management practices. Having transformed a rather loose federation of companies into a highly centralized administrative structure, which Alfred Chandler cites as an antecedent for "modern industrial management," they turned GE back over to Coffin around 1900. The Schenectady plant concentrated on large and nonstandardized machines such as generators and motors, the Lynn factory manufactured smaller, standard items like arc lamps and small motors, and the Harrison plant specialized in incandescent lamps.[3]

In December 1893 the Calculating Department learned that it was to be moved to Schenectady. By then its staff had shrunk to three engineers: Steinmetz, Ernst Berg, and E. B. Raymond. When Horace Parshall, who had been in England on company business, resigned in early 1894 to become a private consultant in London, Steinmetz was put in charge of the department.[4] Having lived in five cities in five years, the twenty-eight-year-old engineer moved to Schenectady in January and settled down. His long *Wanderjahr*, which had begun with a Socialist trial, ironically ended in a corporate office. In fact, he later became so identified with GE that the public knew him as the Wizard of Schenectady.

Founded in the mid-seventeenth century as a Dutch settlement on the Mohawk River in upstate New York, Schenectady became a commercial center in the 1830s, when the Erie Canal and railroad lines connected it to the agricultural and industrial revolutions occurring in the Old Northwest and the Northeast. A locomotive works and the Westinghouse Farm Machinery Company—founded by George Westinghouse, a graduate of Schenectady's Union College—helped double the city's population to over ten thousand in 1870. In 1886 Edison moved his dynamo factory from New York City to Schenectady, partly to avoid labor problems. The plant became the headquarters of Edison General Electric when it was formed in 1889. Within a year the firm's large immigrant work force helped swell the city's population to almost twenty thousand. With the growth of GE and American Locomotive in the early twentieth century, boosters called Schenectady the "city that lights and hauls the world." GE provided these services for its Schenectady workers—in a manner reminiscent of more isolated "company towns"—when it purchased the local gas, electric lighting, and street-railway companies in 1898.[5]

When Steinmetz came to Schenectady he lived for the first few weeks at a hotel and then at a rooming house at 233 Liberty Street. Sometime that fall he invited Ernst Berg to share an apartment within walking distance of the factory at 53 Washington Avenue, thus beginning a close personal and professional relationship that lasted for nearly three decades. At a banquet in 1913 Steinmetz introduced Berg, who was six years younger and his protégé in many respects, as his "alter-ego."[6]

Numerous photographs taken by Steinmetz show that he and the handsome Swedish engineer were nearly inseparable during these early years in Schenectady. Ernst's brother Eskil, who had also graduated from the Royal Polytechnikum in Stockholm, joined them in 1895. Steinmetz had urged him to come over, saying he "never saw a sensible man who had lived a few years in the United States, willing to go back to Europe to stay." In May 1898 the threesome moved from Washington Avenue to a two-story rented house at 243 Liberty Street, the former home of the founder of the Schenectady Locomotive Works. Christened Liberty Hall by the young bachelors, the house became well known in the Steinmetz

Ernst Berg (*left*), Steinmetz (*standing*), and Eskil Berg (*right*) at their Schenectady boarding house on Washington Avenue, about 1895. Source: Schaffer Library Special Collections, Union College, Schenectady, N.Y.

folklore for its quasi-Bohemian atmosphere. In the backyard lived a menagerie of unusual pets, including raccoons, cranes, owls, crows (John and Mary), a monkey, and an alligator that swam in the Erie Canal alongside the house, much to the consternation of the neighbors. Exotic orchids and cacti grew in a conservatory attached to the house and in an octagonal greenhouse in the yard. Steinmetz's half-sister Clara, who was a poet and a painter, arrived from Germany in the winter of 1895–1896 for the first of several visits—adding to the Bohemian flavor of their life. She boarded nearby in 1897, then moved to New York City, apparently annoyed by the practical jokes played on her by the "boys." Steinmetz was closer to the Bergs and called them his "family" in a caption on a photograph taken at Washington Avenue. Other photos show the group at parties, picnics, and on boating, hiking, and bicycle trips (with the hunchbacked Steinmetz pedaling far behind). One juvenile biography that focused on these and other human-interest stories was entitled *The Electrical Genius of Liberty Hall*. Adding to the legend, GE claimed for many years that the laboratory Steinmetz and the Bergs set up in a barn behind the house in the fall of 1898 was the birthplace of the General Electric Research Laboratory and, thus, of industrial research in America.[7]

Some of the folklore was based on fact. Liberty Hall did house a menagerie, but the Bergs, who were avid hunters and had introduced skiing into that part of New York, seem to have paid more attention to the backyard zoo than did the urbane Steinmetz. Ernst purchased Florida alligators, a monkey, and a horse. While on a business trip he wrote Eskil that he was "very sorry to hear that our beloved crow has died, and particularly that it is John and not Mary." He told Eskil how to care for Mary, the owl, the raccoons, and the dogs while he was away. Steinmetz ordered a horned owl and exotic birds, and he had a butterfly collection mounted. He was also fond of the crows. He claimed he could communicate with them and later had them stuffed and kept them on a shelf in his library. (One of the stuffed crows is now part of the Steinmetz collection at the Schenectady Museum.)[8] In regard to the laboratory born in a barn, Willis Whitney, its first director, did work with Steinmetz at Liberty Street, but only until new quarters were built in the factory (see Chapter 6).

The barn was also home to the Society for the Adjustment of Differences in Salaries—a Saturday night poker group organized in 1896. The society approved a constitution in 1898 and elected Steinmetz "permanent president" two years later. Steinmetz, the Bergs, and Broderick were charter members of this American version of the Breslau university clubs. Many GE engineers and officers stopped by on a Saturday night, including Albert Davis, the company's patent attorney; John Dempster, an assistant to Steinmetz; and Edwin Rice. Rice, whose father founded the American Sunday School movement, attended only once, according to the records.[9]

Ernst Berg ran Liberty Hall; he paid the bills, ordered supplies, complained to the landlord, and so forth. Steinmetz had little to do with it, as shown by a letter from Ernst to his sister in late 1899. Berg was considering hiring a housekeeper while he was away and told his sister, "Mr. Steinmetz will be alone in the house and he could not well show her her duties and break her in properly."[10] Berg also ran the administrative side of Steinmetz's department at GE, mainly because Steinmetz took almost no interest in these matters. After moving to the University of

Illinois in 1909, Berg wrote Steinmetz's secretary, "I have written several letters to Steinmetz recently but as usual he does not reply . . . I am afraid that since I left the General Electric Company, Steinmetz has gotten into a bad way. He needs a boss and I am afraid that Hayden [Berg's replacement] is not exercising all his powers."[11]

Steinmetz's Calculating Department

Berg was an effective boss for the absent-minded Steinmetz, especially when he guided engineering at GE from 1894 to 1900. The centralization of management at Schenectady in 1894 made Steinmetz, as head of the Calculating Department, the de facto chief engineer of the company. GE was then organized as a highly centralized, functionally departmentalized corporation, a form of "line and staff" management borrowed from the railroads. President Coffin presided over two vice presidents, a general counsel, and the technical director, who was made a vice president in 1896. Eugene Griffin, a captain in the U.S. Corps of Engineers before heading the Railway Department at Thomson-Houston, was first vice president in charge of sales. Under him were four departments, each covering a specific product area, which sold and installed equipment. Department engineers, like William Le Roy Emmet in the Lighting Department (central station work) and Edward Hewlett in the Railway Department, acted as consultants between their customers and GE, selling mostly customized apparatus to utilities, electric street railways, and factories. J. P. Ord was second vice president in charge of financial matters. Frederick P. Fish headed the Law Department, which included patent attorneys.

Technical director Edwin Rice had charge of the Manufacturing and Electrical Department. The works managers at Schenectady, Harrison, and Lynn formed its manufacturing wing, while the engineering side included the Expert Department (the testing division, which also trained newly hired engineers); the Standardizing Laboratory, formed in 1895; the Publicity Bureau, established in 1897; and the Calculating Department. Although each plant had its own product-development group, Rice's engineers at Schenectady did most of the research and development for GE: inventing, designing, and improving transmission lines and electrical apparatus.

At the heart of the engineering staff was the Calculating Department. Steinmetz later said that before 1900 "all the designing calculations of the company, with the exception of a few special departments, were made by the Calculating Department," while the "manufacturing engineering departments . . . carried out these designs in [the] drafting room and factory."[12] A good example of this division of labor was the design of high-voltage transformers for the long-distance transmission lines in Sacramento, California, and Guadlajara, Mexico, in 1894. Steinmetz made the preliminary calculations and gave general specifications and design recommendations for the transformers to Walter Moody at Lynn. An electrical engineering graduate of MIT, Moody carried out more detailed calculations and completed the final design. After the transformers were built, Steinmetz tested their insulation at Schenectady before they were shipped.[13]

Representing GE's technological frontier of the 1890s, these polyphase systems were the main business of the Calculating Department. Westinghouse had installed the first commercial AC transmission lines in the United States—at Portland, Oregon, in 1890 and at Pomona, California, in 1892—but these were single-phase systems. Following the lead of European companies, GE introduced the first three-phase transmission networks in the United States, as we have seen. Engineers chose three-phase lines for long-distance work because they were more economical, requiring less copper than single- or two-phase lines.[14]

Between 1893 and 1897 GE installed at least nineteen lines over five miles in length, ranging from 2,500 to 30,000 volts. All but three were hydroplants. About one-half of these were on the West Coast, where waterpower was plentiful but rather distant from settled areas. Some installations, like those in Portland, Sacramento, and Salt Lake City, were "universal systems," in which local stations used three-phase current to power factory motors, transformed it into single-phase current to light arc and incandescent lamps, and converted it into direct current to drive streetcars. (Westinghouse introduced the universal system at the Columbian Exposition in 1893, before GE built these plants, and then installed its first universal system at Niagara Falls in 1895.) Six of the GE plants were in other countries (Canada, Mexico, Argentina, Japan, and Italy), an indication of the company's multinational character at this early date. By 1894 GE had also installed at least ten polyphase systems under five miles in length, mostly on the East Coast. All but two were hydroplants that mainly served "dedicated" customers (mines and factories) rather than urban centers.[15]

As GE's expert in alternating current, Steinmetz investigated numerous high-voltage problems on these systems. From 1894 to 1898 he studied the non-arcing properties of lightning arresters, the effect of fog on the breakdown voltage of air, the absorption of moisture by porcelain line insulators, the rise of voltage caused by opening a short circuit with oil switches, and the effect of high-voltage oscillations on cable insulation. The voltages in this work ranged from 25,000 to 95,000 volts, far above that used in practice.[16] He also did more basic research on high-voltage phenomena in this period, which led to his pioneering book on transient analysis ten years later.

The major task for Steinmetz and the Calculating Department was to design the equipment connected to these networks: polyphase alternators, motors, transformers, and rotary converters (for converting AC to DC and vice versa). Steinmetz applied for more than seventy patents on these devices before 1900 (see the Appendix). Most were gigantic, multi-kilowatt, custom-built machines that dwarfed the skilled mechanics who built them in the lofty, cathedral-like shops in Schenectady. In addition to doing the systems analysis of the transformers mentioned earlier, Steinmetz helped design the rotary converters for the Portland system in 1893 and spent a good deal of time on synchronous alternators and motors. He investigated schemes for alternator regulation (keeping the output voltage constant while the load changed), experimented with the parallel operation of alternators, and designed high-speed alternators to be directly connected to steam turbines. Between 1894 and 1900 he took out twelve patents on the regulation of alternators and six on their improvement. In 1901 he reported that he had "controlled the design of . . . some hundred thousand K.W. [kilowatts] of synchronous motors."[17]

The Calculating Department's most innovative work along these lines was on the induction motor. Picking up where Ernst Danielson had left off, Steinmetz made the subject his own and became GE's expert on these motors. In the summer of 1893 he conducted a literature search and found that British and French researchers had anticipated Tesla's work. Elihu Thomson received the news with "great satisfaction" and thought that Tesla's patents, purchased by Westinghouse, now had "less force."[18] The patents eventually withstood these priority claims. In the meantime Louis Bell, Steinmetz, and other engineers at Lynn utilized the work of Danielson to design three-phase induction motors that performed much better than the two-phase Tesla motors produced by Westinghouse.

GE made the first large-scale installation of the motors in a new textile mill in Columbia, South Carolina. Induction motors were favored here over the DC machines proposed by Westinghouse and other firms because of their constant-speed characteristics at full load, needed to drive textile machines smoothly, and because of the safety factor of such sparkless motors in the dust-laden, fire-hazard environment of textile mills. In 1894 GE installed seventeen 65-horsepower three-phase motors at the Columbia Cotton Mills, mounting them on the ceiling to save floor space. The installation set a precedent for polyphase equipment and group-drive in southern textile mills and gave GE the lead in this field. Bell designed a resistor starting device that remained standard on GE induction motors for several years. Steinmetz established the design principles of the motor, calculated the resistor values, and changed the design to one having more slots for the field winding per pole.[19]

He brought this experience to Schenectady. In 1894 he experimented with an induction motor operating at 60 and 125 Hertz, which became the basis for frequency changers. He developed a line of small, inexpensive induction motors the next year, and in 1896 his staff was designing a 100-horsepower single-phase induction motor that would start with considerable torque. As in the case of synchronous motors, Steinmetz recalled in 1901 that he had "controlled the design of some hundred thousand K.W. of induction motors" as head of the Calculating Department.[20]

Many of his patents dealt with the induction motor. Nearly one-third of the sixty patents he filed on electrical equipment between 1890 and 1900 were on improvements to the motor, how to control it, or its use as a frequency changer, regulator, wattmeter, rotary condenser, or phase transformer. Steinmetz filed all but one of these patents while working at the Calculating Department. The most promising one dealt with AC railroading. The induction motor, which did not have a commutator, solved the sparking problem he had encountered with Eickemeyer's AC streetcar motor, but its constant speed was unsuitable for railway work. In 1893 he applied for a patent that provided two speeds by connecting two polyphase induction motors in tandem. After the patent was granted in 1897, GE installed tandem control on a sixteen-mile track in Italy. The system operated satisfactorily, but GE turned to single-phase railroading using an improved AC commutator motor when Westinghouse introduced that system after the turn of the century.[21]

More successful from a commercial standpoint was the "monocyclic" AC distribution system, Steinmetz's most comprehensive invention. Introduced in the fall of 1894 and consisting of alternators, a distribution network, and motors, the system operated from household rather than polyphase current. Before that time only polyphase systems could supply both light and power, because a practical self-starting single-phase motor was not available. The monocyclic system generated single-phase current to light incandescent lamps and had an additional third wire to start polyphase motors. Once started, they ran like single-phase motors and did not draw power from the third wire. Smaller in cross section than the single-phase mains, the third wire was connected to small "teaser" coils placed adjacent to the main coils on the field magnets of the alternator. The teaser coils produced a current out of step with the main current, a difference in phase that enabled the induction motors to start.

Since the teaser coils did not produce as much voltage as the larger field coils, the system was electrically equivalent to a highly unbalanced three-phase network. Steinmetz, however, claimed that it was single-phase, most likely to avoid Tesla's polyphase patents held by Westinghouse. The basis for his claim was that the monocyclic alternator produced pulsating power—as did single-phase systems—instead of the constant power of polyphase generators. Steinmetz did not use this definition of polyphase in his monocyclic patents. He simply said that the system produced polyphase currents to start the motors and then ran in a single-phase manner.[22]

After the patents were issued in January 1895, Louis Bell, then with GE's Chicago sales office, presented a paper to the National Electric Light Association on the monocyclic system. Bell classified it as single-phase, in accordance with Steinmetz's criteria, which had been published the previous fall. This redefinition of polyphase systems—based on output power instead of phase differences—for transparent commercial reasons sparked a lively debate in the pages of *Electrical World*. The journal's editor and such prominent electrical engineers as Reginald Fessenden at the Western University of Pennsylvania (now the University of Pittsburgh) and Dugald Jackson at the University of Wisconsin rejected Steinmetz's definition, while Harris Ryan at Cornell supported him. Steinmetz wrote that his definition was not new, because he had introduced it in an 1892 paper.[23]

Steinmetz's letter brought down the wrath of André Blondel, professor of electrical engineering at the Ecole des Mines and the Ecole des Ponts et Chaussées in France. Blondel wrote *Electrical World* that the "monocyclic system is, in fact, nothing else than a *polyphased system* by means of which two phases are transformed (for what reason we do not know, unless it is to avoid the patents of another company), into three phases or into one . . . To wish to modify the definition of polyphased currents in order to avoid this conclusion is evidently a radical proceeding, and one which will not be countenanced by those having no ulterior interest in the matter." Steinmetz drew on his high reputation as an AC expert to reply, "I consider the *perversion of scientific facts or definitions for commercial purposes as dishonorable*, and I am afraid I would have to denote the action of the writer to impute such purpose without sufficient proof, *with the same name*, if I would not hope that Mr. Blondel has passed these remarks by mere inadvertency, without sufficiently considering their bearing." Blondel apologized. He had not intended his letter to be an "attack upon [Steinmetz's] scientific character, which I admire sincerely, nor upon his personal character, to which I would be the first to render homage." But he did not reverse his technical judgment. At least one American textbook agreed with him and called the monocyclic network a "regular three phase system in bad balance."[24]

Steinmetz stuck to his guns and defined polyphase plants in terms of constant power in an 1895 paper. He did not give way until 1900, when he added a footnote to a revision of his major book on alternating currents. He said that the "unbalanced polyphase system" was "also called [the] 'polyphase monocyclic system,'" since its voltage relationships were similar to that in the "single-phase monocyclic system."[25] He clarified matters somewhat in a 1901 edition of another textbook by characterizing the monocyclic network as one with a "polyphase system of e.m.f.s [electromotive forces] with a single-phase flow of energy."[26]

The Steinmetz archives shed little light on his motives regarding the monocyclic system. I agree with his contemporary critics and historian Harold Passer that Steinmetz probably made the invention in order to avoid Tesla's patents. To a large extent the monocyclic system was an insurance policy that would allow GE to sell AC equipment should the courts uphold Tesla's patents and force the company out of the polyphase business. Yet Steinmetz's personal preference had a bearing as well. At the 1893 meeting of the International Electrical Congress he said, "I do not believe in the polyphase systems very much myself. I consider them only as a state of the art which we have to use now because the single-phase system is not as yet developed sufficiently to place entire reliance on it, but I hope the polyphase system will be gone very soon and we will have the single phase."[27] He liked the simplicity of the single-phase plant with a synchronous motor and criticized the difficulty of keeping the polyphase system in "balance" (i.e., maintaining an equal load between its phases). Not long after the meeting he invented the monocyclic system to solve this problem. Incandescent lamps, the elements most affected by a network out of balance, were connected to the single-phase part of the monocyclic system, which was not thrown out of balance by the small motor load on the third wire.[28]

GE agreed with Steinmetz's assessment and promoted the monocyclic and polyphase systems as its complementary AC lines. Since polyphase units required less copper and operated motors more efficiently, the company sold them for long-distance lines and large power applications. GE sold monocyclic equipment primarily to replace single-phase lighting plants in areas wanting to add a small number of motors. In 1895, the firm installed monocyclic systems in several American cities, including St. Louis, Chicago, Salt Lake City, Middletown, Ohio, and Galesburg, Illinois. The St. Louis alternator was rated at 800 kilowatts and had a rotor sixteen feet in diameter, one of the largest in the country. By 1896 the company had installed 33,500 horsepower of three-phase alternators and 14,800 horsepower of monocyclic alternators, the latter having ten to twelve sizes ranging in power from 50 to 1,500 kilowatts.[29] The system seems to have been easy to keep in balance. The Miami Bicycle Company in Middletown, for example, reported in 1896 that lights on the single-phase mains did not flicker when the system's motors were turned on.[30]

Steinmetz invented a monocyclic induction motor, but GE did not manufacture it and chose instead to sell three-phase motors for monocyclic systems. The approach worked fairly well, despite the mismatch between the polyphase motors and the monocyclic system. This proved to be an advantage in the long run, however. Benjamin Lamme, chief engineer at Westinghouse and arguably the best designer of induction motors in the United States in his day, recalled that Steinmetz's invention advanced the design of induction motors, because a three-phase motor had to perform exceedingly well to operate satisfactorily on the highly unbalanced monocyclic network.[31]

The invention also helped GE in a more indirect manner. In March 1896 GE and Westinghouse signed a patent-pooling agreement because they felt boxed in by the more than three hundred patent suits pending between them. GE, for example, held the controlling patent on the trolley method of connecting streetcars to an overhead electric line, while Westinghouse held Tesla's polyphase patents. Since both companies guaranteed their customers protection against patent infringement suits, they wanted to eliminate this litigation expense and create an administered market that would avoid the "ruinous competition" of laissez-faire capitalism. Negotiations began in the spring of 1895, shortly after GE introduced the monocyclic system, but broke off soon thereafter. In the meantime a court upheld GE's trolley patent and the monocyclic system was beginning to sell, a situation that probably persuaded Westinghouse to come to terms, especially since the courts had not upheld its polyphase patents as yet. The monocyclic system—an inferior but marketable, noninfringing invention—thus gave GE more bargaining power in negotiations leading to the patent-pooling agreement.[32]

GE continued to manufacture the monocyclic system for a few years. In April 1896 the company announced that the monocyclic and three-phase systems were its two major AC lines. An 1897 installation book for GE engineers gave instructions on how to install both systems, and the company sold more monocyclic equipment to the utility in Middletown in 1898 to help supply the growing demand for motors in the area. In the same year GE installed the first long-distance monocyclic plant, a fourteen-mile transmission line in Vermont, and sold monocyclic equipment to the naval base at Mare Island, California.[33] Steinmetz described the system in a 1901 paper, but GE dropped it around 1900 in favor of the technically superior polyphase equipment.[34] The main reason for waiting four years after the patent-pooling agreement to make this decision was probably GE's need to amortize its investment. But engineers also liked the monocyclic system. Berg recommended it to a midwestern utility, Rice listed it in classes of GE apparatus as late as January 1899, and Steinmetz described its advantages in an internal report that February. Thus, the monocyclic system was not merely a stopgap measure like the Westinghouse stopper lamp, invented to light the 1893 Columbia Exposition to get around Edison's patents and then quickly discontinued. Steinmetz's reputation undoubtedly contributed to the system's long life, but GE seems to have been genuinely interested in promoting it as an alternative to a full-fledged polyphase plant.[35]

The monocyclic idea also lived on for a few years in the guise of the monocyclic starting device, which allowed three-phase induction motors to operate from regular single-phase lines. The device simulated a monocyclic condition by means of a phase-splitting technique patented by Steinmetz and Berg. In 1898 Steinmetz reported that the "three-phase motor with monocyclic starting device . . . is the only type of single-phase induction motor which has found an extensive commercial application, in sizes from ½ H.P. to 100 H.P." Steinmetz listed the device in his textbooks as late as 1916.[36]

In 1897 he invented another way to power polyphase motors from single-phase lines: a three-phase motor with a capacitor connected across the third coil. The capacitor shifted the phase of the incoming current, which enabled the motor to start from single-phase lines. GE manufactured this motor in sizes of $1/2$, 1, and 2 horsepower in 1899. Once low-cost capacitors became available in 1925, GE engineers developed a motor with a series capacitor that finally made single-phase induction motors practical. These motors came into widespread use, especially in fractional horsepower sizes, and are prevalent today because they run on household current instead of polyphase or monocyclic power.[37]

Theories of Machines

The Calculating Department prided itself on developing theories of machines to complement this considerable practical experience. The symbiotic relationship between theory and practice enriched Steinmetz's value to GE and made him one of the best AC theorists in the world. Only a few of his rivals had much design experience. Gisbert Kapp designed and invented alternators, motors, and transformers for a manufacturer and then as a private consultant before accepting a university position late in his career. Bernard Behrend, a German engineer, designed induction motors for the Maschinenfabrik Oerlikon in Switzerland and Allis Chalmers in the United States before becoming a private consultant in Boston. Englebert Arnold was the chief electrical engineer of Oerlikon from 1891 to 1894, then a professor of electrical engineering at the polytechnic in Karlsruhe, Germany. But Kapp, Behrend, and Arnold were atypical. Most theorists, those who published research articles and wrote design treatises, were college professors who did some consulting for industry. These included such leading engineers as John Hopkinson, John Ambrose Fleming, and Silvanus Thompson in Britain, André Blondel in France, Erasmus Kittler in Germany, and Arthur Kennelly and Harris Ryan in the United States. Others, like Michael Pupin at Columbia College, had no design experience whatever.[38]

At the other end of the spectrum, few of the master designers in industry published their theories or design techniques. Michael von Dolivio-Dobrowolsky, chief engineer of Allgemeine Elektrizitäts-Gesellschaft in Germany; Charles E. L. Brown, Arnold's predecessor at Oerlikon, who established the Brown Boveri Company in Switzerland; Sebastian Ziani de Ferranti, founder of Ferranti Limited in Britain; and Charles Scott and Benjamin Lamme at Westinghouse published mostly general descriptions of their pioneering work on AC equipment. Lamme, for example, waited more than twenty years to publish a version of his theory of the induction motor that was as detailed as what Steinmetz had revealed. Although Lamme said he withheld the theory on the advice of a colleague who thought it was too mathematical, commercial reasons probably prevailed. Most inventors were content to let others work out the theories. In 1888 Tesla replied to a critic, "Although my motor is the fruit of long labor and careful investigation, I do not wish to claim any other merit beyond that of having invented [the motor], and I leave it to men more competent than myself to determine the true laws of the principle and the best mode of its application."[39]

Steinmetz thus held a unique position among the leading AC theorists. Well trained in science and mathematics, he was chief engineer of one of the largest electrical manufacturing companies in the world. GE provided an able group of assistants and a testing room to compare theory with practice. Just as important, ever since leaving the University of Breslau he had had a burning desire to establish a scholarly reputation. GE permitted him to publish his research, as long as he did not divulge proprietary information. Steinmetz worked this matter out with Rice and published over 150 articles and 12 technical books on AC systems during his thirty years with the company.

Before Steinmetz and Rice reached this agreement GE had severely restricted what its engineers could publish. Steinmetz told Behrend, for example, that he had independently discovered the basis of the "circle diagram" approach to analyzing the induction motor in 1893. But "commercial reasons prevented him from publishing," according to Behrend.[40] GE changed its policy the next year when it permitted Steinmetz to publish his theory of the induction motor. The theory contained results as important as the circle diagram but left out some crucial equations. In 1897 Rice formalized GE's censorship policy by establishing the Publication Bureau, which screened all company bulletins, advertising, and employee articles—including Steinmetz's many technical papers— for proprietary information. Authors could publish most material after it was patented, but other information was kept secret—a major difference between the norms of the scientific and engineering communities. Steinmetz recognized the importance of this distinction, saying in 1897 that GE was "loath to describe much of a knowledge which has cost enormous sums to accumulate."[41]

The bulk of his many papers and books contained theories of apparatus. These consisted of equations that described the operation of a device like a transformer in terms of its input and output characteristics (voltage, current, and power) and such design parameters as resistance, inductance, and capacitance. Other formulae—the design equations mentioned in Chapter 1—related these parameters to the dimensions of the device, such as the number of coils on a transformer winding, and to the physical characteristics of its material, such as the hysteresis of a transformer's iron core. Steinmetz made his reputation by publishing the operational equations, but he rarely revealed the design equations, in order to protect GE's competitive position—the compromise he reached with Rice.

These equations illustrate remarkably well a pervasive relationship between science and technology in electrical engineering. Steinmetz developed operational equations by applying the principles of electrophysics and magnetic-circuit theory to a generalized machine. He based his design equations on experiments with machines and other empirical data. The design equations thus linked the science of Ohm and Faraday with the technology of a specific machine and its materials. Furthermore, mature theories of apparatus typically embodied most of the types of engineering knowledge described in Chapter 1. They were derived from scientific principles, were written in engineering mathematics, and included design equations based on empirical data. They did not contain design rules of thumb, which they tended to replace, or technical skill, which resided in experienced engineers and workers. Successful theories thus fulfilled the research program Gisbert Kapp had laid out in 1885 when he encouraged engineers to develop "connecting links between pure science and practical work."[42]

A good example is the theory of the induction motor. Steinmetz developed his theory between 1894 and 1898, a period Lamme called the "Age of Induction Motor Analysis."[43] And so it was, because engineers in Europe and the United States published numerous papers and treatises on the mathematical theory of this promising motor. They wanted to bring theory up to practice—Tesla had demonstrated a workable motor in 1887—and enable engineers to design improved motors from the drawing board with less reliance on cut-and-try methods.

Most engineers understood that the induction motor worked on the principle of Arago's disk. In 1824 the French physicist François Arago discovered that a magnetic needle suspended above a copper disk would rotate on its axis as the disk was spun around. The effect remained a mystery until Michael Faraday explained it in 1831 on the basis of his theory of electromagnetic induction. Faraday reasoned that the rotating disk cut lines of magnetic force emanating from the needle, thereby inducing eddy currents in the disk on either side of the needle. The currents produced magnetic fields that acted on the needle, causing it to revolve with the disk.

The step from Arago's disk to an electric motor was a lengthy one. More than fifty years later, in 1879 and 1880, the British physicist Walter Baily and the French engineer Marcel Deprez independently discovered how to create a rotating magnetic field electrically. Tesla and the Italian physicist Galileo Ferraris independently picked up this work in the mid-1880s. They used two two-phase currents to create a rotating magnetic field, which caused an armature—a copper cylinder in Ferraris's case—to turn because of the induction of currents in the cylinder. In 1888 Tesla wrapped a closed-circuit winding around an iron-core drum armature in a rotating-field motor, thus creating the modern form of what came to be called the induction motor.

Ferraris published the first theory of the motor in the paper describing his invention in 1888. He reasoned that the rotating field induced eddy currents in the copper cylinder and that the interaction between the magnetic field produced by these currents and the revolving field caused the cylinder to turn. Using this explanation from Faraday and basic definitions of energy loss and output power, he derived a mathematical theory consisting of two fundamental equations. From this theory he calculated that the motor should produce maximum power when the cylinder's speed was one-half of that of the rotating field, resulting in a maximum efficiency of 50 percent.

In 1888, when tests of Tesla's motor with a closed-circuit armature showed an efficiency of over 60 percent, theorists took a closer look at the motor. They first tackled the problem with Maxwell's equations for an induction coil, because the induction motor could be considered as an induction coil with a rotating rather than a stationary secondary winding (the winding receiving induced currents). In 1891 Louis Duncan, professor of applied electricity at Johns Hopkins University, and two French engineers, Maurice Hutin and Maurice Leblanc, independently published theories of the induction motor based on Maxwell's equations. By including the coefficient of self-induction in their equations, which Ferraris had neglected in order to simplify his analysis, they derived equations that described the operation of the induction motor much more accurately than those published by Ferraris.

This application of Maxwell's equations brought theory a quantum jump closer to practice, but it was still too ideal for engineering purposes because it assumed a perfect air-core machine. Like transformers, real motors had magnetic leakage and iron cores. Engineers had known since the late 1880s that magnetic leakage had to be considered for induction motors because it decreased starting torque substantially; they also knew that three phenomena associated with an iron core—hysteresis, eddy currents, and variable permeability—caused the operation of iron-core machines to deviate from theory. Thus, Maxwell's theory, whose constant coefficients depended on constant permeability, gave inaccurate results for iron-core machines, even if hysteresis, eddy currents, and magnetic leakage were somehow taken into account.

To incorporate variable permeability into their theories of the induction motor, engineers turned to the magnetic-circuit method of dynamo design introduced in the mid-1880s. The first theorists to apply this approach to the induction motor were Arnold at Oerlikon and Blondel in France. Their analyses, published independently in 1893, used magnetic-circuit theory to account for variable permeability. They had mixed success with magnetic leakage, hysteresis, and eddy currents, because they used a hybrid approach of combining Maxwell's equations and magnetic-circuit theory to handle these parameters.

One year later Kapp and Steinmetz independently improved and simplified this hybrid approach considerably by developing a pure magnetic-circuit method. They based the technique, which came into general use, on their theories of the transformer. Previous researchers from Duncan to Blondel had also applied a theory of the transformer to the induction motor. While they had employed Maxwell's coefficients in one fashion or another, Kapp and Steinmetz eliminated the coefficients by modifying traditional magnetic-circuit theory with two new parameters borrowed from their transformer theories: leakage inductance, introduced by Kapp in 1888, and primary admittance, introduced by Steinmetz in 1894. Because these terms gave a better representation of magnetic leakage, hysteresis, and eddy currents than previous theories, Steinmetz, who used both terms, could calculate the performance of induction motors more accurately than anyone theretofore.

Although Blondel referred to this approach as the "Kapp-Steinmetz" method, there were significant differences. Kapp employed rotating vectors, while Steinmetz used both vectors and complex quantities, which gave him an edge in analytical capability. Steinmetz's theory was also more comprehensive. Kapp assumed that energy losses due to hysteresis and eddy currents were constant at constant speed, and so he did not include their vectors in his diagram. Steinmetz accounted for these losses by his new term, *primary admittance,* and thus obtained more accurate values for the performance of the motor.[44] The new parameters—leakage inductance and primary admittance—were not empirical coefficients added to make ideal theory correspond more closely to practice. Kapp and Steinmetz defined them by analogy to Maxwell's coefficients. Yet because of the rather unorthodox manner in which they treated the magnetic relationships defined by these terms, their theories appeared to differ sharply from Maxwell's.[45]

How far they had gone became apparent during a heated debate at a meeting of the American Institute of Electrical Engineers in October 1894. At the meeting Michael Pupin read a paper on the motor by Lieutenant Samuel Reber, a member of the Signal Corps who had recently graduated in electrical engineering from Johns Hopkins. Reber based his theory on Maxwell's equations for an ironless transformer, modifying them to include a variable coefficient that depended on the position of the rotating armature. Like other Maxwellians, Reber neglected energy losses due to hysteresis, eddy currents, and magnetic leakage. After hearing Pupin read this paper Steinmetz remarked that its differential equations made it too mathematical for practicing engineers, and that it should have considered energy losses because these determined the performance of the motor. He quickly sketched on the blackboard his complex-number theory, which was much simpler mathematically and accounted for these losses. After Steinmetz finished, Pupin criticized him for not having used the Maxwellian coefficients. Pupin considered two of his terms, *leakage inductance* and *primary admittance*, to be empirical rather than fundamental parameters. Steinmetz replied that these were not empirical, but rather terms he could calculate before the motor was built. He disliked the Maxwellian approach because the coefficient of self-induction included both mutual and leakage flux, which made it impractical as a design parameter.

Two leading engineers quickly supported Steinmetz. Kennelly, often an arbitrator of technical disputes at the AIEE, favored Steinmetz's theory: "I, for one, am opposed to other than engineering methods for treating engineering problems. I believe that the method which has just been discussed by Mr. Steinmetz before us on the blackboard is far more capable of giving us the engineering properties of induction motors than are the methods based on Maxwell's equations." Rankin Kennedy, a prominent British designer of AC machinery, agreed with Kennelly and reprinted Steinmetz's theory in an article on induction motors. Kennedy thought that the Maxwellian technique was too mathematical and that the "results of this elaborate work [of Reber's] are useless to the engineer."

Pupin stubbornly clung to the Maxwellian method as late as 1918, when he criticized electrical engineers for using Steinmetz's approach. Pupin declared that the "attempts of ordinary mortals to do better than Maxwell did must be discouraged. Let us follow Maxwell as long as we can, then when someone is born who is more profound than Maxwell, we will bow to him."[46]

Pupin's criticisms illustrate the intricate relationship between science and technology in electrical engineering at the turn of the century. He directed his sarcasm at engineers who had the effrontery to deviate from the canon of applied science. An inventor of considerable reputation, Pupin had applied Maxwell's electromagnetic theory in his invention of the loading coil for long-distance telephony. For Pupin, Maxwell's theory was the surest basis for electrical engineering. Attempts to modify it to fit the needs of practical work were bound to erode the foundation on which the entire edifice of electrical engineering rested.

Steinmetz and other theorists of the induction motor saw the matter differently. They were not trying to do better than Maxwell in physics. They had no intention of modifying his electromagnetic theory, which they, like Pupin, considered to be the scientific basis of electrical engineering. Rather, they were trying to do something different from what Maxwell had done: they wanted to derive an engineering theory of an electromechanical device, which they accomplished by adapting Maxwell's theory to their needs. And, contrary to Pupin, these "ordinary mortals" did better than the strict followers of Maxwell, like Reber, in devising useful theories of the induction motor. Pupin may not have liked how far they departed from Maxwell, but they were Maxwell's intellectual heirs more than he realized.

The theory of the induction motor owed a considerable debt to science. The motor's scientific principle was Arago's rotations, as explained by Faraday. But a straightforward application of this science by Ferraris led to a woefully inadequate theory of the motor. Duncan, Hutin, and Leblanc improved matters by turning to Maxwell's equations of the induction coil, but these were not useful for engineering work either until they were substantially modified by Kapp and Steinmetz. It was this transformation that alarmed Pupin, who need not have been that concerned. The theory's debt to electrophysics was hidden by several "translations" of information between science and technology. Its mathematical structure came from Maxwell, whose differential equations were translated into forms more useful for design work: graphical analysis and complex algebra. Its magnetic-circuit approach came from Rowland's work on the magnetic properties of materials, which was then applied to dynamo design by Kapp and Hopkinson. And its non-Maxwellian parameters—primary admittance and leakage inductance, whose seemingly empirical and unscientific character so incensed Pupin—were in fact defined analogously to Maxwell's coefficients.

It is no wonder that Pupin did not recognize the theory as applied science. It had gone through so many metamorphoses that it bore little resemblance to classical electrophysics. It resembled instead what Edwin Layton and others have described as "engineering science." That is, engineering theorists transformed scientific and technological information about the induction motor into a systematic body of knowledge, a set of equations, that could be used to design these motors. The theory's derivation from experiment and mathematics, its cumulative nature, general applicability, and so forth also demonstrated its qualities as an engineering science. But the theory differed from Layton's model in two important respects: the status accorded to theorists and the nature of physical forces. Contrary to Layton, theory was highly regarded by electrical engineers, who held Kapp, Steinmetz, and their colleagues in high esteem precisely because they published useful theories of apparatus. *Electrical World,* for example, published a biographical sketch of Steinmetz as early as 1894 in which it praised his theoretical work.[47]

Pupin was fighting a rear-guard action in 1918; engineers have used the induction motor theories of Kapp, Steinmetz, and others, rather than the Maxwellian approach, since the mid-1890s. A major reason is that engineering theorists continued to improve their equations, which were more accurate than the Maxwellian method to begin with. Steinmetz made his theory more useful for design work by deriving an equivalent circuit for the induction motor in 1897. A diagram of connected components which translated his mathematics into an electrical language, the circuit modeled a complicated electromechanical device just as simpler circuits modeled the electrical wiring of a house. That is, the circuit reduced a device that worked on both electrical and mechanical principles to a purely electrical model. Since electrical engineers were familiar with AC circuits, the equivalent circuit gave them a visual understanding of the physical relationships hidden by Steinmetz's rather abstruse complex-number equations. By reducing the operation of rotating machinery to a single circuit subject to well-understood AC circuit analysis, Steinmetz greatly simplified the theory of the induction motor. After he applied this concept to other apparatus, engineers could draw equivalent circuits of generators, transformers, transmission lines, and motors on one diagram and analyze the entire interconnected system by AC circuit theory. As noted earlier, the equivalent circuit became a powerful engineering tool, especially in electronics, where engineers applied it to vacuum tubes in radio receivers as early as 1919. Since then electronics engineers have derived equivalent circuits for all types of vacuum tubes, transistors, and even the laser.[48]

The power and flexibility of the Steinmetz approach—a complex-number theory containing energy-loss parameters in an equivalent circuit—are shown in his theory of a general induction machine. He was able to apply it to a transformer, an induction motor, or an induction generator simply by changing the value of the term *slip* (difference between the speed of the rotating field and that of the armature) in a general set of equations. He summarized these results in a single curve on a diagram of torque versus slip. For a slip greater than one, the curve described the performance of a backwards-driven induction generator; for a slip of one, a transformer; for a slip between one and zero, an induction motor; and for a slip less than one, an induction generator running above synchronism. Few of his operational equations were so universal in nature, but they all applied to an entire class of machines instead of to a particular device.[49]

The formulae that tied these operational equations to specific machines, that linked the science embedded in Steinmetz's theory to the technology of the artifact, were the design equations mentioned earlier. Most of these fell under the category of proprietary information at GE. Steinmetz published equations for the primary admittance of a transformer, which could be adapted to the induction motor, but he did not divulge his design equations for leakage inductance, a more difficult parameter to determine. Instead, he gave general rules for the relative values of primary and secondary leakage inductance in 1897 to supplement the tests he had described in 1894. He indicated the commercial importance of these equations for the induction motor in 1897 when he said that GE engineers had to be able to calculate leakage inductance beforehand, because they could not easily change its value after building the motor. Precisely for these commercial reasons he maintained that these formulae "belong[ed] in the field of induction motor design, and therefore [fell] outside the scope" of his published theories, which consisted mostly of operational equations.[50]

It was left to theorists outside of industry to publish design equations for leakage inductance. Kapp published these formulae in the *Elektrotechnische Zeitschrift* in 1898 when he edited that journal. Behrend derived equations for the leakage coefficient of induction motors from tests he did at Oerlikon between 1896 and 1897. He left the company a year later and published the equations in 1900, shortly after joining the Bullock Manufacturing Company in the United States. Comfort Adams, professor of electrical engineering at Harvard University, published equations for leakage inductance in the AIEE *Transactions* in 1905.

After Adams read his paper to the AIEE a design engineer complained that the professor's equations were "too complicated for practical work." Steinmetz rose to Adams's defense and replied that the method was not too complex and that GE engineers used a similar technique. As in 1897, Steinmetz did not divulge GE's equations, but he revealed that the company's practice differed from Adams's method only in that "every operation is not gone through separately, but [is] inserted in a formula, which is still further abbreviated when fixed types of apparatus are under construction. But the calculation is actually done by the principles stated here."[51] Steinmetz described more fully how he tested and calculated primary admittance and leakage inductance after 1905, but he never published design equations as detailed as those reported by Kapp and Adams.[52]

By 1897 Steinmetz had developed his induction motor theory to such a mature stage that as late as the mid-1930s textbooks and handbooks saw no reason to modify it. He introduced four variables to simplify his equations and referred all secondary-winding values to the primary winding, which enabled him to convert his equations to a form suited for the equivalent circuit. He also introduced three constants of the motor—power factor of admittance, power factor of impedance, and the ratio of exciting current to short-circuit current at standstill—that enabled him to compare the operating characteristics of induction motors.[53]

From 1898 to 1900 Steinmetz extended his theory to cover the increasingly important single-phase motor. He did not follow Ferraris's 1894 method of splitting the motor's revolving field into two oppositely rotating vectors. Instead, he recognized that at, or near, synchronism the rotating field resembled that of the polyphase motor. He improved this theory at low speeds in 1900 by taking into account secondary effects neglected in his earlier polyphase papers. He then expressed primary admittance and leakage inductance as second-order functions of slip, but he still did not publish their design equations.[54] The simplifying power of Steinmetz's theory was evident for the single-phase motor, because he could treat it and its starting device as one equivalent circuit, subject to AC analysis, rather than having to deal with the complicated problem of analyzing an electromechanical motor connected to an electric circuit.

During the next few years Steinmetz used his complex-number approach to develop theories of all AC apparatus. These included other induction machines (frequency converters, induction generators, and the tandem connection of induction motors), synchronous alternators and motors, commutator motors, and rotary converters. He published most of the theories in the AIEE *Transactions* and then collected them in two books: *Theory and Calculation of Alternating Current Phenomena* (1897) and *Theoretical Elements of Electrical Engineering* (1900). Each went through several editions.

Steinmetz wrote the first draft of *AC Phenomena* in 1894, based on his IEC paper of the previous year. W. J. Johnston and Company, publisher of *Electrical World* and the predecessor of McGraw-Hill, agreed to publish his manuscript under the title *Complex Quantities and Alternating Currents*. Kennelly, one of the reviewers of the manuscript, praised the book and thought it would "take its place in the literature of alternating currents as a classical work." His only suggestion was to use the "international system of mathematical notation." Steinmetz resisted making the wholesale changes and asked Kennelly and the editor of *Electrical World* for assistance.[55] Berg gave Steinmetz a hand in publishing an expanded manuscript in 1897, but they did not make the changes requested by Kennelly—a decision that came back to haunt them a decade later.

Steinmetz's many articles and books bear witness to the flexibility of his engineering theory. To analyze the tandem connection of induction motors, for example, he merely inserted the proper value for slip into his general transformer equations and considered the connected motors as one equivalent circuit. In the case of the synchronous motor, he defined a new term, synchronous reactance—analogous to leakage reactance in induction motors—as part of the machine's characteristic impedance. In keeping with GE's policy, Steinmetz described tests to measure synchronous reactance but did not publish its design equations. His theory produced complete power curves of the synchronous motor, where only partial curves had been obtained experimentally before. It also served as the basis for theories of the synchronous alternator and the parallel operation of alternators that were more comprehensive than any to date. The key to its success, as in all his AC machinery analysis, was a set of equations that described the relationship between the internal and external impedances of the machine. For the AC commutator motor, which he had worked on at Eickemeyer's, Steinmetz divided it into three classes— the repulsion, shunt, and series motor. He derived the theory of the repulsion motor from his equations for induction machines and developed separate theories for the shunt and series commutator motors, the ones most widely used in practice.[56]

Having a different theory for each motor was intellectually unsatisfying to GE's chief scientist-engineer. Like "pure" physicists at the time, Steinmetz was searching for a "general theory," not of relativity but of the electric motor. In 1904 he carried this quest as far as he would take it by publishing a general theory of the induction and AC commutator motors. Gabriel Kron, a talented mathematician-engineer whom Philip Alger, Steinmetz's successor, brought to GE in the 1930s, reached Steinmetz's goal by developing a general theory of electrical machinery applicable to all motors and generators, whether they were AC or DC.[57]

Steinmetz's most advanced work in AC machinery considered how secondary phenomena distorted the shape of alternating currents, a factor he had neglected in earlier theories. These phenomena included such esoteric matters as nonuniform rotation of an armature, pulsation of the magnetic field, and the variation in reactance caused by magnetic saturation of the core, hysteresis, and synchronous motion. Some effects of the higher harmonic frequencies produced by these phenomena were increased output of alternators, decreased energy lost by hysteresis in transformers, and insulation breakdown. Steinmetz accounted for these distortions to the perfect sine waves in his theories by expressing them with a Fourier series (a sum of weighted sinusoids of different frequencies) or by an "equivalent" sine wave.[58]

At the other end of the spectrum, he developed design rules to complement theories of apparatus. Like other industrial engineers, Steinmetz knew that empirical rules of thumb still had a place in the inexact science of electrical engineering. Companies often considered design rules, like design equations, to be proprietary information, but he did reveal one for alternators. The rule, which had no scientific basis, simply stated that the product of the length and diameter of an alternator's armature was proportional to its output power. Silvanus Thompson, author of widely used textbooks on electrical machinery, published the rule in 1901. Thompson called the proportionality constant in the rule "Steinmetz's coefficient" and gave its values for single-phase and three-phase alternators of different sizes. By providing initial numbers to plug into Steinmetz's equations, the rule helped engineers design alternators from scratch, rather than simply analyze machines already built. During a visit to Schenectady in 1986 I discovered that today's engineers use similar methods—albeit with theories of machines stored in a computer program—to design custom-made motors and generators. Many of the theories in the GE computer were improved versions of the ones Steinmetz had developed at the turn of the century. Company engineers thus bestowed a form of immortality on the Wizard of Schenectady, even though high-powered public relations men failed to keep the legend alive after World War II.[59]

Steinmetz did not write a treatise on his philosophy of engineering science, but he made enough scattered comments on the subject for us to piece together his views. The basis of his philosophy was a distinction between rational and empirical equations. The former were derived from first principles, typically from electrophysics. Empirical equations came from experimental research and technological knowledge. While Steinmetz preferred rational to empirical equations because the former were more universal,[60] he recognized that the nonideal world of engineering often required empirical equations. In 1895, for example, he criticized a paper on AC theory by a mechanical engineer, noting that in "mechanical engineering, and sometimes even in electrical engineering, we are obliged to use [an] empirical formula occasionally, wherever a rational equation is not available or is not known. However, I cannot see any reason for attempting to represent by [an] empirical formula, phenomena of which a rational equation is in evidence," in this case a Fourier series.[61] Other examples of rational formulae are what I have called the operational equations in theories of apparatus. His design equations and design rules are empirical equations, as is the law of hysteresis, although Steinmetz initially considered it to be a rational equation, that is, a scientific principle.

Rational and empirical equations might look good on paper, but they had to be useful to the design engineer. Steinmetz made this clear during a discussion at the AIEE in 1894: "A theory of a machine is very nice and of high scientific interest, but to be useful, the *first* condition is that you can determine all the quantities which enter into the theory beforehand by calculation, the *second* condition is, that you can easily determine these quantities by experiment to check the correctness of your calculation, and the *third* condition is that you can easily recognize the limits of the practical application of your theory."[62] Steinmetz often criticized theories for being too "scientific," that is, too ideal. In 1893 he complained that the Maxwellian theory of the transformer described a device that "does not exist in practice, but merely haunts as a phantom transformer the text-books and mathematical treatises on transformers."[63] In early 1894 he lamented that most theories of the induction motor were written "only by theorists who never constructed a motor themselves and who have never seen a motor taken apart" (a criticism that could also have been made of his first theory of the transformer). Such theories perpetrated the "wildest confusion" about the performance of the motor, a state of affairs he thought to be "thoroughly discouraging."[64] It was this situation, also shown in Reber's theory of the motor, that prompted Steinmetz to publish the theory he had developed at GE. He made similar comments in later years, saying in 1908 that the "phantom [transmission line] circuit of [uniformly] distributed capacity and inductance" was "very different from the circuit existing in practice."[65]

We have seen how Steinmetz overcame these problems by developing operational and design equations, experiments to determine design parameters, and design rules for AC apparatus in the 1890s. In 1911 he described this process in general: "All engineering theory consists of three parts: first to find out that a thing exists; second to get the general theory, which is really the phenomenon as it would exist under ideal conditions, conditions which never exactly but only approximately exist in practice . . . and the third part is the adaptation of those general theories to the specific conditions under investigation."[66] This evolutionary process was evident in the case of the induction motor. Engineers tested and improved the motor invented by Tesla and Ferraris. Reber and others derived a theory for an ideal, ironless motor based on Maxwell's equations for the induction coil, which neglected hysteresis, eddy currents, and magnetic leakage. Then Kapp and Steinmetz modified this approach considerably by taking these secondary phenomena into account, which in this case determined the actual performance of the device.

The initial theory usually followed practice in his work, but Steinmetz occasionally tried to deduce new technology from a general theory. In 1916 he stated that electrical engineering problems had historically been solved using the "synthetic" method: proceeding from the specific artifact to the general theory, thus obtaining a "complete structure of the engineering science." With the growth of scientific knowledge, the analytic approach was possible: from "general (differential) equations of the science based on the fundamental underlying laws" to specific cases. In a paper on the general equations of the electric circuit, Steinmetz used the latter technique to "check whether any class of [electric] current of industrial importance has escaped recognition."[67] Although unsuccessful in this case, he noted that the synthetic and analytic methods were complementary and could be used to check each other. Practice often followed theory in his work in this manner. He used the theory of AC circuits to invent the monocyclic system and the theory of the induction motor to invent the tandem connection of these motors, for example. He and his staff also employed his theories to improve AC motors and ancillary devices.

The symbiotic relationship between theory and practice is quite evident in the case of the induction motor. Experimentation played a key role in the development of the motor's theory, and Steinmetz incorporated his research on hysteresis and eddy currents into the theory. He built an experimental "general alternating current transformer" that could be operated as a transformer, frequency changer, induction motor, or phase transformer. Because of this flexibility, he probably used it to develop his first theory of the induction motor, especially the proprietary equations for leakage reactance.[68]

After developing a theory through a combination of experimentation and applied science, Steinmetz and his staff designed motors and calculated their characteristics. In 1895, for example, Eskil Berg employed Steinmetz's theory to calculate the efficiency, power factor, running current, and other characteristics of an induction motor, given the primary admittance, impedance, and other parameters of the motor. He then plotted these results as a function of output power.[69]

The next step was to build and test a machine in order to compare theory with practice. Steinmetz reported some of these results in his AIEE papers to show the value of his theories. In 1894 he gave the speed-torque curves of a 100-horsepower induction motor, similar in design to those installed at the Columbia Mills textile plant. He stated, "These curves were calculated theoretically originally, but after the motor was built and tested I had no reason to change the curve because the observed values" were very close to those desired.[70] In 1897 he published the load and speed curves for the motor calculated by Eskil Berg, remarking that the theoretical and tested results were in "exceedingly close agreement."[71]

Two associates give further evidence of the practical value of his theory of the induction motor. Henry Hobart, his colleague at Lynn, recalled that "during 1893 and 1894 Steinmetz introduced much greater (and, doubtless, desirable) precision into the designing of polyphase induction motors, and fine motors of many sizes and various types and speeds were built in accordance with his methods." Hobart credited the good characteristics of GE's pre-1893 induction motors to Ernst Danielson and the "fine quality" of the later ones to Steinmetz.[72] Horace Parshall recalled that in 1893 the "general theory of induction motors was understood but how to determine the specific performance was another matter. [Steinmetz developed] the diagrams and formulae until finally the induction motor design assumed the same definite form as had previously existed in the case of the [well-understood] commutating machine."[73]

The last step in developing an engineering theory was to modify it to accord with practice. In the case of the induction motor Steinmetz and his staff noted a slight discrepancy between the calculated and tested impedance of a motor due to eddy currents induced in its mechanical structure. Steinmetz reported that after part of the structure was removed, theory and practice agreed. He also noted that a nonuniform air gap would make theory deviate from practice, because primary admittance was a function of the length of the air gap. Since his equations assumed a uniform gap, he modified his theory to take the realities of construction into account. Walter Slichter, an engineer in the Calculating Department, described similar modifications in the spring of 1899. Slichter discovered that calculated values of performance parameters for twenty-one induction motors above 5 horsepower differed from tested values, mostly because of vagaries in construction, and suggested changing several constants in the design equations for these motors.[74]

Comparing theory with practice in this manner was a normal procedure in the Calculating Department. In early 1899 the department consisted of Eskil Berg (design of alternators and converters), John Dempster (assistant to Eskil), A. E. Averrett (design of induction motors and regulators), H. S. Meyer (assistant to Averrett), Slichter (final specifications on induction motors and regulators), Ernst Berg, and Steinmetz. Each engineer followed the progress of machines he had designed through the drafting room, the construction shops, and the testing room. Ernst Berg, who administered the department, instructed them to comment "on the relation between calculated and observed values and the reasons for the discrepancies" as the machines were tested.[75] The calculated values came from Steinmetz's theories, as modified by design equations and other constants to fit theory to practice.

End of an Era

The department was much different in 1899 from what it had been five years before, when it did all the design calculations for the company. Steinmetz recalled that as GE's business grew in the mid-1890s and as engineering became standardized, design work was moved out of the

Calculating Department. The first to be transferred was that on "intermediate sizes of standard lines," and "this gradually extended, until most of the design work was done by the manufacturing engineering" departments under Rice. The Calculating Department designed only new lines of equipment and special types and sizes. It also developed new classes of machines, such as the AC commutator motor, and did "general investigations." But this condition was "unstable," because the department could not accept full responsibility for the work. Thus, in the latter part of the 1890s the engineering departments designed their own machines, and in many cases the Calculating Department engineer who had been in charge of a machine went over to the "designing engineering department, together with the transfer of the work."[76]

Steinmetz noted that "this reorganization was completed about 1900." The "engineering department then comprised . . . a number of designing engineering departments, the calculating department had ceased," and he, "as Consulting Engineer, devoted his time to the new fields in electrical engineering, which then opened: electrochemistry, the new developments in electric lighting, the higher potential [voltage] problems, etc."[77] Although many of these changes did occur around 1900, there was more to the demise of the Calculating Department than Steinmetz recalled.

The department began to shrink in mid-1899. The group was together during a planned rearrangement of offices in May, when Steinmetz asked Rice to move his department to a corner office, which had more windows. His assistants needed more light than engineers in other departments because they were "expected to spend the whole office hours in calculating work, using mostly slide rule."[78] The proposed move went by the boards during the summer. In late September Steinmetz prepared for J. P. Ord, second vice president in charge of financial maters, a feasibility study of storage-battery propulsion for railway engines in the New York Central tunnel. Other letters indicate that Steinmetz had turned mostly to consulting work by then. In early November he was "informed that an attempt is being made to reduce the size of the office promised to us to an area less even than my present office." It is likely that this occurred and that his staff was reduced to one or two assistants.[79]

Other events in 1899 support this conclusion. Early that year Rice was redefining the Calculating Department's responsibilities. In April he asked design engineers to send requests for special tests on commutating, synchronous, and induction machines to Steinmetz. And in May

Rice gave him responsibility for approving all design changes from the original specifications on AC equipment requested by the design engineers. The result of the last order was that much engineering was held up, waiting for Steinmetz and his small six-person staff to process what must have been a deluge of paperwork. In May, for example, the machining of parts for New York Edison's rotary converters was held up awaiting Steinmetz's approval. Other departments also asked him to do some calculations on reactance coils and high-frequency effects on cables in May and August.[80] Rice soon concluded that a centralized Calculating Department had become a relic of the past—when the company was much smaller—and that Steinmetz's talents, which did not include the ability to direct a large staff effectively, were better suited to consulting work.

The growth of design engineering outside the Calculating Department corroborates this view. In 1898 design engineers worked in eleven departments: railway, power and mining, lighting, supply (small electrical items), alternating current, direct current, transformer, switchboard, rheostat, and engine. By 1899 these departments were no longer under the first vice president for sales and had been transformed into engineering departments and placed under Rice, third vice president for manufacturing and engineering.[81] These groups took over the Calculating Department's work, along with some of its engineers. In 1905, for example, A. E. Averrett, former head of induction motor design in the Calculating Department, was a member of the A.C. Engineering Department.[82]

Another reason GE shifted engineering away from Steinmetz was that some engineers thought his designs were too conservative. It is rather curious that the politically radical Steinmetz was a conservative designer, but the evidence supports this view. Emmet, chief engineer for the Lighting Department and an "intrapreneur" inside GE in such areas as the Niagara Falls and steam-turbine projects, drew up plans for generators in the second powerhouse at Niagara Falls when bids were invited in May 1900. GE won the contract for eleven alternators at 5,500 horsepower each, but not because of Steinmetz. Emmet recalled that "Steinmetz had refused to approve the radical procedure [for the alternators] which I proposed and I took the matter entirely out of the hands of the calculating department which he directed and had the work done by H. G. Reist's men [in the A.C. Department], who worked out all the details and made the designs."[83]

Steinmetz was also conservative as an in-house consultant. He recommended in at least three cases not to pursue technologies that other companies later turned into profitable businesses. In June 1895 he advised Rice not to purchase the patents on a steam turbine invented by Charles Parsons in England. Although Steinmetz considered this field to be "very promising" and thought GE should pursue work in this area, he did "not think much of Parsons' steam turbine." He thought "that a simpler design, even if not quite as efficient in steam consumption, would be desirable." He recommended that GE "delay the closing of any agreement with Parsons until we have satisfied ourselves with regard to our own design" by W. H. Knight.[84] Rice followed his advice, with the result that Westinghouse purchased Parsons's patents later that year and forged ahead of GE. GE then leapfrogged the competition—a characteristic of its approach to innovation noted by George Wise—by developing a turbine invented by the American Charles Curtis in 1896. After much experimentation GE installed a 5,000-kilowatt Curtis turbine for Commonwealth Edison in Chicago in 1903. The machine started a profitable turbine business that helped revolutionize the utility industry by providing a more economical means to generate electricity. Steinmetz played only a minor role in this work, developing a theory of the turbine and participating in design studies with Curtis.[85]

The second example of Steinmetz's conservatism involves electric lighting. In October 1895 Steinmetz advised Rice not to purchase the rights to the Moore tube, an electric-discharge lamp invented by D. McFarlan Moore, a former Edison employee who had left GE in 1894. Steinmetz studied the descriptions of the device in the technical press and concluded that it was of "*no practical or commercial value whatever.*" It was less efficient than similar lamps invented by Elihu Thomson and Tesla and "had no chance to compete in efficiency with ordinary incandescent lighting." Steinmetz further noted that "ever since Tesla started his high frequency experiments [in 1892,] dreamers all over the country have gone wild with revolutionizing the method of electric lighting, but have produced thus far nothing whatever, and not even shown the possibility of success."[86] While many of these "dreamers" produced little of value, by 1904 Moore had improved his gas-filled tube to the point where it began to compete successfully with incandescent lighting before the invention of the tungsten filament. GE then purchased Moore's companies and patents in 1912, and Moore rejoined the firm. Although unable to compete with tungsten lamps, the Moore tube formed a basis for the much later development of neon tubing and fluorescent lighting.[87]

The last example involves high-voltage transmission. In October

1899 Steinmetz advised Rice not to develop commercial 60,000-volt lines or to guarantee them at this voltage, which the Stanley Electric Company proposed to do for a plant near San Francisco. Steinmetz, who had performed experiments on the effect of fog on lines at 95,000 volts, was worried about fog on the proposed line. He had predicted in 1891 that future systems would be in the hundreds of kilovolts and had experimented with transformers up to 160,000 volts. But he now argued that 60,000-volt lines had not been proved in practice and noted that a 50,000-volt line in Colorado had not been reliable. He would "rather be glad to let somebody else try his luck [with 60,000 volts] but should recommend to follow the progress of this plant carefully."[88] The Stanley company had good luck with 60,000 volts and higher, especially in the West. Consequently, it led the way in high-voltage engineering in the late 1890s until GE purchased the firm in 1903—another example of GE leapfrogging the competition, this time by buying it out.[89]

In fairness to Steinmetz it should be noted that these three cases formed only a small percentage of the consulting work he did for GE up to 1900. Records survive for more than forty of his technical opinions during this period, covering the entire field of electrical engineering: electrical machinery, electric lighting, and high-voltage phenomena. In the late 1890s Rice relied on Thomson at Lynn and Steinmetz at Schenectady as his main technical consultants. In most instances Rice or another vice president needed a quick answer, which gave Steinmetz a week or two to formulate an opinion. Most of his recommendations appear sound and some, like that to form a research laboratory, made a lasting contribution to the company (see Chapter 6). We should also note that the advantage of hindsight makes the above cases seem obviously wrongheaded, while many of his contemporaries in large electrical firms shared his relatively cautious approach to engineering.

Steinmetz also reported on engineering at GE and abroad, especially as the work of the Calculating Department wound down in the late 1890s. In 1897 he visited electrical engineering firms and major installations in Europe. The tour included London, Paris, Milan, Zurich, Vienna, Berlin, and his home in Breslau. He saw Dobrowolsky at AEG in Berlin and probably visited the Oerlikon and Brown Boveri companies in Switzerland, the Siemens company in Berlin, and the Ganz company in Austria—the top firms in the field. Steinmetz was not much impressed, though. European alternators were much larger than American ones of the same capacity because of a "faulty electrical and magnetic design," which he attributed to an unwillingness to spend money on experimentation and low pay for the companies' chief engineers. Induction motors were up to snuff because of their very small air gaps, but rotary converters were poorly designed. He concluded that, in the area of AC engineering, the "European Companies are far more backward than even I anticipated."[90] In a report on GE engineering in 1898 he surveyed the company's status in all major lines of electrical apparatus and electric lighting and made recommendations for new types of equipment and lines of research to follow.[91]

Indicative of his distance from day-to-day engineering at this time was his status vis-à-vis the Engineering Committee. Rice established the committee in 1901, chaired by Emmet, to consider engineering that affected the general policy of the company or the work of more than one department. Rice did not appoint Steinmetz to the committee, because he was not very interested in these details or in purely commercial matters. But Rice said to Steinmetz that he would write Emmet that "no action affecting the policy of the Company, or engineering, should be taken without your full knowledge."[92]

By this time Steinmetz was concentrating on electric lighting research in the laboratory behind Liberty Hall. While head of the Calculating Department, he along with Ernst Berg had supervised the design of most of GE's advanced machinery, particularly induction and synchronous motors. He developed mathematical theories of AC apparatus that became standard not only at GE but in the international electrical engineering community. Although not as proficient a designer as Lamme at Westinghouse, nor as innovative an engineer as Emmet at GE, Steinmetz left his stamp of scientific technology on the company. Alger later called him the "patron saint of the GE motor business." Emmet, who criticized his conservative design approach, acknowledged, "To him, I owed, as many others did, much of my knowledge of electric principles . . . Steinmetz's greatest usefulness was as a teacher, and apart from the books which he wrote, his greatest work was to start the General Electric engineers upon the use of proper methods of calculation."[93] Emmet was referring, of course, to Steinmetz's complex-number method of AC analysis—the basis of which was a fruitful combination of theory and practice.

Steinmetz helped GE maintain its position as the largest electrical manufacturer in the country, particularly after it weathered the business depression of the mid-1890s. In 1898 GE employed over four thousand hands at the Schenectady Works, while Westinghouse Electric had about the same number in all its plants. After purchasing the Fort Wayne

Electric Company in 1898 and the American branch of Siemens & Halske in 1900, GE numbered over twelve thousand workers in all its plants in 1900. That same year the Schenectady Works employed seventy-five hundred workers, placing it at the top of the second tier of the largest manufacturing plants in the United States. It was not far behind such huge steel companies as Carnegie's Homestead plant, which employed between eight thousand and ten thousand workers in 1900. GE's growth was made possible by the expansion of the Schenectady Works to cover 130 acres, on which 114 buildings provided over a million square feet of floor space. The company made $6.2 million in profits in 1900, had orders of $27.9 million, and increased its cash on hand to $6.6 million—an indication of Coffin's "born-again" financial conservatism. Driving the city's population to nearly thirty-two thousand in 1900, GE's employment of about one-fourth of them put some truth into the local boosters' slogan that Schenectady was the "Electric City."[94]

Although GE and Westinghouse grew into industrial giants during these years, the electrification of America proceeded at a slow pace. One journal reported only twenty-four hundred central stations at the end of 1896, fifteen hundred of which operated only at night. It was not until 1921 that more than one-half of all urban residences had electricity, 1925 for urban and rural households. The electrification of factories did not take off until after World War I, mainly because of the large amount of capital invested in the older system of steam engines, belts, shafts, and pulleys. For residential use the unreliability and high cost of early electrical service help explain why it took over forty years for electricity to become common in the home. The frequent breakdown of dynamos and the breakage of light bulb filaments due to voltage fluctuations, for example, led to light fixtures with both gas and electric mantles. In regard to cost, although Edison had priced electricity the same as gas for the Pearl Street station in 1882, he did not match the gas companies in slashing their prices. Electricity cost almost twice as much as gas lighting in 1890 and was priced at the steep figure of 20 cents per kilowatt-hour in Chicago in 1892. (The average residential rate nationwide was a more affordable 10 cents per kilowatt-hour in 1907.) Gas became even more attractive in the mid-1890s with the introduction of the incandescent Welsbach mantle, which was efficient and did not burn as an open flame.[95] GE met these challenges from gas and from improved European light bulbs by turning from engineering product development, which had been successful for electrical machines, to research based more on science in new laboratories established after the turn of the century.

6 New Settings for Research

To improve its position in electric lighting, the company established the General Electric Research Laboratory, an event historians generally regard as the founding of "science-based" industrial research in the United States. Other companies had hired physicists and chemists before then, some of whom worked in well-equipped laboratories. What set the GE endeavor apart and made it a model for other corporations was that—eventually—it gave highly talented scientists, like Nobel Prize–winner Irving Langmuir, a fair degree of freedom to investigate the scientific basis of industrial problems, as well as to use existing science and scientific methods to develop new products. The industrial research laboratory was thus a major route by which GE and other companies "incorporated" science-based innovation in-house, as they had earlier incorporated engineering inside the firm.[1]

GE traced its roots to the most famous laboratories in America: Thomas Edison's facilities at Menlo Park (1876) and West Orange, New Jersey (1887). Contrary to popular belief, Edison regularly employed scientists and engineers, including mathematical physicist Francis Upton, Arthur Kennelly, chemist Jonas Aylsworth, and other college-educated chemists. Kennelly published first-rate engineering research in hysteresis and alternating-current circuit theory, as we have seen, but Edison hired chemists mainly for their skill in performing experiments he laid out beforehand. Both laboratories focused on the *development* part of the later expression *research and development*. The Menlo Park "gang" invented a practical incandescent lamp and lighting system; the West Orange group produced improved phonographs, the nickel-iron storage battery, and motion-picture equipment in Edison's "invention factory." They focused on innovation instead of discovering the scientific principles of existing products or new product lines—a goal the GE Research Laboratory only partially attained after many years. As noted by Andre

Millard, the early GE laboratory, staffed by Ph.D. physicists and chemists, had much in common with the West Orange facility run by the Old Man and his "muckers" (round-the-clock experimenters) in regard to product development. West Orange, though, was more autonomous and was not tied as closely to the needs of the corporation.[2]

GE had its share of the materials testing, standardizing, and engineering development laboratories that American manufacturing firms had established since the 1870s. The company had at least six of these laboratories in 1900. At Schenectady the Standardizing Laboratory calibrated electrical instruments under the direction of engineer Lewis Robinson and a staff of fifteen. The Physical and Chemistry Laboratories, as part of the Purchasing Department, tested materials used in the company's products. Product development occurred at Elihu Thomson's laboratory in Lynn, at the Lamp Works in Harrison, New Jersey, and at the Testing Department in Schenectady.[3] The procedure at the latter facility was well established by 1900. In 1896, for example, Steinmetz filled out a shop order requesting the manufacture of an induction motor with different types of windings for his investigation in the testing room. Thomson and Edison followed a similar procedure at Lynn, Menlo Park, and West Orange.[4] Westinghouse Electric had comparable engineering laboratories for materials testing and product development that dated to the late 1880s. Yet the company waited until 1916 to establish a science-based laboratory as a "theoretical group" to support the works laboratories.[5]

Steinmetz and Whitney

GE created its science-based laboratory essentially from scratch in 1900 to improve its position in electric lighting. As the Calculating Department disbanded in the late 1890s, Steinmetz—the company's best-trained employee in science and mathematics—turned from engineering to chemical research and played a crucial role in establishing the GE laboratory and guiding its early years. In a letter to Edwin Rice in September 1900 Steinmetz proposed that the company set up an "electro-chemical experimental laboratory," insulate it from the demands of the factory, and place it in the charge of a "good and well paid practical chemist of considerable originality." The lab should concentrate on five areas: mercury-vapor lamps, arc-light electrodes, metallic filaments for incandescent lamps, the Nernst lamp, and the physical constants of chemical compounds. Steinmetz said he would "be very glad to direct such an experimental laboratory, provided that the administration is put in the hands of the chemist."[6] We can better understand the significance of this important letter if we consider Steinmetz's research in electric lighting prior to 1900.

GE had been interested in the product development proposed in the letter ever since inventors began to challenge the firm's dominance in electric lighting after Edison's patents expired in 1893. The Edison companies had doubled the efficiency of the original bamboo lamp in 1888 by coating the filament with asphalt. But there things stood for nearly a decade, mainly because of the monopoly enjoyed by U.S. lamp companies. Europeans led the assault on this position in 1893 with the widespread introduction of an incandescent gas mantle invented by the Austrian Carl Auer von Welsbach. Impregnated with metallic oxides that glowed brightly when heated, the mantle reduced the cost of gas lighting by two-thirds and threatened to overtake the incandescent lighting market. Another challenge came from the enclosed arc light, which gave off less glare than street arc lamps. Electric-discharge devices like the Moore tube and the mercury-vapor lamp also promised to be much more efficient than the incandescent light.

The most worrisome threat came from two incandescent bulbs. The Nernst lamp, invented in 1897 by Walther Nernst, a German professor of electrochemistry, had a metallic-oxide filament and a heater coil to start it. Although rather expensive and cumbersome, the Nernst lamp was almost twice as efficient as the Edison light bulb and found a good market in Europe and the United States after 1902. The second innovation was a lamp invented by Welsbach in 1898 that used the metal osmium as its filament. Sold only in Europe, the osmium lamp was expensive because of the rarity of osmium, yet its high efficiency convinced researchers (correctly) that the future of incandescent lighting lay with metal filaments.[7]

Steinmetz was able to suggest promising lines of research in electric lighting in 1900 because he had thoroughly investigated the subject. In late 1894 he advised Rice to purchase the patents on the Welsbach gas mantle, but he pointed out its disadvantages and recommended an alternative system using calcium carbide. He suggested these same carbides for incandescent lamps in 1896 and later criticized osmium because it had some dangerous properties and could not reach as high a temperature as carbon. Although he considered carbon to be the best material for incandescent filaments, he suggested that the Lamp Works experiment with osmium, metallic oxides like thoria, and carbides of metals like

titanium. He was wrong about the high temperatures possible with osmium filaments, but he had a good knowledge of contemporary research
in American physics journals. Steinmetz did not foresee the bright future
of the Nernst lamp when he said in 1899 that the mercury-vapor lamp
was more promising because it was much more efficient.[8]

The higher efficiency of electric-discharge tubes and arc lamps persuaded him to concentrate on these devices. After advising GE not to
purchase the Moore tube in 1895, he began to reconsider vacuum-tube
lighting two years later on the suggestion of chief engineer John Howell at
the Lamp Works. Steinmetz recommended that Howell work along the
lines Edison had tried with a Crookes tube coated with calcium tungstate
and use an automatic vacuum adjuster to avoid the loss of vacuum noted
by Edison.[9] The prospects of a different type of electric-discharge
tube—the mercury-vapor lamp—brought Steinmetz into the fold in
1899. In March he went with Howell and Rice to visit the private laboratory of Peter Cooper-Hewitt in New York City to see his mercury-arc
lamp. Following the lead of experimenters in the late 1870s, Hewitt
employed an inverted U-shaped tube filled with mercury and connected
to a high voltage. The mercury vapors emitted a bright bluish green light
once the discharge began.[10]

The visit stimulated Steinmetz to experiment vigorously on mercury-
vapor lamps at his laboratory in the Liberty Street barn. In a short time he
prepared a detailed report on the lamp, which indicates that Steinmetz,
the mathematical physicist turned engineer, had greatly increased his
knowledge of chemistry since taking two courses on the subject at
Breslau fifteen years before. He found that the phenomenon was a
Geissler tube glow with the current conducted by mercury vapors. He
analyzed the spectrum of the light with a spectroscope and saw that it had
only one color line, which accounted for the lamp's "ghastly" color. While
adding sodium, potassium, and lithium to supply the missing yellow,
orange, and red lines in order to achieve a white light, he discovered that
these chemicals left a dark deposit on the glass unless introduced in an
ionized state. He concluded that "while the results derived thus far do
not offer prospect for immediate commercial application they are sufficiently encouraging to warrant the expenditure of considerable sums"
toward that end. He pointed out how his work differed from Hewitt's,
patented the color-correction method about a year later, and suggested
that GE secure Hewitt's "cooperation" and establish a laboratory at the
Lamp Works to develop the mercury-vapor lamp. Steinmetz recom-

mended his assistant "Mr. Kjeldson" for the new laboratory and said he would visit it once a month to check on its progress.[11]

GE did not establish this laboratory, nor did it secure Hewitt's cooperation, since Westinghouse had begun funding his experiments at about this time.[12] Instead, GE organized the Research Laboratory at Schenectady a year later on the basis of Steinmetz's letter to Rice, which listed his "confidential report" on the mercury-vapor lamp as the first line of research. Steinmetz had made similar suggestions before. In the summer of 1897 he renewed a recommendation he had made "some time ago" to "have some experimental work done in electro-metallurgy." He suggested that Rice place the laboratory in the Engineering Department, where it could address topics like the production of calcium tungstate and the development of condensers. In 1899 he changed his terminology and wanted to "resuscitate" his recommendations "made some years ago" to establish an "electro-chemical laboratory" to take advantage of the surplus power from the GE hydroplant at nearby Mechanicville, New York. Earlier letters indicate that he thought the company should use this power to produce titanium and other chemicals needed for the development of electric lighting.[13]

Steinmetz also wanted to hire a chemist. In January 1899 he thought that the proposed laboratory to use the power from Mechanicville should be "put in charge of a good chemist." That April Elihu Thomson told Rice that the Hewitt lamp was "of such importance that it may warrant us getting a skilled chemist" and have him produce the noble gases needed for this work. Rice thought the suggestion was one "we should certainly take up at an early date."[14] GE officials did not act on the matter until receiving Steinmetz's letter in September 1900. His proposal listed the order of research as the mercury-vapor lamp, luminous arc electrodes like titanium carbide, metal and metallic oxide incandescent filaments, and the Nernst lamp. Although the Nernst lamp was last on the list, Westinghouse's purchase of its U.S. patent rights probably spurred GE to bring in outside help to improve the company's position in all areas of electric lighting, a less expensive proposition than buying patent rights. The company wanted a trained scientist because it could no longer count on Edison, who had failed to develop an improved lamp filament in the late 1890s.[15]

Steinmetz recommended that the researcher should be a "practical chemist of considerable originality, able to follow and work out independently any suggestions made to him." He should have a "fair knowledge" of electricity and general physics, be very familiar with glass blowing, have good administrative skills, and so forth.[16] Steinmetz wanted a scientist with knowledge and abilities that complemented his own strong points. He probably made the offer to "direct" the lab himself but to leave it in the charge of the chemist, because he wanted the better-trained chemist to follow up the research he had begun at Liberty Street. He did not necessarily want someone to do "basic" research; Elihu Thomson was the only one of the lab's founders to mention this need. Regarded as a scientist as well as an inventor, Thomson said that GE should have a "research laboratory for commercial applications of scientific principles, and even for the discovery of those principles."[17]

In one of those bursts of bureaucratic work that Steinmetz could maintain (briefly) for some projects, he carried through his recommendation to establish a research laboratory. He asked patent attorney Albert Davis to write Charles Cross, professor of physics at MIT and founder of its electrical engineering program, to recommend a chemist. Cross suggested Willis Whitney, an instructor at MIT who had received his doctorate under the electrochemist Wilhelm Ostwald at the University of Leipzig in 1896. With Rice's approval, Steinmetz asked Thomson to meet Whitney to see "whether he appears to you suitable for the intended work."[18] Whitney's lucrative consulting work for a manufacturer of photographic paper probably convinced Thomson that he had the proper combination of theory and practice for the job. After much negotiation Whitney struck a bargain with GE: he would take the position and spend only two days a week at Schenectady during the school year.[19]

Whitney started work in mid-December 1900 at the laboratory on Liberty Street. Steinmetz had requested $20,000 to set up Whitney's laboratory and to pay for operating expenses for the first year,[20] but quarters were apparently not ready when Whitney arrived. He first worked with Steinmetz on the project to improve the color of the mercury-vapor lamp, probably by the method of ionized salts that Steinmetz had patented the previous year. Whitney apparently had as much trouble with Steinmetz as he did with the fickle mercury lamp. By all accounts the two men were often at loggerheads. George Wise, Whitney's biographer, has aptly described their differences: "It was a cigar-smoking socialist meeting a non-smoking Republican; an abrasive practical joker meeting a natural diplomat; a theoretician meeting an experimenter."[21]

With Whitney absent from Schenectady for a good part of the week, Steinmetz took an active part in running the new lab, especially after a fire destroyed the roof of the Liberty Street barn in January. Although the only equipment damaged was "about a dozen pairs of skis," Steinmetz "had to give up the former laboratory and have transferred all things to Dr. Whitney's laboratory in the factory."[22] Throughout February and March Steinmetz handled many administrative details, including ordering equipment; setting up a factory shop order for Whitney; giving Whitney's staff recommendations to Rice; delegating equipment requisitions to John Dempster, his former assistant at the Calculating Department, now working for Whitney; and requesting that the works manager fix the road to the laboratory, a one-story wooden structure known as Building 19.[23]

The flurry of administrative work ended around April, when workmen completed a new laboratory on a property Steinmetz had purchased on Wendell Avenue in the exclusive General Electric Realty Plot. A seventy-five-acre tract of woods near Union College purchased by GE in 1899, the plot was reserved for employees who could afford to build a house costing at least $4,000, as stipulated by deeds on the lots. Edwin Rice, president of the realty association, built the first house. Steinmetz probably used the $2,500 GE had paid him for the mercury-lamp patents—the patents were his property because he had done the work outside the plant at the Liberty Street lab—to construct the Wendell Avenue lab, and GE paid its upkeep. Contemporary photographs show a two-story wooden structure with the laboratory on the first floor and living quarters above it.[24]

Steinmetz still lived at Liberty Hall, but the group had broken up. Ernst Berg, who had been thinking about leaving since late 1899 because of problems with servants and the landlord, moved to the Edison Hotel in the spring of 1900. His brother Eskil joined him in the fall, leaving Steinmetz alone in the house with his sister Clara, who had come up from New York City to take care of him. He and Ernst were apparently still close, though, because Ernst purchased a lot next to his in the GE Realty Plot in the summer, kept up a burial plot that he, Eskil, and Steinmetz had bought earlier, and sent Steinmetz photographs he had purchased in Europe. Steinmetz then befriended Joseph LeRoy Hayden, a young engineer from the Lynn Works who worked with him on an arc-lamp project at Wendell Avenue after the laboratory part of the building was completed in the spring of 1901. Fifteen years younger than Steinmetz, Hayden had no formal schooling and had left his home town of Haydenville in western Massachusetts at the age of sixteen to find work in Boston. He learned the electrical trade at the Boston Electric Company

Steinmetz in the Wendell Avenue laboratory, 1912. Source: Hall of History Foundation, Schenectady, N.Y.

and then joined GE at Lynn in 1900. As the friendship between Hayden and Steinmetz grew, Ernst Berg moved from the Edison Hotel in the summer of 1901 to board in a private house and asked the GE realty company if he could exchange his lot for one down the street from Steinmetz. In August Steinmetz, Clara, and Hayden moved into the house-laboratory on Wendell Avenue, where they lived and worked for two years while Steinmetz's elegant, three-story Elizabethan brick house was being built on the property.[25]

It is not clear how Steinmetz split his time between his home laboratory and the factory at the turn of the century. He gave Whitney advice on the mercury-vapor and Nernst lamps in 1901 and delivered an "impromptu" talk on the induction motor in the lab colloquia series in early 1902. He probably spent less time at the Research Laboratory once Whitney became its full-time director in the fall of 1901. Dempster recalled that Whitney and Steinmetz argued frequently and that Steinmetz often left for his home laboratory. "Then we didn't see him often," Dempster said. Both laboratories focused on the projects outlined by

Steinmetz in 1900. Whitney placed the mercury-vapor lamp in the charge of Ezekiel Weintraub, a Ph.D. physicist hired in the summer of 1901 who later invented a practical mercury-arc rectifier. Whitney also made resistance rods for lightning arresters and meters from materials having high negative-temperature coefficients, which Steinmetz had called "pyro-electrolytes" in his original list of research projects. William Weedon, a Ph.D. chemist, took up the titanium-carbide arc lamp in 1903—another research area suggested by Steinmetz. Whitney also broadened his research agenda and had more than thirty product-development projects under way by the fall of 1902 with a research staff of fourteen.[26]

Working mainly with Hayden, Steinmetz took up the subject of luminous arc lighting, the second project on his original list. As the largest U.S. producer of arc lamps in the 1890s, GE was concerned about the German development of the luminous arc lamp by Hugo Bremer in 1900. Bremer used carbon rods treated with nonconducting salts that evaporated into the arc between the rods, producing a bright light with the best efficiency of any arc lamp. This work stimulated Steinmetz to develop a different type of luminous lamp based on the iron oxide magnetite. He had previously worked with magnetite at Eickemeyer's, where he tested its hysteresis properties, and at GE, where he patented its use as a "pyro-electrolyte" material for lightning arresters, to start induction motors, and to form joints in arc lamps.[27]

Patented in early 1902, the magnetite lamp had an upper positive electrode of copper enclosed in an iron shell and a lower negative electrode of magnetite in an iron tube. The magnetite provided the luminous material for the arc. Steinmetz mixed it with titanium oxide to increase the lamp's output and chromium oxide to prolong its life. The reliable lamp operated on direct current, required a chimney to carry off the fumes, and produced a strong white light at a good efficiency. Steinmetz made occasional improvements after GE placed the lamp in production in 1903, and it was used for street lighting as late as 1922. Arthur Bright, a leading historian of electric lighting, noted in the 1940s that the "flame arc and the magnetite arc were the last fundamental advances of broad commercial importance in arc-lighting technique." The magnetite lamp was the only major invention in arc or electric-discharge lighting not made by an independent inventor.[28]

Working in his home laboratory, Steinmetz was a quasi-independent inventor who drew on GE's resources to develop the lamp. Whitney's laboratory helped him determine the correct proportions of materials in the electrodes, and Lynn engineers solved many design problems after Steinmetz turned the lamp over to them in the spring of 1902. C. A. B. Halvorson, Jr., and Richard Fleming at Lynn developed a mechanism to keep the electrodes separated when the lamp was off and to bring them together when it was on. They placed the copper electrode in a steel casing to prevent oxidation, added the chimney, and made many other improvements to make it marketable. Steinmetz, Hayden, and Lynn engineers also did the commercial development of the mercury-arc rectifier invented by Weintraub in Whitney's laboratory, which enabled the magnetite lamp to operate on alternating current.[29]

Although Steinmetz and Hayden spent most of their time at Wendell Avenue, Steinmetz did not abandon the laboratory he had fought for so many years to establish at the factory. In 1903 Rice appointed him to the laboratory advisory council along with Thomson, Davis, and Howell. One of his first suggestions was to purchase a "liquid-air" plant to exhaust mercury-vapor lamps. He was also included in a photograph of laboratory personnel dated "about 1904," which must have been taken near the end of his working relationship with Whitney. From then on he preferred the more pleasant environment of the Wendell Avenue laboratory—which GE supported financially and with a few laboratory assistants—to the large bureaucratic structure Whitney was building at the plant.[30]

Wendell Avenue, of course, was more than a workplace. When Hayden moved out to marry a local woman, Corinne Rost, in May 1903, Steinmetz induced the couple to move back in when the big house was finished that June (Clara had moved to another part of the city that year). The arrangement became permanent. Steinmetz legally adopted Hayden when the Haydens had their first baby in 1906 and relished his role as the doting "grandfather" to the Hayden children. By all accounts Steinmetz was proud of the family he had inherited and cared for under his roof. In one of the few personal remarks in his private papers, Steinmetz, who had left his homeland many years before and had called the Bergs his "family" in the 1890s, wrote in a notebook in 1911 about his new life:

20 years have passed since I started this book; 11 years since I wrote in it again.

Great changes have since taken place. I have a family, my adopted son Joseph L. R. Hayden and his wife Corinne Rost, and 3 grand children: Joseph Steinmetz Hayden, Marjorie Hayden, William Steinmetz Hayden. I

Steinmetz and the Haydens with the Detroit Electric, 1914. Source: Hall of History Foundation, Schenectady, N.Y.

own house and property, with laboratory and green house, just as I always wished.

Symbolic of these changes is that Steinmetz sold the Bergs his share of their cemetery plot; he wanted to be buried with the Haydens.[31]

Near the end of 1902 Steinmetz moved further away from daily work at the factory, having accepted a part-time appointment as head of a new electrical engineering department at Union College. Detailed records of the negotiations that led to GE's funding one-half of the department, which Steinmetz directed while working for the company, have not survived. But the evidence indicates that the desire he had felt as a student in Germany to become a college professor was still strong nearly two decades later.

His path to academia was rather circuitous. By 1900 Steinmetz had brought out two revised editions of his first book, *Theory and Calculation of Alternating Current Phenomena* (1897), which many colleges were using as an advanced textbook. He wrote a more general textbook, *Theoretical Elements of Electrical Engineering,* in 1900 and published a second edition two years later. He taught in GE's postgraduate program for newly hired engineers for several years and gave a course on AC machinery to seniors at Columbia University in the spring of 1902. In June of that year Harvard College awarded him an honorary M.A. In presenting the honor president Charles Eliot called him the "foremost expert in applied electricity of this country and therefore the world." During the same month Steinmetz gave his presidential address to the American Institute of Electrical Engineers on the subject of engineering education and called for close cooperation between industry and academia. Those aware of these activities—and Rice and Coffin, the president of GE, surely were—could see that he was seriously considering a university position in 1902. He had almost finished the magnetite project and was still sparring with Whitney about the Research Laboratory. That summer the president of Union College approached him to reorganize its electrical engineering program, a proposal he discussed with Coffin and Rice in the fall. It thus seems likely that GE arranged the position with Union later that year in order to keep Steinmetz in Schenectady.

The arrangement, which lasted a decade, suited him well. He taught mornings at Union College, a short walk from his house on Wendell Avenue, worked afternoons in his home laboratory or at the plant, and spent most of the summer at Camp Mohawk, a rustic cabin he built on a creek feeding the nearby Mohawk River in 1899.[32] Steinmetz thrived in this comfortable environment, free from most of the bureaucratic demands of the corporation. At Union he developed a strong electrical engineering curriculum, strengthened the college's relationship with GE, and wrote several textbooks based on his lectures (see Chapter 7).

Transient Phenomena

At home and at the plant Steinmetz began a lengthy investigation of transient phenomena in power systems. The project proved successful and led to a body of work that he later ranked as one of his three major technical accomplishments (the other two were his research on magnetic hysteresis and the development of the complex-number method).[33] His previous theories of machines and transmission lines had dealt with the steady-state condition of AC circuits: the sinusoidal patterns of voltage and current that existed after disturbances in these quantities had died out. Now he took up the more general case of transient phenomena (surges) caused by electrical changes in the circuit.

Electrical manufacturers recognized the industrial importance of these esoteric phenomena with the rapid growth of long-distance, high-voltage AC systems described in Chapter 5. Engineers had known for some time that transients occurred in every electric circuit. They now took more notice of them because the unprecedented voltages and power of these systems meant that surges on their lines were potentially destructive. Switching motors, lights, and other transmission lines on and off these systems set up voltage surges that spread throughout the network, often destroying line insulators, transformers, and generators. Lightning also caused problems in the power lines that were spreading across the country after the turn of the century. Direct lightning strokes or nearby flashes created surges on the lines that were usually more destructive than those caused by switching operations. Engineers designed several protective devices—including line insulators, lightning arresters, and power-limiting reactors—and based most of their work on trial and error rather than on scientific or engineering knowledge.

Steinmetz and his GE assistants developed a substantial portion of this engineering knowledge in the early days of this new field by doing experimental and theoretical research. As we have seen, Steinmetz had investigated insulating materials at Eickemeyer's and tested high-voltage equipment at GE in the late 1890s. For the latter research he connected twelve transformers to an alternator at the Schenectady plant in 1896 and produced an electric discharge of 150,000 volts at 35.5 kilowatts across a fourteen-inch air gap.[34] Producing one of the most powerful electric discharges so far created, this research preceded his famous "lightning generator" by twenty-five years.

Research with the 1896 equipment produced two important papers. In the first one Steinmetz investigated the wide disagreement over the dielectric strength of air (a measure of its electrical insulating property). He found that the size and shape of the discharge electrodes were factors and used sharp points to obtain more stable data. Although the data indicated a variable dielectric strength of air, he held that if it was constant, it was between 60,000 and 80,000 volts per inch—within the range of the modern value of 75,000 volts per inch. He derived an empirical equation for dielectric strength and did not repeat the mistake he had made with magnetic hysteresis: he recognized that this equation did not represent a physical law. Steinmetz mentioned two phenomena in this paper that became important research topics in later years: corona and the influence of atmospheric conditions, especially fog, on the dielectric strength of air. He had observed the former in 1893, when he used the term *corona* for the blue electrostatic aura surrounding his electrodes.[35]

In the second paper Steinmetz examined the nature of the discharges by means of high-speed photography. Much to his surprise, the photographs showed an oscillating discharge at a higher frequency than that of the alternator producing the fireworks and the frequency predicted by an analysis of the discharge circuit. He had expected that the first discharge would reduce the dielectric strength of air enough to start an arc at the generator frequency. Faced with the photographic evidence, he concluded that a number of oscillations were required to "weaken the disruptive strength of air sufficiently to maintain an arc."[36]

While closer to physics than to engineering, this research had practical implications. Like most scientists and engineers at the time, Steinmetz accepted the theory of English physicist Oliver Lodge that lightning was a high-voltage oscillatory discharge, similar to what he was producing in the laboratory. His research would thus lead to better designs for lightning arresters, which he had tested at Eickemeyer's and GE. In addition, Steinmetz knew that switching off the load from a high-voltage system resulted in an arc discharge between the contacts of the switch. Extinguishing that arc, usually by means of drawing apart the contacts of a switch or immersing them in oil, often caused destructive surges, which he hoped to understand better with his study of laboratory discharges. After calculating the frequency of lightning discharges and switching transients on a transmission line in a theoretical paper in 1898, he put aside the lightning question for a few years to concentrate on the problem of high-voltage switching.[37]

This problem was of immediate concern because GE and Westinghouse had locked horns in the lucrative circuit-breaker market for high-voltage systems. Westinghouse developed several types of switches that broke the circuit in air by separating the breaker contacts relatively slowly. In contrast, GE developed a device in 1900 that quickly extinguished the arc in oil: the Type-H metal-tank breaker with brick partitions. This breaker was safer than Westinghouse's switch, whose sustained arcing in the open air was a fire hazard, and replaced it within a decade. But many engineers in 1898 feared that the quick action of the oil switch would set up destructive surges on transmission lines.[38]

Steinmetz published a paper on this question in 1901, based on research with an artificial transmission line at GE. The set-up consisted of a 10,000-volt alternator, one-half mile of high-voltage cable, a circuit breaker, and a spark gap to measure voltages. While measuring the surges caused by opening the line with different types of circuit breakers, he noticed that if he brought the metal balls of the spark gap close together to start a discharge, the system produced an extremely high voltage of 300,000 volts. He attributed this dangerous value to oscillations produced by the self-interrupting arc that formed across the spark gap (similar to that produced across the open-air switch). Once the arc started, its high current heated the air, which blew out the arc explosively, whereupon it restarted, blew out again, restarted, and so forth.[39]

Steinmetz mathematically studied the rise of voltage caused by this arc by obtaining a "complete solution" (steady-state and transient) of the differential equations for the transmission line. This was his first transient analysis. He had made his reputation in the 1890s by inventing the complex-number method to *avoid* solving differential equations, a technique that worked only for the steady-state solution. He simplified the problem considerably in this paper by modeling the line as having its capacitance lumped across the midpoint of the line, rather than regarding it as distributed along the line (the more accurate but mathematically more complicated approach he had taken in 1898). Steinmetz calculated that the frequency of the line oscillations was a function of the line's parameters, rather than the alternator, and that the maximum voltage rise when turning on the line was less than twice the generator voltage and, thus, relatively harmless. But switching open a loaded line could produce up to a fourfold increase, and rupturing a short circuit gave a destructive tenfold rise of voltage.[40]

Steinmetz tested this theory in the field in the summer of 1901 on the forty-four-mile 25,000-volt line at Kalamazoo, Michigan, which had been installed just two years before.[41] Experimenting with different types of circuit breakers, he found that GE oil switches worked the best and that his theory was largely correct. Low-frequency oscillations of less than twice the generator voltage were produced when the oil switch opened a loaded line, as compared with voltages three to four times that of the generator for the Westinghouse open-air switch, which produced a self-interrupting short-circuit arc. The oscillograph (an electromechanical forerunner to the cathode-ray tube oscilloscope) showed that his prediction of the frequency of these oscillations was also correct and that the oil switch opened the circuit at the zero point of the AC wave, making the quick action of this switch harmless. The only discrepancy between theory and practice occurred in the case of closing an open-ended line, where low-frequency surges (80 to 90 Hertz) of up to three times the generator voltage occurred no matter what switch he used.[42] These unexpected surges and those caused by self-interrupting arcs grabbed Steinmetz's attention. They were potentially more dangerous than higher-frequency surges, because they lasted a longer time before dying out.

Their power caught GE's attention as well, during a dramatic accident involving the company's oil switches in 1903. The accident occurred at the power station of the Manhattan Elevated Railway Company in New York City, at the time the "largest steam-driven electrical generating plant in the world."[43] Eight 5,000-kilowatt Westinghouse alternators supplied three-phase power at 11,000 volts, which was transmitted to substations by thirty-four three-inch, lead-covered underground cables manufactured by GE. From the substations Westinghouse transformers stepped this voltage down to 390 volts, which operated Westinghouse rotary converters to supply 625 volts DC to the elevated railway cars. To protect the equipment GE engineers had designed an elaborate system of circuit breakers: 11,000-volt Type-H oil switches at both the power-house and substation ends of each cable. The breakers at the powerhouse end worked after a three-second time delay. Upon installing its part of the system in 1901, GE declared that it was so designed that "any failure of a cable, a switch or other device shall not result in a general disruption of service."[44]

Nevertheless, a cable failure did shut down the entire system during evening rush hour on July 23, 1903, and again six hours later around midnight. The first shutdown stranded hundreds of commuters for near-ly an hour on the El and made front-page news in the *New York Times*.[45] Henry Stott, superintendent of motive power for the railway, provided the following details about the accident. During rush hour operators ob-served sparks on static dischargers for cables at substations seven and eight. Within a few minutes an explosion in the underground-cable manhole nearest the powerhouse raised the pavement around the man-hole and blew out one foot of a cable for substation eight. Alternator number four then short-circuited, which broke, bent, or badly burned twenty-eight armature bars and eleven terminal connections of the larg-est engine-type generator ever built. Four other cables short-circuited before a combination of automatic circuit breakers and operator action shut the entire system down to prevent further damage. After the acci-dent, inspectors noticed no damage to the substations, but they discov-ered slightly bent armature bars and burned insulation on the other five alternators on line. Shortly after midnight, after bringing the undamaged part of the system back up, operators observed a ground on the 11,000-volt line, followed by a short-circuited cable to substation four, which shut the system down again.[46]

Stott, a former GE engineer in the Buffalo office, ordered a full investigation by Steinmetz and Percy Thomas, Westinghouse's high-voltage expert. Westinghouse had guaranteed that its alternators could withstand 25,000 volts for thirty minutes, 30,000 volts for one minute, or 35,000 volts for one second without damage. The company had also sold the static dischargers, which shunted high-frequency surges to ground.[47] GE had the greater stake in the matter, because its engineers had designed the cables and circuit breakers. Although Steinmetz had shown at Kalamazoo that oil switches did not cause abnormally high voltage rises upon opening a circuit, Westinghouse engineers, who preferred their open-air switch, were understandably skeptical of the role played by the GE oil switches in the accident.

Thomas and Steinmetz published their analyses of the accident two years later in 1905. By that time Thomas had conducted tests on other high-voltage systems and found that oil switches did not cause dangerous oscillations. He therefore laid the blame for the accident on a resonance condition that produced a voltage rise at least 5,000 volts above Westinghouse's safe voltage limit. Steinmetz blamed the accident on the failure of a cable, which produced an "arcing ground" similar to the self-interrupting arc he had observed in 1901. He calculated that an arcing ground in the Manhattan system produced a low-frequency surge of 120,000 volts. Neither Thomas nor Steinmetz faulted the equipment of GE or Westinghouse. The absence of the usual finger-pointing prompted Stott to commend them publicly for laying aside "mere commercial considerations."[48]

Steinmetz investigated the accident using the knowledge of high-voltage phenomena he had gained at GE. In his view the trouble began with a spark discharge between one conductor and the grounded metal sheathing of the defective cable that later exploded. This was an arcing ground that did not produce a rise in current sufficient to trip the oil circuit breakers. Using the equations in his 1898 paper, he calculated the fundamental frequency of the line oscillation produced by the discharge to be nearly 2,700 Hertz. Because the discharge was self-interrupting, it set up two wave trains, which corresponded to "traveling waves" with a frequency of 13,000 Hertz and steep wave fronts. These high-frequency waves created the static disturbances noticed at the substations. Before operators could locate the ground, high voltages produced by the traveling waves weakened the insulation of the cable and caused it to short-circuit between two conductors. The short unleashed the entire power of the alternator, which destroyed the cable, whereby a second, more powerful arcing ground at the manhole created a low-frequency surge throughout the entire system. Steinmetz used the equations of his 1901 paper to calculate that this surge had a frequency of 270 Hertz and that it produced 120,000 volts between the lines with a short-circuit current of 9,000 amperes. This enormous power caused the remainder of the damage to cables, transformers, alternators, and so forth. He believed that the second shutdown was due to the partial breakdown of cable insulation during this surge. Then a small switching transient completed the cable failure and shut down the system. Stott accepted Steinmetz's explanation, rather than Thomas's, and adopted his suggestions to protect the system in the future, which included keeping GE's oil switches.[49]

The success of his theory in explaining the accident convinced Steinmetz that he was on the right track when he returned to his study of lightning. At the request of Rice he had made improvements to the inexpensive spark-gap arrester in 1901.[50] After the Manhattan Railway accident Steinmetz realized that arresters also had to protect against internal lightning: disturbances caused by arcing grounds and switching transients.

GE improved its position in this area in 1904 by transferring Elmer Creighton from the Stanley Electric Company in Pittsfield, Massachusetts, which it had just purchased, to the Schenectady plant and placed him in charge of lightning arrester development as an assistant to Steinmetz. Creighton was well qualified for this work. After graduating from Stanford University with the E.E. degree in 1897, he helped the French engineering professor André Blondel develop the oscillograph—an important tool in high-frequency research—in 1901 and then worked in high-voltage engineering for Stanley. Upon joining GE Creighton became an assistant to Steinmetz at Union College as well. As an assistant professor, Creighton used the GE-funded Protective Apparatus Laboratory at the college to carry out research on the basis of Steinmetz's theories of arresters and lightning. When he went to work full-time at GE in 1908, he continued to work in the laboratory at Union, which gave him the title of consulting professor.[51]

One of Steinmetz's theories was that of the multigap lightning arrester. Invented by Alexander Wurts of Westinghouse in 1892, the multigap arrester consisted of a vertical string of metal spheres, separated by gaps, with the top one connected to the transmission line and the bottom one to ground. The widely used arrester discharged the lightning stroke to ground and did not ground the follow-on generator current for more than one-half cycle. Steinmetz considered the arrester to be a distributed-parameter circuit and used a nonmathematical form of his theory of this circuit to analyze the arrester's proved ability to discharge high-frequency oscillations.[52]

This analysis led to a theory of lightning that was more in the realm of natural science than engineering. In 1906 Steinmetz stated that lightning flashes were probably similar to a series of discharges from sphere to sphere in the multigap arrester. He expanded this idea considerably in a 1907 paper, where he suggested that lightning was not a single discharge between oppositely charged bodies (for example, from cloud to ground) but a series of small discharges along the lightning path. Each discharge equalized a voltage gradient that had reached the breakdown voltage of air. He explained the existence of these gradients within a cloud, from cloud to cloud, or from cloud to ground on the basis of uneven condensation that occurred during the formation of thunderclouds. He supported his theory by alluding to recent work in Germany by Bernhard Walter and in the United States by Alex Larsen, who had photographed up to forty successive lightning discharges in six-tenths of a second using a rotating camera.[53] In the early 1930s experimenters using an improved camera showed that the lightning stroke consisted of a "stepped leader" formed by successive discharges, which was not an oscillatory discharge. Before this work Steinmetz and most other researchers had believed that the discharges seen by Walter and Larson were oscillatory. Estimating the energy of a stroke and the characteristics of the discharge path, he calculated the stroke's current to be 10,000 amperes, with a frequency of 500,000 Hertz. The main purpose of this part of the theory was to provide useful estimates with which to design lightning arresters.[54]

Steinmetz made a clearer distinction between his scientific musings and engineering knowledge in a second paper published in 1907, where he discussed "external" and "internal" lightning. The first category was a scientific one, consisting of disturbances due to atmospheric electricity. The second category fell into the realm of engineering, since it consisted of transients due to hardware failures and switching operations on transmission lines. These forms of lightning produced four types of disturbances—steady stresses, traveling waves, standing waves, and surges—which were often interrelated. For example, a thundercloud could set up a bound charge (steady stress) on an electric line below it. If the cloud was discharged by a lightning stroke, the bound charge would break up into two traveling waves, moving in opposite directions. After reflecting from each end of the line, the waves could produce standing waves at a dangerously high voltage that could destroy an insulator. A damaged insulator, in turn, could produce a short-circuit destructive surge, such as the one that had caused the Manhattan Railway accident.[55]

Steinmetz recommended several devices to protect against external and internal lightning. The overhead ground wire protected against direct lightning strokes, and the multigap arrester shunted traveling waves of short duration to ground. The latter could not protect the system against the surges caused by an arcing ground, because the heat built up by the current from the short-circuited generator passing through the arrester with every discharge rapidly destroyed the arrester. To overcome this defect he called for an arrester that would discharge continuously without damaging itself.[56]

Using these theoretical guidelines, which Steinmetz had developed as proprietary knowledge before publishing them,[57] Creighton developed a lightning arrester in 1906 that would meet these criteria. The aluminum-cell arrester consisted of several conical aluminum dishes filled with an electrolyte and connected in series from the power line to ground. Since the insulating film, formed on each dish by a chemical reaction set up by the generator current, punctured at voltages higher than that of the line, the arrester shunted lightning and other electrical transients to ground. As the voltage across the arrester decreased to the value of the line voltage, the film resealed itself and did not short the generator current to ground, as did the multigap arrester. Consequently, the aluminum cell could discharge surges for up to thirty minutes. This gave operators time to locate the arcing ground or other failure without bringing down the entire system—a design specification springing from Steinmetz's analysis of the Manhattan Railway accident.

Creighton's research put GE in the forefront of high-voltage engineering in the United States. By 1908 the company had placed an aluminum-cell arrester on the market—as did Westinghouse independently at the same time—and Creighton had published four lengthy articles on lightning and protective apparatus. One of the papers reported field tests he made on a high-voltage line in Colorado that recorded lightning discharges across a spark gap by means of moving photographic film through the gap. The equipment (correctly) confirmed Steinmetz's hypothesis that multiple lightning discharges often occurred, but (wrongly) showed that the discharges were oscillatory at a frequency close to Steinmetz's estimate of 500,000 Hertz.[58] In 1908 Steinmetz told Rice, "Our engineer, Mr. Creighton, is now pretty generally recognized as the highest authority on lightning and lightning protection in the United States, while heretofore this position has always been held by Westinghouse engineers: Wurtz [sic] and later Percy Thomas."[59]

The Manhattan Railway accident had shown that Steinmetz led

Thomas in theory, but he lagged in experimentation and product development, the areas in which Creighton excelled. The accident had also convinced Steinmetz that a much better theoretical knowledge of traveling waves on transmission lines was needed in order to understand the destructive effects of internal and external lightning. Thomas had published a nonmathematical treatment of these waves before the accident,[60] and Steinmetz had given equations for them afterwards. But Steinmetz found this work wanting and attacked the problem in a much more rigorous mathematical fashion with a lengthy paper in 1908.

The analysis proved to be a turning point. Steinmetz had analyzed the distributed-parameter line in 1893 by using complex numbers to convert the line's partial differential equations involving two independent variables (time and distance) into ordinary differential equations having only distance as the independent variable. This method produced the steady-state solution of voltage and current as a function of distance along the line.[61] We have seen that he did a transient analysis of a transmission line in 1898 in which he simplified matters considerably by treating the line as having "lumped" rather than distributed parameters. In 1908 he faced the problem head-on, returned to the partial differential equations, and derived the steady-state and transient solutions of voltage and current as a function of time and distance. The resulting equations described two transverse electromagnetic waves traveling down the line in opposite directions, whose propagation constant was a function of the distributed line parameters. He then studied the propagation of energy through a power line in terms of these waves traveling along the line, mostly outside it, in contrast to the popular analogy of electricity flowing like water through a pipe. The steady state resulted from the combination of main and reflected waves; transients (surges and oscillations) were caused by the initial traveling waves produced by circuit disturbances; and the rise of voltage came from traveling waves combining to form standing waves.[62]

Oliver Heaviside had obtained these results for telephone and telegraph lines almost thirty years before. Steinmetz went beyond him to consider problems peculiar to power transmission. As we have seen, Heaviside adapted Maxwell's electromagnetic theory and applied his own operational calculus to analyze the propagation of energy along wires. But since Heaviside was mainly concerned with the telephone and telegraph, he had analyzed only systems consisting of a source, line, and load (for example, a telephone transmitter, line, and telephone receiver). Steinmetz, however, investigated AC power systems, which consisted of a source (generator), step-up transformer, line, step-down transformer, and load (motors and lights). He introduced a distance variable that allowed him to describe the complex circuit with one set of differential equations, even though the distributed parameters differed widely in each section of the circuit. Next, he derived equations for the power dissipated in, or supplied from, each section of the circuit as the traveling wave moved through it. Finally, he derived equations for the voltage rise at the transition points of the circuit (for example, between a transformer and the line) due to the reflection and refraction of the traveling waves at these points.[63]

When Steinmetz did not refer to Heaviside or other theorists, Dugald Jackson, head of the electrical engineering department at MIT, put his work in context during the discussion of the paper at the AIEE. Jackson observed that Steinmetz's equations were "finely developed, comprehensively organized offspring" of Lord Kelvin's and Heaviside's equations for telegraph and telephone lines. Jackson further noted that "Dr. Steinmetz"—Union College had awarded him an honorary doctorate in 1903—"has taken a third great and important step [beyond Kelvin and Heaviside] in developing equations which are applicable to complete circuits composed of diverse characters and which also take adequate account of the terminal conditions or apparatus. The paper deals with the offspring of old friends in electrical science, but the offspring are much more comprehensive and fully developed than their parents."[64]

Steinmetz's paper was the first rigorous application of traveling-wave analysis to practical power systems. Although Heaviside's theory had proved essential for the invention of the loading coil for long-distance telephony at the turn of the century,[65] most engineers did not consider such abstruse mathematical investigations necessary for AC engineering. Percy Thomas, for example, described transients in AC transmission lines on the basis of traveling waves, but he did not publish the mathematical theory of the propagation of these waves. Steinmetz's 1908 paper showed engineers that the mathematical investigation of these waves was possible and that it was necessary in order to calculate the increasingly important phenomena of transients in the design of high-voltage transmission systems. His paper thus marked the beginning of a new field in electrical engineering, one that by the 1930s had grown into a vital subdiscipline.

In the latter part of 1908 Steinmetz, with the help of assistant professor Olin Ferguson at Union College, incorporated this paper, his previous papers on high-voltage surges, and further research into *Theory and Calculation of Transient Phenomena and Oscillations* (1909).[66] This was his magnum opus and the most advanced electrical engineering book to date. While his *Alternating Current Phenomena* was the classic work on steady-state theory, *Transient Phenomena* became the classic on surges in AC circuits and machines. Steinmetz wrote the book by applying Maxwell's theory of electromagnetic fields and waves to large-scale AC networks, but he also developed engineering methods for analyzing these systems.

In 1916 Ernst Berg called it the first "really advanced book on practical electrical engineering problems," an opinion shared by reviewers. *Electrical World* predicted that it would be the "standard treatise on the subject of transient phenomena for many years to come." The *Elektrotechnische Zeitschrift* thought Steinmetz treated the subject in an "exhaustive manner" (*erschopfender Weise*), and *Electrical Engineering* considered it a "veritable classic." Perhaps the most propitious comment came from an unlikely source, the London *Electrician*, which had criticized the use of higher mathematics in electrical engineering after publishing Heaviside's articles. In 1909 a reviewer for the journal simply wrote, "This book opens a new chapter in [electrical] engineering."[67]

The new chapter was far too mathematical for many, though, prompting Steinmetz to publish *Elementary Lectures on Electric Discharges, Waves, and Impulses* in 1911. Based on tutorial articles for GE engineers and on lectures at Union College, the book was intended to provide a "translation from mathematics to English" of *Transient Phenomena*. Steinmetz dropped the higher mathematics and presented the subject using the physical image of the surging of energy between electric and magnetic fields surrounding the line. He also included many oscillograms to enable engineers to visualize the oscillations and wave trains described by the lengthy equations (up to four pages long) in *Transient Phenomena*. As an elementary textbook, *Waves and Impulses* put his work into a readable but mathematically correct form, which helped engineers understand the equations necessary to design protective devices for power systems.[68]

The Consulting Engineering Department

While developing the theory of transient phenomena, Steinmetz spent little time at the plant. He later admitted that he "had practically retired from active work for the Company [and] visited the factory only rarely, sometimes not for weeks" during this period. Except for the recession of 1907 when he returned to full-time work at the plant, he spent his time teaching at Union College, working in his home laboratory, and writing papers and textbooks at Camp Mohawk during the summer.[69] GE was not shortchanged on its investment. Between 1903 and 1913 Steinmetz took out sixty-three patents and wrote many of the textbooks that brought prestige to the company (see the Appendix).

Besides the obvious personal benefits of having such freedom, Steinmetz had other reasons for working at home. In 1911 he wrote in a private notebook that Whitney's laboratory "has not been, what I hoped for it, and I have practically no relations with it any more."[70] He had hoped that it would have been more responsive to the needs of GE engineers. In 1908 he told Rice that the older labs at the plant—the Testing Room and the Standardizing Laboratory—were adequate for developing electrical equipment. "Entirely unsatisfactory however are the facilities for pioneer work in the field of electrochemistry." He thought that Whitney's laboratory, which by early 1907 had grown to 150 employees, "has done very good work in developing the suggestions which it has taken up, but its organization is entirely unsuitable for carrying out any smaller preliminary investigation, or such pioneer work [in metal filament lamps], or to assist the engineers in their work where it is of [a] chemical nature. This as you remember was the reason why I re-established my private laboratory."[71]

Steinmetz reminded Rice that his home laboratory had been at the disposal of GE engineers or the Patent Department for electrochemical work. The lab had developed the magnetite arc lamp, the constant-current mercury-arc rectifier, the aluminum-cell lightning arrester, and a silicon-carbon alloy for lightning arrester rods. Work in progress in 1908 included that on multigap lightning arresters, high-voltage DC systems, vacuum-tube lighting, and the extraction of nitrogen from the air to make fertilizers or explosives. Steinmetz did not have facilities to do research on other matters, including "ductile tungsten" for lamp filaments, which were "dragging [on] for years." European researchers were far ahead in this area. In 1906 GE paid $350,000 for two metal-filament processes, an expense Whitney's laboratory had been established to avoid. Steinmetz thought it "shameful . . . that the successors of Edison should cease to be considered as the leaders in the development of incandescent lighting, and European engineers get the credit which our Company should have retained, if its engineers had not been asleep."[72]

As Steinmetz wrote this letter in October 1908, Whitney lay in a Schenectady hospital, recovering from a relapse of a physical and mental breakdown that had occurred the previous winter. His spirits revived near the end of 1908 when physicist William Coolidge, assistant director of the laboratory, began work on a hot swaging process for making brittle tungsten into a flexible filament wire, a process he patented in 1910. The invention helped return control of the incandescent lamp business to GE and placed Whitney and his laboratory in an unassailable position at the firm.[73]

In the fall of 1908, without the benefit of hindsight, Steinmetz thought that an engineering laboratory modeled after the one at Wendell Avenue was better suited than Whitney's group for such projects as ductile tungsten. He thus recommended to Rice that "a facility and organization should be provided to carry out such pioneer investigations in fields requiring chemical knowledge and facilities."[74] In the same letter Steinmetz complained about the lack of originality of GE engineers and that neither he, Rice, nor Berg had been able to keep tabs on product development since engineering had been decentralized some years before. One way to coordinate it would be to establish an engineering council similar to the laboratory advisory council for Whitney's group.

Steinmetz revived many of these recommendations in a proposal he made a year later to establish a Consulting Engineering Department. It would develop new lines of apparatus, conduct basic engineering research, standardize design methods, and provide consulting services both within and outside the company. The department would not supervise engineering at GE but would help coordinate it by staying informed of design work throughout the firm and making suggestions. In carrying out these tasks the group would educate engineers with broad experience and help them gain a national reputation, thus halting the decline of GE engineering Steinmetz had observed in 1908.[75]

GE followed Steinmetz's recommendation—as it had in 1900 for Whitney's laboratory—and established the Consulting Engineering Department in May 1910, under the administration of C. W. Stone with Steinmetz as its technical director.[76] Stone performed the role Berg had played in the Calculating Department, a post Berg could no longer fill because he had left the year before to head the electrical engineering department at the University of Illinois. Although Steinmetz had turned primarily to electrochemical and high-voltage research, his relatively small staff—most of whom had engineering degrees—covered the entire field of electrical engineering. Ernst F. W. Alexanderson, a Swedish engineer who had come to Schenectady in 1902 to work with Steinmetz, specialized in railway motors and radio equipment. (In 1904 Steinmetz assigned him the project of designing a radio-frequency alternator for Reginald Fessenden, which led to Alexanderson's developing an improved alternator that formed the basis for the establishment of the Radio Corporation of America fifteen years later.)[77] Engineer John Taylor concentrated on transformers. Walter Slichter, who had designed induction motors at the Calculating Department, continued similar work in the new group before leaving to become professor of electrical engineering at Columbia University in the fall of 1910. He was replaced by Henry Hobart, an experienced designer who had worked with Steinmetz at Lynn before becoming a private consultant with Horace Parshall in England. Hobart returned to GE in early 1911 to design AC and DC machines in the new department. On the high-voltage side were Frank W. Peek, Jr., in corona research, C. M. Davis in high-voltage transformers, and, of course, Steinmetz. Assisting them were L. A. Simmons and W. S. Andrews. Steinmetz also took in student engineers from the GE test course as assistants.[78] Although not listed in the department in early 1911, Creighton was a de facto member who wrote reports to Stone and had an office in the same building. He had joined the department by January 1912.[79]

In keeping with his conception of the proper inventive environment—which James Brittain has aptly called "creative engineering in a corporate setting"—Steinmetz gave his engineers a rather free hand. In view of the innovative and wide-ranging nature of their work, he considered an "autocratic" organization undesirable. (He also hated administration, as we have seen.) He permitted his engineers to choose their own tasks as long as they were along lines useful to the company or to the electrical industry in general, which also helped GE since it was the largest electrical manufacturer. The goal was to provide "practically complete independence of the engineers of our department."[80] His supervision of the staff was as light as it had been at the Calculating Department in the 1890s. In 1911 he told Whitney that he would be at the plant two hours a day, from 10 a.m. to noon, four days a week for "shorter consultation," and at the Wendell Avenue laboratory from 3 p.m. to 5 p.m. for longer conferences. He kept up with the work in the department by requiring biweekly and then monthly reports from his staff. This management style contrasted sharply with that of Edison at West Orange and Whitney at GE, who closely monitored their researchers.[81]

In May 1912 Steinmetz proposed to Rice that the Consulting Engineering Department provide a "centralized informal supervision of development" throughout the company. The department was already receiving reports from two engineers at Pittsfield, one at Lynn, and two at Schenectady in the areas of high-voltage engineering and arc lighting, in addition to reports received from A. McKay Gifford's chemical laboratory at Pittsfield and Whitney's laboratory since liaison engineer Laurence Hawkins joined it in 1912. Yet Steinmetz thought that reports were needed from all engineers and scientists doing developmental work for GE in order to coordinate this vast decentralized activity. He asked Rice to request heads of departments and managers of factories to keep the C.E. Department informed of all development work so that it could give advice and act as a clearinghouse. This function never materialized, but he expected his engineers to keep up with R&D throughout the company.[82]

Steinmetz's group did a good deal of R&D on its own: experimental research and the development of products to the point where they could be turned over to an engineering department. For the first few years the C.E. Department did not have its own laboratory. Alexanderson and Hobart developed their machines in the Testing Room, as GE engineers had done since the 1890s. Creighton and Peek experimented on high-voltage transmission lines in the field and at Schenectady, where Creighton worked in the Protective Apparatus Laboratory at Union College and Peek did experiments at an outdoor high-voltage laboratory set up at the factory in 1910.[83] Steinmetz and Hayden worked at Wendell Avenue on such projects as electric insulation, lightning arresters, tungsten lamps, arc lighting, and high-voltage phenomena. Some projects they turned over to the C.E. Department, while others were cooperative efforts with engineers at the Pittsfield and Lynn Works. In addition to transient phenomena, Steinmetz investigated the energy produced by the tungsten filament, which resulted in empirical equations similar in form to the law of hysteresis.[84]

Steinmetz expanded the Wendell Avenue laboratory in 1911 by adding three rooms to the two-story facility, which was conveniently connected to the main house. Well-equipped electrical, physical, and chemical laboratories, a darkroom, and storerooms occupied the first floor. The second floor contained a glass-blowing room, a high-voltage laboratory, and a technical library with a conference area. In many respects it was a smaller version of Edison's West Orange laboratory without the machine shops. GE assigned Hayden and one or two assistants to the lab, lent it equipment, and paid the bills of what was, in effect, one of the company's research facilities.[85] GE also paid its director handsomely. When the laboratory was enlarged in 1911 Steinmetz's salary was raised from $12,000 to $18,000 per year, making him one of the highest-paid employed engineers in the United States.[86]

The Consulting Engineering Department obtained its own laboratory in 1914, a year after Steinmetz resigned his post at Union College in favor of Ernst Berg and returned to full-time work at GE. The Consulting Engineering Laboratory became the organization he had petitioned Rice for six years earlier. As director, Steinmetz divided the lab into sections with a "consulting engineer" in charge of each. In the high-voltage branch, Creighton and members of the Protective Apparatus Laboratory, which had moved from Union College to the plant, developed lightning arresters and other protective devices. Peek studied dielectric phenomena, and G. B. Shanklin tested cable insulation. On the machinery side, Alexanderson worked on AC apparatus, magnetic hysteresis, and radio-frequency alternators. The electrochemist-engineers did research on arc lighting, ultraviolet radiation, and nitrogen fixation by the electric arc. Since Steinmetz had done research on all of these topics, the C.E. Department's lab, like Edison's at West Orange, bore the unmistakable stamp of its creator.[87]

We can better understand how Steinmetz's laboratories at home and at the plant compared with others in industry by considering two case studies. The first case—the development of condensite, an early plastic—began with a conversation between Steinmetz and Edison at a trade association meeting in September 1910. Steinmetz learned that Edison's laboratory had invented an improved process for making Bakelite, a plastic invented by chemist Leo Baekeland in 1907. A formaldehyde condensation product, Bakelite was not flammable like celluloid, the first plastic, but it was much more brittle. Edison wanted to make an improved Bakelite for his phonograph cylinders, while GE and other electrical companies liked its electrical insulation properties. Steinmetz had experimented with the material in 1909 in the form of carbonized Bakelite filaments for electric lamps and had mixed Bakelite with paraffin oil, hoping to make it less brittle and thus a better insulating material. Neither project was successful. Edison's process for making Bakelite, developed primarily by his chemist Jonas Aylsworth, was reportedly simpler and quicker than Baekeland's, produced a material that could be applied as an insulating coating to wires and coils, which he later called "condensite," and appeared not to infringe Baekeland's patents. Steinmetz recommended to Rice that Edison "be approached by some of his former assistants in the company, to induce him to give us the advantage of his investigation."[88]

Steinmetz asked Whitney for information on condensation products and tried to persuade him to follow up Edison's lead. Whitney replied that his researchers had experimented with Bakelite, that Baekeland's representatives had shown them how to work with it, and that they had visited Baekeland's laboratory and planned to do so again. Bakelite had not been used thus far because it was too expensive and had some manufacturing problems. Whitney was interested in condensite but did not offer to investigate Edison's work. Steinmetz replied that Whitney misunderstood the situation and was not "working in the direction where we are badly in need of an insulating material practically regardless of its cost." Steinmetz wanted someone from Whitney's laboratory to learn about factory methods of insulating high-voltage armature coils to see if a Bakelite product could be used in this application.[89]

When Whitney did not oblige, Steinmetz decided to direct the work himself. He asked Gifford, the chemist of the Pittsfield Works, to coat coils with Bakelite that October, wrote Edison for more information on condensite, and visited Edison's laboratory with Rice and Stone in January. The visit prompted him to send Hayden and Gifford to the lab to learn more about the process. They left high-voltage coils with Edison to be coated with condensite, which were then tested at GE. L. E. Barringer, an engineer working on Bakelite in Whitney's laboratory who was appointed head of the Insulation Engineering Department in January, helped Steinmetz on the project.[90] While GE used Bakelite in arc lamps and transformers in early 1911, Gifford could not make condensite flexible enough for insulating purposes. When he did achieve some success in May 1912, Steinmetz recommended that GE hire "several chemists" to do research on condensite and related artificial resins (plastics), including the investigation of their "chemical structure." He believed that "this work should be carried out in close cooperation with our Research Laboratory [i.e., Whitney's group], but should be done in Pittsfield." This suggestion eventually resulted in a cooperative effort between Whitney's group, Barringer, and Gifford's Pittsfield laboratory in developing Glyptal, an alkyd synthetic resin that helped start GE's industrial-plastics division at Pittsfield.[91]

The contrast between Steinmetz and Whitney in this case is revealing. Steinmetz had admired Edison since the late 1880s when he wrote a popular article on the wax-cylinder phonograph for a Zurich newspaper and applied for a job at the Edison Machine Works. He later praised Edison as one of the founders of his profession and had faith in Edison's type of research, whose methodology was not that far removed from his own empirical investigations of hysteresis and high-voltage phenomena. Although Steinmetz's mathematical style of invention and research contrasted sharply with Edison's more visual, shop-culture approach, he had more in common with Edison than with Whitney in many respects. He and Edison were highly regarded in the community of electrical inventors and engineers, and the public saw him as a "wizard" in the mode of Edison. Steinmetz was much more active than Edison in organizations like the AIEE, yet he saw Edison at meetings of the Association of Edison Illuminating Companies. Whitney, on the other hand, belonged more to the scientific community and was active in groups like the American Chemical Society and the American Electrochemical Society. Furthermore, Steinmetz and Edison were self-taught chemists (Steinmetz's formal training in chemistry was rather meager), while Whitney, the Ph.D. chemist, put more faith in the university-trained Baekeland than in Edison or Aylsworth. Critical of Whitney in his letter to Rice in 1908, Steinmetz apparently used Edison's condensite project as another means to criticize Whitney. Conversely, Whitney probably thought that the purpose of his science-based lab—which cultivated the school culture of its many German-trained Ph.D.'s with university-like colloquia—was to replace the outdated Edisonian tradition at GE, not to treat it as an equal in cooperative ventures.

Steinmetz knew something about Whitney's work. He sat on the laboratory's advisory council, borrowed equipment from Whitney, had a storeroom in the Research Laboratory, and followed the work of Weintraub on metalized carbon filaments. But he knew nothing about Whitney's research on Bakelite, and Whitney apparently knew little about the high-voltage engineering problems Steinmetz's group was trying to solve. The tension evident in their relationship from the beginning increased during the condensite project, which carried over in 1910 to the issue of the life of GE lamps. As in the case of condensite, Steinmetz became interested in the matter at a meeting of the Association of Edison Illuminating Companies, where he learned that many light bulbs burned 25–60 percent longer than GE's lamps. He then recommended ways to improve the manufacture and design of the lamps. Whitney agreed with some of the suggestions and talked to Irving Langmuir about Steinmetz's idea of a reflected focus on a filament, but nothing seems to have come of it.[92]

The second case study—engineering research by Peek on high-voltage coronas—was more successful than the condensite project. The presence of this luminous discharge along power lines prevented raising transmission voltages, because it caused a substantial loss of power above the critical voltage where the corona first appeared. Steinmetz had acquired much experience in this area since first observing coronas in 1893. Ten years later he designed a means for measuring energy lost by corona and directed corona tests on the Kalamazoo line in 1908. During the same year he suggested that GE spend more money on high-voltage research, including the corona problem, instead of developing high-voltage equipment without doing basic engineering research. And in 1910 he laid out and monitored corona tests as a function of altitude and humidity conducted by GE engineer Giuseppe Faccioli on power lines in Colorado. Since Faccioli's results were inconclusive, Steinmetz recommended that Peek conduct further research on GE's experimental transmission line in Schenectady in order to provide a basis for interpreting Faccioli's data.[93]

Peek had excellent qualifications for this work. After receiving the E.E. degree from Stanford University in 1905, he joined GE and spent two years in the Testing Room, a common practice in those days. He then worked for two summers helping Creighton conduct high-voltage tests on power lines in Colorado. Peek was a charter member of the Consulting Engineering Department and, upon Steinmetz's recommendation, received the M.E.E. degree from Union College in 1911. In that year he also taught a course for seniors on the operation and protection of high-voltage transmission lines.[94]

Peek performed his corona research from 1910 to 1912 on experimental lines at the plant and in the laboratory. The outdoor set-up consisted of five-hundred- and one-thousand-foot lines with variable spacing of conductors. Three full-size boxcars on a railway track running beneath the overhead lines carried instrumentation, equipment to raise the voltage to 200,000 volts, and a darkroom to develop photographs of corona on the lines. These facilities were vastly superior to those Steinmetz had used in his research on the dielectric strength of air at Eickemeyer's and at Wendell Avenue. In a letter to Dugald Jackson, then president of the AIEE, Steinmetz stated that Peek's research was "not some laboratory experiments, made with limited facilities, and therefore [of] limited weight, but that practically unlimited facilities have been made available to finally settle all the important questions on the subject of corona."[95]

Peek published three papers between 1911 and 1913 that answered most of these questions. A mathematical analysis of his data revealed that the loss of power due to corona was proportional to the line frequency times the square of the difference between the line voltage and the critical voltage at which corona first appeared. The proportionality constant depended on the radius of the wire, the distance between wires, and the density of air. The critical voltage was a function of those parameters, the surface condition of the wire, and the disruptive strength of air. Peek found the latter to be 76,000 volts per inch at standard temperature and pressure—thus confirming the estimate Steinmetz had made in 1898. Peek also discovered that humidity and strong winds had no effect on the power consumed by a corona or its critical voltage. But smoke, fog, rain, and snow lowered the critical voltage and increased the power loss.[96]

Previous experimenters, such as Harris Ryan at Stanford University, Ralph Mershon at Westinghouse, and J. B. Whitehead at Johns Hopkins University, had reached some of these conclusions,[97] but Peek's exhaustive analysis settled many points. Ryan informed AIEE members in 1911 that Peek, who had graduated from Stanford the year Ryan joined the faculty, had determined the best values to date for the disruptive strength of air, the critical distance from the wire for a corona to form, and the relationship between critical voltage and air density, in addition to pointing out the need to consider the irregularities of the surface of the wire. He also praised Peek's equations for power loss due to corona, the disruptive critical voltage, and the visual critical voltage for their practical value to designers of high-voltage systems. In 1915 Peek incorporated these papers and further research into *Dielectric Phenomena in High-Voltage Engineering,* which became the standard book on the subject for many years.[98]

Steinmetz influenced this research in several respects. Peek started from Steinmetz's work on the disruptive strength of air. He disagreed with Steinmetz's explanation of the role of a condensed layer of air surrounding the wire in increasing the breakdown gradient, yet accepted his advice on methodology and other theoretical matters. He used Steinmetz's wattmeter method to measure energy lost by corona (reminiscent of the wattmeter method to measure hysteresis), a curve-fitting technique favored by Steinmetz to derive his empirical equations, and Steinmetz's theory of the dielectric strength of air to explain the formation of corona. To a great degree Peek followed the style of engineering research Steinmetz had developed during his hysteresis experiments at Eickemeyer's twenty years before. The titles of Peek's papers were "The Law of Corona and the Dielectric Strength of Air (Parts I, II, & III)," which mimicked his mentor's "The Law of Hysteresis (Parts I, II, & III)." Both men marshaled massive amounts of experimental data and used curve-fitting techniques to derive empirical "laws" of value to design engineers. Steinmetz did not direct Peek's work every step of the way. He provided a model for engineering research, set the corona program before him, and persuaded GE to fund it, even though immediate commercial benefits were not assured. One reason Peek's work was more successful than that on condensite was because it fit squarely in the tradition of engineering research pioneered by Steinmetz.[99]

We will see that after the United States entered World War I, Steinmetz joined the ongoing military research in the Consulting Engineering Laboratory. In 1919 the laboratory merged with Lewis Robinson's Standardizing Laboratory to form the General Engineering Laboratory. Robinson became the director of the new organization, whose bureaucratic structure lacked the creative environment of the original department. Steinmetz took charge of the high-voltage section, where his assistants built the celebrated lightning generator upon which so much of his popular fame rests. During this reorganization Alexanderson became the first chief engineer of RCA, which GE helped establish, and worked part-time as head of the Radio Consulting Department at GE. Eventually these various departments were merged into one engineering laboratory. In 1965 GE recognized the increasingly close connection between the scientific research performed in the laboratory founded by Whitney and the engineering research in the laboratory started by Steinmetz and merged them into the present Corporate Research and Development Center at Schenectady.[100]

In the Consulting and General Engineering Laboratories, Steinmetz institutionalized a different type of research from that performed at the Research Laboratory. Whitney's group did a good deal of research on the scientific basis of new products. Irving Langmuir, for example, studied atomic reactions at high temperatures and low pressures during his investigation of why light bulbs darkened. The work led to the controlling patents on the gas-filled incandescent lamp in 1912, as well as the Nobel Prize in chemistry in 1932. Although Langmuir's ability to merge high-quality scientific research with product development was exceptional, several of Whitney's researchers did some science-based research, as would be expected from a group of Ph.D. physicists and chemists.[101] Steinmetz's group, on the other hand, did engineering research of the type pioneered by its director. Much of the work, like Creighton's on lightning arresters, was product development. Yet they also studied the operation of machines, including transmission lines struck by lightning, and the characteristics of their materials in an attempt to derive new engineering knowledge. Alexanderson investigated the magnetic hysteresis of iron at high frequencies for his radio-frequency alternator, Peek developed a "law" of corona formation on transmission lines, and Steinmetz further developed his theory of arcing grounds. Some science was involved in these investigations, and the research, like Steinmetz's on hysteresis in the 1890s, was carried out by a scientific method. But since the engineers did not have advanced training in science and focused on improving technological knowledge, they did not perform the type of science-based work done at Whitney's lab.

Steinmetz's laboratories were more like those at AT&T and Westinghouse during this period. In 1911 AT&T established a Research Branch in its Engineering Department to develop a transcontinental telephone system by the target date of 1915. Although the group had many Ph.D. physicists, its early work was directed toward engineering research. Physicist H. J. van der Bijl, for example, developed mathematical theories of vacuum tubes that guided the design of these devices. These "technological theories," to use Leonard Reich's phrase, resembled the theories of apparatus Steinmetz and his assistants had derived for AC machines and transmission lines. Unlike GE, however, AT&T initially did not permit its researchers to publish their theories. The situation at Westinghouse was somewhat different. In 1916 the company established a new laboratory in its Engineering Department, set it apart from the plant at nearby Forest Hills, and staffed it with Ph.D. physicists and chemists to do scientific research for the older materials and product-development laboratories at the factory. In most respects, the works labs at Westinghouse were comparable to Steinmetz's engineering laboratories and the Forest Hills laboratory was like Whitney's group at GE. Some investigators like Tyrgve Yensen at Forest Hills, however, did research on magnetic materials that was similar to the type of work done in Steinmetz's labs. The Forest Hills group turned away from science-based research to develop radio equipment in the 1920s, then began a new scientific research program in nuclear physics in the 1930s.[102]

Steinmetz's laboratories stood midway between Edison's product-development laboratory and the science-based laboratory pioneered by Whitney. Edison's group used science and a scientific methodology more than is commonly assumed, but Whitney's lab paid much more attention to the production of scientific knowledge. Steinmetz's laboratories spanned the spectrum from product development to scientific research, yet they concentrated on what I have called engineering research: experimental and theoretical investigations of technological artifacts and their materials, with the object of providing the theories of devices, design equations, and other engineering knowledge discussed in Chapter 1.

A scientist-engineer who bridged the two communities, Steinmetz made his reputation primarily in the engineering world and usually saw himself as an engineer, not a physicist. In 1908 he complained to an editor at McGraw-Hill that "physicists still use the term coefficient of self-induction to a considerable extent [instead of the term *inductance* used in electrical engineering], in conformity with their habit of lagging [a] quarter of a century behind."[103] He established the Consulting Engineering Department and signed himself as chief consulting engineer because of his strong belief in the value of the engineering variety of industrial research.

He was not unknown in physics circles. James McKeen Cattell, editor of *American Men of Science,* asked him in 1909 to "arrange our physicists in the approximate order of merit," since he had been highly ranked in Cattell's previous study. As Steinmetz became better known in the popular press, which before 1920 usually made little distinction among scientists, engineers, and inventors, he saw himself more and more as a scientist. He chose the title of professor of electrophysics when he retired from Union College in 1913 and wrote an article on "scientific research" in industry in 1916. He apparently chose the physics title in a game of one-upmanship with his old rival Michael Pupin at Columbia University, who was professor of electrodynamics. While purporting to describe scientific research, the 1916 article focused on Peek's corona investigations.[104] Like his mentor Steinmetz, Peek produced equations of use to design engineers, not a scientific understanding of the natural phenomenon of luminous electric discharges. The difference was more than one of semantics, because of the higher status of scientists after World War I. But it mattered little to General Electric, which was able to incorporate both science and engineering in a highly profitable manner.

Part 3

≡ **Engineering Society**

7 Reforming a Profession

$\equiv$ **W**hile working to incorporate science and engineering at General Electric, Steinmetz stepped onto the wider stage of public affairs as a college educator, professional society leader, political economist, politician, and cultural hero. What bound these roles together was the attempt, common in the Progressive Era, to "engineer society," to apply the methods and values of science and technology to reform education, the professions, business, and government. As observed by Robert Wiebe and others, many reformers came from a "new middle class" of doctors, lawyers, social workers, journalists, and other professions emerging at the turn of the century. Although the goals of professionalization and social reform were often not identical, the new professionals did seek to impose a bureaucratic order, run by experts, on a national culture being shaped by the forces of urbanization and industrialization.[1]

They also wanted to improve their status by raising the standards of their professions. As part of the new middle class, electrical engineers shared the belief that a "professional" was formally trained in a systematic body of knowledge, had a sense of social responsibility, and belonged to a group that was self-regulating.[2] A leading engineer and a Socialist, Steinmetz worked for reforms along these lines. He called for better cooperation between industry and academia while teaching at Union College and as a leader of the American Institute of Electrical Engineers campaigned for a code of ethics, technical standards, and professional autonomy.

Union College

As a GE engineer and educator, Steinmetz represents the type of corporate influence on engineering education analyzed by David F. Noble. Noble argued that "corporate reformers" in industry, government, and academia cooperated to make industrial research, patent laws, modern management, and engineering education the servants of corporate capitalism. Electrical manufacturers like GE provided the leadership for the corporate reform movement. In the area of education GE reformers set up cooperative courses with colleges and in-house postgraduate programs that trained engineers to corporate specifications, thus helping stabilize capitalism against the liberating force of scientific technology. There is little doubt that GE helped shape electrical engineering to benefit corporate capitalism, yet Steinmetz's role in this endeavor has been relatively unexplored.[3] On the surface, his socialism conflicts with the ideology of most corporate reformers, but his politics meshed well with their philosophy of "corporate liberalism," as we will see in Chapter 8. Steinmetz's experience at Union College provided a basis for his later "corporate socialism" and helps us better understand GE's role in electrical engineering education.

When Union appointed Steinmetz professor of electrical engineering in 1902, it was not breaking with its liberal arts tradition. Founded in 1795, the college had introduced several curriculum innovations in the early 1800s under the sixty-year presidency of Eliphalet Nott. A noted inventor of stoves and steamship boilers, Nott established one of the earliest science degree programs in the country in 1828 and set up the first engineering program not funded by outside business interests in a liberal arts college in 1845. Union became a respected civil engineering school, graduating forty engineers from 1850 to 1865, while its neighbor, Rensselaer Polytechnic Institute, graduated nearly ninety in the same period. Union also held a prominent place among American colleges, being third in enrollment in 1829 and fifth behind Yale, North Carolina, Virginia, and Harvard thirty years later.[4]

But the college fell on hard financial times after the Civil War. It failed to keep pace with the growth of other liberal arts schools and lost its place in engineering education to MIT, Cornell, and the midwestern land-grant colleges. Union finally emerged from its financial wilderness when several economic reforms instituted by president Andrew Raymond brought the college out of debt in 1901.

Another factor in Union's revitalization, according to one historian of the college, emerged when GE moved its headquarters to Schenectady. In response, Union established an electrical engineering course as a senior option in the civil engineering program in 1895. There were seven graduates in 1898 under Bryon Brackett, who held a doctorate in physics from Johns Hopkins University, and three to six graduates per year from 1898 to 1902 under Horace Eddy of the University of Minnesota. President Raymond had approached GE to aid the struggling program, but he was unsuccessful until Steinmetz agreed to chair a new department of electrical engineering in 1902.

The appointment helped both Union College and GE. Union was recovering from its economic woes and welcomed the financial support from GE. In a college publication that heralded Steinmetz's arrival as the beginning of "Union's New Era," president Raymond predicted that the college would soon have the "best electrical engineering department in the country," thanks to GE's support and Steinmetz's reputation attracting a large number of students.[5] GE's motivations were less clear. The company did not support Union primarily to bolster electrical engineering education; in 1902 more than three thousand undergraduate E.E. students were already enrolled in more than one hundred American colleges and universities. Rather, GE benefited from the favorable publicity of aiding higher education, the opportunity for its engineers to obtain advanced training nearby, and the prospect that Steinmetz would stay in Schenectady.

Supporting Union College also complemented the company's other educational programs. GE's de facto policy in the early twentieth century was that engineering colleges should teach basic principles and leave practical training to corporations. This philosophy had its roots in the early days of the electrical industry, when GE set up an in-house training program because of the lack of college-educated engineers. In 1885 Thomson-Houston organized an "expert's course" to train design engineers and salesmen. When the firm merged with the Edison companies to form General Electric in 1892, GE set up a new in-house training program: the student engineering course, or "test course," as it came to be known. The company placed newly hired graduates in the two-year course, where they learned to coordinate theory with practice by running factory-acceptance tests on equipment and then working in other engineering departments. Through the test course GE selected the best men for the company—few women graduated as engineers at the time—and indoctrinated them with the corporate team spirit. The company usually hired only one-half of the graduates and placed the other half with potential customers, thus educating a sizable group of nonsalaried salesmen for its products. Ernst Berg told the head of the program in 1912 that its students were "probably as much or even of more use [to GE] outside than in the company."[6] Following GE's example, Westinghouse, Western Electric, and Allis-Chalmers established similar internship programs.

GE also used the test course to improve its relations with colleges by hiring professors for the program in the summer. Steinmetz told a company manager in 1909, "This is one of the most efficient and useful—and cheapest—ways of effectively advertising our apparatus." Carl Magnusson, head of the electrical engineering department at the University of Washington, for example, spent the summer of 1911 and the next two academic years working at GE, first in the test course and then with Steinmetz in the Consulting Engineering Department.[7]

As a lecturer in the test course, a member of GE's Education Committee, and an engineering theorist, Steinmetz held an educational philosophy that meshed well with company programs. In his AIEE presidential address in 1902 he stated that colleges should ground students in the fundamentals, rather than try to turn out immediately useful engineers. This view met with the approval of both GE and Union College. The company used the test course as an internship program to train design engineers and managers, while Union maintained its liberal arts tradition by teaching almost no commercial practice to regular students. In serving both GE and Union College, Steinmetz believed that each could retain autonomy in its sphere of influence. The rather paradoxical idea of autonomy through cooperation was his guiding philosophy during his decade at Union and helped determine GE's involvement in other educational programs.

The cooperation between the company and the college covered three areas: financial and material support from the firm, GE engineers teaching at Union, and cooperative educational programs. Union benefited substantially from GE's donations of money and equipment. When Steinmetz joined the college, GE's board of directors agreed to donate $2,000 worth of laboratory equipment and $2,000 per year for five years, provided that Union matched one-half of the latter sum. Alumni and friends raised this amount and Union continued its previous allocation of $1,000 per year to electrical engineering. GE thus paid one-half of the new department's budget of $4,000 per year. Steinmetz agreed to organize the department, give two lectures per week, and direct a laboratory and postgraduate work without pay. But he was paid well by GE, earning an annual salary of $18,000 in 1911. When he retired from Union in 1913, GE continued to fund his professorship. The chair went to Ernst Berg, who had left GE in 1909 to head the electrical engineering department at the University of Illinois. Berg's latest three-year contract at GE was up that summer, which left him free to drive a hard bargain.[8] His income at Illinois was $10,000 per year, of which $6,000 came from consulting contracts with GE and Illinois utility companies, including Commonwealth Edison in Chicago. President Charles Coffin of GE matched this sum and guaranteed him $10,000 per year to head Union's E.E. department and do consulting work for the company.

GE continued to donate equipment to the college, so that Union had "considerable [sic] more than the average institution in America" in 1914, according to Berg.[9] Most of the machinery was obsolete, though, and did not give the type of up-to-date experience available at the test course. In 1905 GE gave Union an engineering building on campus worth $25,000. Andrew Carnegie promised $100,000 to equip it, provided that the college endowed a matching sum. Once the friends of the college and GE contributed their share in 1909, Union had a modern electrical engineering laboratory and classroom building. In return for these and other gifts, Union allotted GE space for a campus laboratory in 1904. Headed by Elmer Creighton, the lab performed the high-voltage research and development noted in Chapter 6 until it moved into the plant in 1913 to make way for a new graduate laboratory.

During the chairmanships of Steinmetz and Berg, GE engineers often taught at Union—the second area of cooperation. In the fall of 1903 eleven leading engineers gave postgraduate lectures on state-of-the-art design techniques to special students, who were most likely other GE engineers. The lectures covered the entire range of apparatus manufactured by the company. From 1904 to 1906 Creighton was assistant professor to Steinmetz as well as head of GE's laboratory at the college. After Creighton returned to full-time work at the plant in 1906, Steinmetz asked Union to appoint him "consulting professor," since he still gave lectures at the college. Steinmetz asked that Berg be given the same title for teaching the course, E.E. Practice, to seniors. In 1912 eight GE engineers taught segments of this course, which had become a regular part of the curriculum. Henry Hobart, who had just joined the Consulting Engineering Department, was scheduled to teach the undergraduate design course that year. Union also conferred honorary doctorates on prominent GE engineers, including Berg and William Le Roy Emmet.[10]

Steinmetz had a strong voice in determining the college's E.E. curriculum for nearly a decade. In 1903 he established the undergraduate course according to his ideas of a broad engineering education. Students took "general culture" studies, mathematics, and science in their freshman and sophomore years. They studied some electrical engineering as juniors but concentrated that year on mechanical engineering. The senior year was devoted to electrical engineering. Steinmetz thought this curriculum "will give a thorough grounding in the fundamental principles and their application to all branches of electrical engineering, such as will enable the graduate rapidly to acquire the practical experience necessary to electrical engineering success."[11] This philosophy held up remarkably well during his decade at Union. The undergraduate curriculum of 1912, for example, included only two practical courses—Design and E.E. Practice—both taught by GE engineers. Although Berg taught the latter course when he first took over the department, he appears to have followed Steinmetz's educational policy at least until the mid-1920s. Steinmetz felt confident that he would. In recommending him for the chair at Union, Steinmetz told the president of GE, "Dr. Berg has been connected with me for so many years, that his engineering views are practically identical with mine, and his success in his present position [at the University of Illinois] assures success here, if sufficiently supported."[12]

Despite GE's support and Steinmetz's reputation, Union's electrical engineering department did not become the "best in the country," as Raymond had predicted in 1903. But it did fairly well considering its size and facilities. The number of baccalaureate graduates grew from three in 1903 to about twenty by 1913, while the number of faculty increased from four to five in the same period. These were respectable numbers for a school Union's size, but much lower than figures at the large midwestern land-grant colleges and at private schools like MIT and Cornell, which already had between thirty-five and seventy E.E. graduates at the turn of the century.

Union's graduate program in electrical engineering had a mixed record. In 1903 Steinmetz established two postgraduate courses. The first was a one-year program, open to regular students, that led to the M.E.E., a professional degree. The second program was a special course for "nonstudents," probably GE engineers, that did not lead to a degree. A total of twenty-five students enrolled in the two courses in 1903. In 1910 Steinmetz established a two-year doctoral program following the completion of the one-year M.E.E. degree. Several GE engineers and some Union faculty completed the M.E.E. degree and a few, like F. W. Peek and Ernst F. W. Alexanderson, entered the Ph.D. program, but the college had no "resident" (i.e., non-GE) graduate students in 1912. The department had trouble attracting full-time graduate students because of the lack of laboratory facilities and Steinmetz's part-time role at the college. Although he taught his classes on a regular basis until 1912 or so, he was only there in the mornings and spent even less time on campus after establishing the Consulting Engineering Department. He admitted in 1910 that he had "turned over the administrative and executive work of the department . . . already some years ago" to assistant professor Olin Ferguson.[13]

Union also considered establishing cooperative engineering courses with GE—the third area of interaction between the company and the college. Yet despite Steinmetz's insistence on a close relationship between academia and industry, he disliked the cooperative movement in engineering education started by Hermann Schneider of the University of Cincinnati in 1906. Schneider's plan was that students spend alternate weeks in the classroom and factory in order to coordinate theory with practice. In 1907 Magnus Alexander, a director of the test course at the GE plant in Lynn, and Dugald Jackson, head of the electrical engineering department at MIT, proposed a cooperative program between the Lynn Works and MIT. The MIT faculty approved the course, but it was put on hold in the spring of 1908 because of a "lack of enthusiasm on the part of some of the General Electric officials," probably due to the business recession of the previous year.[14] In the fall of 1908 Alexander and Jackson went to Schenectady to try to win Steinmetz over to the cooperative program. Alexander told him that president Coffin of GE had said that "if Union College and the Schenectady works would agree to such a plan simultaneously with MIT and the Lynn works, he [Coffin] might be very pleased to give his approval."[15] Steinmetz apparently opposed the plan, probably on the same basis he had given to the AIEE that summer: cooperative courses interfered with the college's function of grounding students well in the basics before handing them over to the corporation for practical training. Consequently, GE did not go along with Alexander's plan, nor did it set up cooperative programs with Union College during Steinmetz's reign.

When Berg took over in 1913 Union was not committed to Steinmetz's policies. Shortly before Berg returned to Schenectady assistant professor Walter Upson recommended that the college establish a cooperative program that differed markedly from that of Alexander and Jackson. Twenty entering graduate students would alternate six-month periods between the Schenectady test course and Union College, thereby receiving both postgraduate corporate training and the M.S. degree in less than two years. The head of the test course and other GE engineers approved the plan, provided that it did not disrupt the course by with-drawing too many students from it at once. Upson argued for the plan on the basis that Union had not been able to attract full-time graduate students despite the reputation of Steinmetz and that the program would provide Berg with full-time graduate students right away.

Instead of adopting Upson's plan, Berg tried an advanced cooperative program in the fall of 1913 in which Schenectady GE engineers worked toward the master's degree. This proved unsatisfactory to Berg, though, because only nine of fifteen engineers completed their work that term. Thenceforth Berg put little faith in cooperative programs, even though he was still a paid GE consultant. Instead, he made visits to the Schenec-tady Works part of his courses, as was common under Steinmetz, and encouraged students to work summers at the plant. Berg tried another cooperative program with GE in 1919 and was dissatisfied with that one as well.

Steinmetz's idea of autonomy through cooperation is apparent in the three areas of interaction between GE and Union College. Although GE funded the college and lent it instructors to teach commercial subjects to company engineers, Steinmetz and Berg created a curriculum that gave autonomy to both organizations. Union taught the scientific principles of electrical engineering to its regular students and left their commercial, postgraduate training to GE's test course, in which Steinmetz and Berg also taught. Union's opposition to cooperative programs further illus-trates this division of labor. Steinmetz thought they destroyed the col-lege's pedagogical autonomy by tying instruction too closely to business interests. Berg shared his opinion after trying two cooperative plans.

The ideal of autonomy through cooperation was common among engineers and educators in the early twentieth century. At three national conferences held during Steinmetz's tenure at Union, practicing engi-neers and college professors discussed how to make graduates better suited for industrial work. The two groups agreed that industry should set specifications for graduates and leave teaching methods to the col-leges. A very popular idea was that colleges should teach the fundamen-tals of electrical engineering and leave commercial training to postgradu-ate courses like those at GE, Westinghouse, Allis-Chalmers, and Western Electric. Steinmetz also preached this gospel before the Nation-al Association of Corporation Schools (NACS) in 1913 and 1914, while president of that group.[16]

The ideal of autonomy through cooperation indicates that GE's relationship to engineering education was not one of master and slave. Cooperative plans and the test course—the cornerstones of the corporate reform movement in engineering education—trained engineers to corporate specifications. But the means to this end were not always driven by the demands of the corporation, especially in the area of cooperative programs. In a study of the GE-MIT plan established in 1917, W. Bernard Carlson showed that Dugald Jackson of MIT viewed it as a means of empire building and that GE officials were reluctant to fund it during difficult financial times.[17] Part of the reason may have been that Steinmetz opposed such plans because they interfered with the established relationship between colleges and corporations. Berg discontinued two cooperative programs for much the same reason. Thus, GE was not the prime mover behind cooperative courses and did not force them on colleges—even at Union, which it financed for nearly forty years.

GE's relationship with Union College adds another dimension to the influence of the test course. Noble shows how it provided technical and management training, habituated students to corporate employment and ideology, and rated, selected, and distributed engineers within and outside the firm. These results came from GE's direct control over the most prestigious avenue of postgraduate electrical engineering education in the United States. The test course also exerted a more indirect and broader influence. When such a prominent engineer as Steinmetz promoted the idea that colleges should teach the fundamentals and leave practical training to corporations, many educators agreed and changed their curricula. In 1907 John P. Jackson, professor of electrical engineering at Pennsylvania State University and brother of Dugald Jackson, said that in the past decade "industrial companies were developing the student engineering courses and the colleges were modifying their courses so as to meet this development." Jackson had firsthand knowledge of GE's program because he had visited the Schenectady Works in early 1907 with his seniors, several of whom had been selected for the test course.[18]

Although Steinmetz assisted the corporate reformers in some areas, they worked toward different ends. Dugald Jackson and his colleagues wanted to strengthen corporate capitalism. Steinmetz, as we will see, believed that universal electrification would help usher in a Socialist society based on the model of an enlightened corporation. Ironically, Steinmetz worked for many of the reforms that Noble believes helped stabilize corporate capitalism—like teaching at Union College while employed at GE, lecturing in the test course, and serving as president of the NACS—but later interpreted them as ways to bring about the Cooperative Commonwealth.

The American Institute of Electrical Engineers

While at Union College Steinmetz helped shape an integral part of the electrical engineering profession: the education of its practitioners. As a corporate engineer, he experienced the characteristic dilemma of engineering professionalism in the early twentieth century: the engineer's divided loyalty between the emerging ideals of his profession and the commercial interests of his employer. Edwin Layton has characterized the engineer's response to these conflicts as a "patchwork of compromises between professionalism and organizational loyalty."[19] Steinmetz's positions on the issues of AIEE autonomy, technical standards, ethics, and social responsibility certainly resembled such a patchwork.[20]

The AIEE was deeply concerned about its autonomy, especially its independence from business interests, when Steinmetz was elected president in 1901. Corporate managers and inventor-entrepreneurs— like Norvin Green, president of Western Union, and Alexander Graham Bell—accounted for 30 percent of its leadership when the institute was founded in 1884, but by the turn of the century college-educated corporate engineers, such as Steinmetz and Charles Scott at Westinghouse, were leading the AIEE. Like his predecessors, Steinmetz worked his way up the bureaucratic ladder, as a manager and vice president, until he reached the presidency. He also served relatively long terms on numerous committees such as standards, ethics, and education.[21]

Although college-educated engineers were in the ascendancy, the AIEE elections were not always gentlemanly. In early 1900 Steinmetz wrote former president Elihu Thomson that he and other prominent members had signed a circular against the "escreeable electioneering" by the friends of Carl Hering, a consulting engineer in New York City. (At the time members could directly nominate a presidential candidate in what amounted to a primary by mailing a nomination form to the board of directors.) Hering won the nomination (and the presidency) from Michael Pupin, whom Steinmetz favored despite their constant rivalry, yet Steinmetz thought that the "stir made will put an effectual stop to such electioneering methods for many years to come."[22] The "stir" did not change things, nor did it prevent a fierce battle between Steinmetz and

Pupin for the next year's presidency. Steinmetz won, as he had most of the heated technical debates with Pupin at the AIEE.

As president-elect, Steinmetz supported a constitutional amendment to increase the professional status of the institute by defining more precisely the existing membership grades of "member" and "associate." The old rules gave the board of examiners a rather free hand. Members were defined simply as "professional electrical engineers and electricians," while associates were those "practically and officially engaged in electrical enterprises." Under the new rules a member had to be "qualified to design as well as direct electrical engineering works," whereas the grade of associate was open to anyone "interested in or connected with the study or application of electricity."[23] Since the amendment also limited the offices of president and vice president to members, it confined the AIEE leadership to technically qualified engineers—an important step in preserving the institute's independence from business executives. The occupations of the nearly one thousand persons who joined the AIEE in the academic year 1902–1903, almost doubling its size, reflect this orientation toward professionalism as well: only 16 percent of them were managers or superintendents without technical expertise.[24]

The 1901 amendment was the beginning of what Layton has called the "first wave of engineering reform" in American engineering societies. Lasting from 1900 to 1912, the wave, led by the "electricals" and the American Institute of Mining Engineers, brought about new election methods, a codes of ethics, and more attention to public policy. Layton argues that the AIEE was in the vanguard of this movement because its close ties with science made it the most professionally oriented group, it was threatened more by bureaucratic control because most electrical engineers worked for corporations, and it had the most democratic and open election procedures. The 1901 amendment also made the institute one of the most professional groups by another measure: by 1905 full members accounted for only 10 percent of the membership. Unlike the case with the civil and mechanical engineers who felt the second wave of reform after 1915, the AIEE was not changed by the type of "patrician reformers" identified by Peter Meiksins. Instead, the "engineering establishment" acquiesced to the relatively mild reforms put forward by prominent members like Steinmetz.[25]

The heightened sense of professionalism was evident during the presidential campaign of 1906. Using the open nomination procedure, Edison, Elihu Thomson, and other friends of GE signed a circular for

Edwin Rice as president. Rice received more nominating votes than Samuel Sheldon, professor of physics and electrical engineering at the Polytechnic Institute of Brooklyn. Instead of placing only Rice's name on the ballot, the AIEE board of directors, on the motion of consulting engineer C. O. Mailloux, broke with precedent and put both names on the ballot. The action stirred a lively debate in the pages of *Electrical World*. Ralph Mershon, a consultant who had worked for Westinghouse, thought that if Rice was elected "it will be mainly due, I believe, to the fact that his friends have at their command the commercial organization of the immense manufacturing interest with which he is connected . . . [which] . . . would constitute a most regrettable precedent." Mershon and Mailloux objected to Rice mainly because he was not active in the AIEE. Mailloux thought it was no excuse that Rice lived outside the city, for his colleague Steinmetz was "at least as great *inside* the Institute as he is *outside* it." But Louis Bell, who worked for Rice in the early 1890s, thought the AIEE would become the "subject of international merriment" if, for example, it would turn down "Edison as president on the ground that he had not been active in its debates." Steinmetz was silent on the matter. Although Sheldon won the election, it did not end the debate over business control of the institute.[26]

Since the higher membership standards of 1901 relegated some prominent businessmen—mostly utility executives—to the same status as junior engineers, the board of directors established a committee to consider an intermediate grade of membership. The chairman of the committee was C. W. Stone, manager of Steinmetz's Consulting Engineering Department.[27] Steinmetz opposed the new membership grade. He told Stone in 1911 that if the institute desired a third category, it should upgrade the admission requirements for members and associates and create a third class of nonvoting affiliates, composed of nonengineers interested in electrical engineering. His definition of *engineer* was rather broad: "everybody who wishes to do engineering work and believes himself to be capable of it, and all that would be required [for admission as an associate] would probably be a statement of education and previous experience by the applicant."[28] On a related question about the nomination procedure for officers, Steinmetz advised past president Henry Stott that the board of directors should be restricted to "*engineers*, who are taking [an] active part in the engineering work of the Institute as represented by papers and discussions" and who could attend the board's meetings in New York City.[29] Steinmetz probably made this suggestion in response to the Rice campaign controversy five years before. Adoption of these proposals would have helped "professionals" maintain control of the institute. Prominent engineers would run the AIEE, engineering managers and junior engineers would vote as associates, and businessmen would participate as a nonvoting class.

Instead of following Steinmetz's advice, the institute amended its constitution in 1912 to create three membership grades: fellow, member, and associate. If all went according to plan, prominent engineers would be elevated to the rank of fellow, executives with some technical expertise would join qualified engineers as members, and other businessmen would be relegated to the status of associates, who could vote. The amendment created a furor when businessmen lacking an engineering background began transferring to the grade of member without the usual examination under a temporary section of the constitution. Faced with the specter of a large number of nonengineers becoming members, president Ralph Mershon led a campaign to restore the rights of the board of examiners during this temporary period. Steinmetz, Elihu Thomson, Michael Pupin, and other prominent members supported Mershon and urged the board of directors to have "eminent counsel" consider the question.[30] The matter was brought before the New York State Supreme Court in 1913, which refused, however, to grant an injunction against the membership transfers.

On at least two other occasions Steinmetz took positions in favor of the AIEE's independence from business interests. In 1910 he opposed establishing a prize for the best student paper on the economics involved in the design of electric plants, because this would defeat the purpose of colleges teaching fundamental principles rather than "routine" engineering. He did "not believe there is any justification in the constitution, for diverting funds of the Institute to reward commercialism."[31] In 1919 he opposed a move to include more advertising in the AIEE *Proceedings* to defray publication costs. Steinmetz thought that there was "also the danger, as soon as we depend to a considerable extent on the income from advertisers, of losing our independence because of commercial interests."[32]

He did recognize the importance of the institute to the commercial interests of his employer. He helped establish and actively participated in the Schenectady section of the AIEE, which was composed almost entirely of GE engineers,[33] and thought that GE should increase its influence in the AIEE at the national level. He advised Rice in 1910 that the "Institute has become so important in the work which it has undertaken on standardization, and in many other directions, that it appears to me wise for the company to pay considerable attention to it, and energetically participate in the running of the Institute in proportion to our importance. This, however, requires appropriate representation [vis-à-vis Westinghouse engineers] at the [annual] convention which has rarely been the case."[34] Matters improved when the midwinter AIEE convention was held jointly in Schenectady and Pittsfield, Massachusetts, in 1911. Several GE engineers gave their first AIEE papers at the meeting, prompting Steinmetz to inform Rice that if "we consider, that the national reputation of its engineers is a valuable asset of the Company, the convention has been very successful."[35]

These statements reveal the dilemma of the corporate engineer. Steinmetz wanted to protect the institute from business control, but he also wanted GE to have a say in its technical rulings and to profit from its reputation. As in the case of publishing technical papers—where he compromised between the scientific community's ideal of the free exchange of information and the business practice of keeping trade secrets—Steinmetz compromised between professional ideals and commercial demands. He thought that engineers rather than businessmen should run the AIEE, but qualified GE engineers should serve the institute for the good of the company.

Steinmetz was not averse to working for his candidates. Despite criticizing the electioneering for Hering in 1900, he told a GE colleague in 1911, "I have always made out a complete slate [of AIEE officers] since many years, and circulated it."[36] When asked that year about improving the open procedure for nominating officers, he complained that it was "practically impossible to bring new men into the Board [of Directors], without circularizing, and as a result, entrance into the Board depends altogether too much on the accidental feature of some friends starting a campaign."[37] He started his own campaign during that year's presidential election between Mershon and Gano Dunn, chief engineer of Crocker-Wheeler. Steinmetz prepared a circular for Dunn and telegraphed Edison, whom he had just visited about the condensite project, to get Edison's support.[38] Accused of supporting Dunn out of "revenge" because Mershon had opposed Rice's presidential bid five years before, Steinmetz reminded a colleague that Mailloux, Dunn's campaign manager, had also opposed Rice.[39] Steinmetz probably campaigned for Dunn because he and Dunn's boss were trying to get a code of ethics through the institute (see below). Steinmetz was not completely opposed to Mershon. He supported him for president the next year and came to his aid during the fight over membership transfers.[40]

More serious backstage maneuvers occurred in 1913. When Berg asked whom he should support for AIEE president that year, Steinmetz's secretary replied, "It was agreed to let Mr. Mailloux run this year, as he really deserves it, and Mr. Lamme [of Westinghouse] next year, and a General Electric man the year after. This brings a General Electric man in at the time of the [International] Electrical Congress" in San Francisco in 1915.[41] The elections did not follow this plan exactly, but they did not miss by much. Mailloux was president in 1913, Paul Lincoln of Westinghouse in 1914, J. J. Carty of AT&T in 1915, H. W. Buck, a GE engineer turned consultant in 1916, and Rice, then president of GE, in 1917. A GE man was not president at the time of the San Francisco congress, but Steinmetz was appointed its honorary president. He apparently reconciled this type of insider politics with his ideal of AIEE autonomy by reflecting that all of these men, except for Rice, had the qualifications he wanted for AIEE officers. Although Rice did not actively participate in institute affairs—as noted in the 1906 election—Steinmetz respected his technical expertise and understandably made an exception in the case of a colleague and boss of nearly twenty-five years.[42]

The issue of who controlled the institute was especially important when it concerned a commercially vital issue like technical standards. Engineering societies began to have some influence in this field when the American Society of Mechanical Engineers (ASME) established voluntary standards for screw threads and steam boilers in the late nineteenth century. Although not legally binding, the ASME and AIEE standards were followed by most manufacturers. The National Bureau of Standards (1901) complemented rather than supplanted the codes of the professional societies, which are still active in this area.[43]

As GE's main representative in the AIEE on standards from the start of that work in 1898 until 1913, Steinmetz stood for the manufacturers' interests against those of the consulting and utility company engineers. A major dispute between them at the turn of the century was over custom apparatus. Steinmetz strongly disliked special designs, because they decreased manufacturing efficiencies inherent in the "American system" of mass production and, in some cases, led to poorer systems. In early 1898, for example, he complained to Rice that a hydroplant in Minneapolis "would be a thoroughly modern" system "if unfortunately so many heterogeneous and contradictory elements and appliances had not been insisted on by the consulting engineer[;] this makes the plant a ridiculous combination of elements."[44] On the other hand, many consultants and utility company engineers thought custom orders advanced engineering practice.

This was a clash between professional ideals and business demands, for consultants and utility company engineers believed the institute would compromise its integrity if manufacturing engineers helped set standards. Yet the institute needed their cooperation in order to enact realistic standards—a dilemma that engineering societies continue to face.[45] A professor of electrical engineering at the University of Michigan summed up the AIEE's dilemma in 1913: "Many of our members are representatives of manufacturers. Some may consider this a difficulty. I do not look at it this way . . . Imagine a strictly professional body establishing standardization rules. It would be much more difficult."[46]

The conflict over who would set industrial standards surfaced during an AIEE conference in 1898. Called by president Francis Crocker, professor of electrical engineering at Columbia University and co-founder of the electrical manufacturing firm Crocker-Wheeler, the conference focused its attention on standards for generators, motors, and transformers. The manufacturers, including Rice of GE, consultants, and utility company engineers agreed that they should cooperate in setting standards. But there was considerable opposition to having manufacturing engineers on the committee—an understandable position to take when GE, Westinghouse, and a host of smaller companies were fiercely competing to standardize their own product lines.

Consulting engineer Cary Hutchinson did not think manufacturing engineers should be on the committee. John W. Lieb, Jr., general manager of the Edison Electric Illuminating Company in New York City, agreed and thought the committee should be "untrammeled by any direct relations with the manufacturing companies."[47] Both men thought the manufacturers should simply provide data to the committee. Steinmetz argued from a technocratic viewpoint: "If the Institute intends to produce something of lasting value, which will be accepted by the whole continent, the committee doing the work must be composed of men of such standing and reputation that, regardless of whether they are connected with manufacturing concerns or not, there can be no question that they will be impartial."[48] The council of the AIEE attempted to balance these rival interests by appointing Steinmetz, Lieb, Hutchinson, and L. B. Stillwell, a Westinghouse engineer, to the first standards committee, along with Arthur Kennelly, Elihu Thomson, and Crocker (chairman) as the "neutral" representatives of science.

The committee followed the recommendations of Rice and others to standardize performance definitions and general testing methods, rather than design or manufacturing details. The extent of Rice's influence can be seen in the fact that the categories of the first Standardization Rules followed the suggestions he had made at the 1898 conference.[49] Although the decision not to standardize types of apparatus pleased the consultants and utility company engineers, the AIEE standards eventually helped Steinmetz's campaign against custom designs. For example, the recommended insulation tests were those typically used by manufacturers, instead of the stricter insulation tests often specified by consultants. Steinmetz summed up a decade of standards work in 1910 by noting that the "procedure which has been followed is never to standardize anything until best practice has already crystallized upon some definite form, and not to create definitions, but [to] accept those definitions toward which good practice tends and which therefore can easily be accepted."[50] The AIEE thus standardized performance requirements, voltages, and frequencies that were in general use throughout the industry—a development Steinmetz encouraged in order to increase manufacturing efficiencies.

The cooperation between the AIEE and industry was proably closer than outsiders suspected. In 1907 Crocker, who was still chairman of the standards committee, told the institute president that the committee would have submitted its latest report for final approval "were it not for the fact that Dr. Steinmetz and the officers of the General Electric Company have not had an opportunity to go over the last proof of the rules"; Steinmetz, who was on the committee, was out of town. Crocker did not think they would make any substantive changes, but he thought they should review it again.[51] The committee's published report mentioned "extensive correspondence with manufacturing, consulting, and operating engineers." If the last two groups had seen Crocker's letter, they undoubtedly would have thought the correspondence was too extensive![52]

GE did not dictate its commercial practices to the institute, however. In many cases Steinmetz encouraged the company to follow the AIEE standards. He interpreted a standard on voltage regulation, advised the GE Engineering Committee against introducing a voltage not standardized by the institute, and asked if the company was conforming to the AIEE standards on efficiency measurements for commutating machines and ratings for autotransformers.[53] He also fought corporate parochialism—the "not invented here" syndrome. In 1910 Rice complained that the institute preferred the Westinghouse term *auto-transformer* over GE's *compensator*. Steinmetz replied that standardization "means, that some concession must be made, and even a company like ours can not expect to have all its private terms standardized, especially terms which are as indefensible as 'compensator.'" He thought that the institute chose GE's nomenclature in most cases because "we had a better engineering staff" in the 1890s, which "adopted more appropriate terms." Westinghouse gradually accepted the AIEE rulings, but GE had not in the case of *compensator*, the only time the decision went against the firm. Steinmetz thought the company's intransigence was "rather unfortunate and [was] setting a bad example."[54] GE took his advice and changed to *autotransformer*, probably because he had been on the standards committee since its inception.[55]

Some outsiders, however, thought the manufacturers had too much control over standards. At an AIEE meeting in 1913 Bernard Behrend, a prominent consulting engineer, observed that the "manufacturer is responsible for the old code and for the last edition of the Standardization Rules, and the application of these rules to actual practice favors the manufacturer." He did not think the manufacturer had done this with "evil intent"; the company simply had to stand by the product.[56] Steinmetz preferred this state of affairs because he believed that large manufacturers like GE and Westinghouse had the best engineering staffs in the country and thus would institute the best practice.

He put this technocratic ideal to work by cooperating with Westinghouse engineers in making a "radical revision" of the AIEE standards in 1914.[57] The subcommittee on machinery ratings—Steinmetz, Benjamin Lamme, and two engineers from their respective companies—met at local AIEE meetings in Schenectady and Pittsburgh from 1912 to 1913 in an unprecedented spirit of cooperation to hammer out the new rules. The experience proved so valuable that the engineering staffs extended the discussions beyond the subject of ratings. Steinmetz and Lamme, the top electrical engineers in American industry, spoke to each other's engineering apprenticeship classes and co-authored a paper on temperature ratings.[58] Lamme wrote Steinmetz that "this growth of friendship between rival engineering organizations is a great step forward in our progress." Steinmetz agreed and thought that the "engineers of the corporation should impress upon the world, and upon the men in the organization, the solidarity of the engineering interests, even if the commercial interests are competitors."[59] Out of experiences like this—of engineering, rather than worker, solidarity—grew Steinmetz's theory of technocratic, corporate socialism described in Chapter 8.

This cooperation also led to a personal friendship. The quiet, self-confident Lamme, a midwestern political conservative who stayed clear of AIEE affairs, became fast friends with the outspoken Steinmetz, a German Socialist and master of AIEE politics. In addition to their similar technical interests and corporate positions, they shared a passion for science fiction by H. G. Wells and other "weird" books. Steinmetz once apologized for not sending the minutes of a standardization meeting to Lamme, saying that he had "been so busy reading Dracula and the other stories which you sent me, that I have really neglected everything else."[60] Cooperating with Lamme and his colleagues at Westinghouse, thus bypassing the business executives of the rival firms, was a novel and satisfying way for Steinmetz to resolve the clash between professional ideals and commercial demands in the area of standardization.

He faced a similar dilemma when he served on the AIEE's first ethics committees. Engineering societies considered adopting codes of ethics at the turn of the century in order to improve the status of engineering vis-à-vis the older professions of law and medicine, which had ethical codes. The AIEE was the first of the major engineering societies to adopt a code, largely through the efforts of Steinmetz and Schulyer Skaats Wheeler. A former Edison employee who established the electrical manufacturing firm of Crocker-Wheeler, Wheeler donated a large collection of historical books on electricity and magnetism to the institute in 1901. The collection formed the nucleus of the present Engineering Societies Library and led to Andrew Carnegie's gift of a building to house the engineering societies in 1903.[61] In his AIEE presidential address in 1906 Wheeler stated that a code of ethics would raise the professional status of electrical engineering. In his view a code should prescribe the engineer's duty to his client, the public, and the engineering fraternity—in that order.[62]

In the discussion of Wheeler's address Steinmetz observed that the "high standing of the medical profession and of the profession of law is, in my mind, undoubtedly due to their strict code of ethics." He noted that electrical engineers faced ethical problems quite different from those of other engineers because a "very large percentage of the prominent electrical engineers are more or less closely associated with large manufacturing or large operating companies." He then moved that the institute appoint an ethics committee to look into the matter, draw up a code if advisable, and report back at the next annual convention. Dugald Jackson spoke in favor of the motion, which passed unanimously, saying that it was a good opportunity "to do some more standardizing."[63]

The first ethics committee—Wheeler, Steinmetz, and H. W. Buck, chief engineer of Niagara Falls Power—drew up a code and presented it at the Niagara Falls convention in June 1907. The code followed the guidelines in Wheeler's talk and contained sections on general principles, the ownership of records, and duties to the client, the public, and fellow engineers. The main criticism of the 1907 code came from president-elect Henry Stott, the engineering manager we met in connection with the Manhattan Railway accident. Stott objected to the provisions that engineers should resign if their employers failed to correct defective machinery, that employed engineers should not accept more responsibility than originally agreed to upon hiring, and that the newly hired engineer should inquire into the conditions of his predecessor's departure before joining the firm. Stott probably agreed with the *American Machinist* that an engineer was the "servant of his employer, and no code of ethics can be enforced which conflicts with the will of the employer."[64]

The question facing the AIEE was who should prescribe the engineer's relations with his client or employer: a professional society, or the business community? This dispute prevented the adoption of a code of ethics at that time. The Niagara convention approved the code, but its opponents used technical grounds to refer the matter back to the board of directors. The ethics committee removed the offending sections, and the board sent the revised code to the membership for suggestions in the fall of 1907. Stott appointed an expanded ethics committee to consider the topic, but there the matter rested, largely because much of the AIEE leadership opposed the code.[65]

Stott's successors were more hostile to a code of ethics. In the fall of 1908 Wheeler complained to Steinmetz that he had heard nothing from the new president, Louis Ferguson, chief engineer for Commonwealth Edison and a member of the expanded ethics committee, about reappointing the committee. Wheeler asked Steinmetz to urge him to do so.[66] Steinmetz replied that he was trying to find out from third parties why Ferguson had not appointed a committee; he himself had "not seen Mr. Ferguson in this matter as I do not like to do so, for personal reasons."[67] Lewis Stillwell, Ferguson's successor and a former Westinghouse engineer, did not favor a code of ethics either.[68]

Nor did the staff at AIEE headquarters. Ralph Pope, secretary of the Institute since its founding in 1884, recalled that "some of the objections were on the grounds that such an action would not be compatible with the dignity of the Institute, and that the only code necessary was that the engineer should be a gentleman."[69] The AIEE enforced this Victorian philosophy in at least one case when the board voted not to transfer George Forbes, the designer of the Niagara Falls dynamos, to the grade of member in 1895 "on account of unprofessional conduct."[70] Steinmetz had implicitly argued in 1906 that a policy of relying on engineers to be "gentlemen" was becoming increasingly anachronistic because most electrical engineers now worked for corporations. They were not private consultants drawn from the "old school" elite known to Pope in the early days of the institute.

When it appeared that Dugald Jackson, who had supported Stein-metz's motion to form an ethics committee in 1906, would be elected president for 1910–1911,[71] Wheeler renewed his campaign. He wrote Jackson throughout the spring and summer of 1910, urging him to appoint a new ethics committee. As a former engineer for Edison General Electric who did considerable consulting work for industry, Jackson was cautious. He had criticized the 1907 code for being too specific regarding the duties of the consulting engineer and now thought the institute should proceed via the Committee on the Relations of Consulting Engineers established by Stillwell. Jackson argued that this committee, chaired by Frank Sprague, should determine rules for the professional conduct of engineers and that it, or an ethics committee, should develop general principles suitable for a code of ethics. After his election Jackson, a consummate diplomat who headed the E.E. department at MIT for over thirty years, asked all sides for advice, including the AIEE staff and six former presidents. In what must have seemed like stonewalling to Wheeler, Jackson finally became convinced near the end of his term that Sprague's committee was not going to accomplish much and that the institute needed an ethics committee.[72]

When Jackson asked his advice in late 1910, Steinmetz recommended a form of cultural relativism: "While the fundamental principles, which properly constitute the code are practically unchangeable, their application may greatly change with the change of industrial conditions. For instance, here in the States, it has been considered improper for a designing engineer to leave a manufacturing company, and immediately go to a competitor, while this has been the custom in Europe. Either way is quite proper under the different industrial conditions of the two countries."[73] To codify this "best practice," Steinmetz advised Jackson not to utilize Sprague's committee, which he thought was, "brutally expressed, 'a committee of consulting engineers to fix minimum wages and eliminate scab labor.'" Instead, Jackson should appoint a small ethics committee that would revise the 1907 code and present it to an advisory group of prominent members. Steinmetz felt this procedure would quell objections that the previous code had not been considered from all sides and give the new code the appearance of a permanent document. In keeping with his belief that engineers should run the institute, Steinmetz suggested that the new committee "should consist only of engineers [who] do engineering work, but not of men, who have more or less left engineering to accept administrative positions."[74] To represent the spectrum of electrical engineers, he recommended that the committee consist of a consulting engineer, a utility company engineer, a design engineer, and a professor with consulting experience. By limiting the committee to professional engineers who would be inclined to codify industrial customs, Steinmetz believed, as he had in the case of technical standards, that the AIEE could accommodate commercial interests and remain autonomous.

The AIEE adopted most of his proposals. Jackson appointed a small ethics committee in June 1911, with the expectation that president-elect Dunn would reappoint it in August. In November the institute established a large advisory committee to review their work. But Jackson bowed to the consulting engineers and called for a code "for the guidance of electrical engineers, and more particularly of the practice of consulting electrical engineers."[75] President Dunn reappointed the ethics committee, consisting of Buck, Samuel Reber, Steinmetz, Stott, Wheeler, and chairman George Sever, professor of electrical engineering at Columbia University and a consultant.

Steinmetz played a significant role in writing the new code. A first draft was completed in December 1911 and sent to the advisory committee. In early January Sever, who was delayed with business in the West, asked Steinmetz to preside over the ethics committee. During this crucial period Steinmetz chaired the committee meetings at AIEE headquarters, handled suggestions from the advisory committee in place of Stott, who had to take care of business matters, and revised the code. Steinmetz submitted the final draft to the board of directors in February 1912,[76] which gave its tentative approval over the strong objections of past president Stillwell. In March the board approved a code that "was considerably modified" to meet Stillwell's criticisms, but not enough to secure his approval, nor that of Ralph Mershon, his colleague on the advisory committee. The board also substituted the phrase *Professional Conduct* for *Ethics* in the name of the committee and the title of the code to describe more accurately the nature of that code. The revised code met with the approval of Stott, who had opposed the 1907 version along with Stillwell.[77]

One reason Stott changed his mind was that the 1912 code dealt more favorably with employers.[78] It reinstated the provision that the engineer should not let nonengineers overrule his technical judgments, but added the qualification "on purely engineering grounds," which opened the door for him to be overruled for nontechnical reasons. Strong language about being "personally" and "morally responsible" for the "character of the enterprises with which he is associated professionally" was dropped in favor of simply requiring the engineer to resign if he found their character questionable. The code also deleted the admonition that it "should not be considered an excuse that his connection extends only to legitimate engineering work" in such enterprises. The 1907 code made the engineer "responsible for defects in apparatus" and required him to report them to his employer or client and "insist upon the removal of the causes of danger as soon as possible." The 1912 version said it was the duty of the engineer merely to report the defect, and it dropped a long clause about the limits of the engineer's responsibilities to his client or employer. In short, the 1912 code favored the employer over the engineer. The engineer's first obligation was no longer to a client, which could be interpreted as the public for employed engineers. Now only consultants owed their first duty to a client. The "client" for employed engineers was their employer, an interpretation American engineers did not overturn until they renewed their concern about ethics during the political turmoil of the 1970s.[79]

The AIEE did reject one of Steinmetz's pet ideas: design standards. The 1907 code had said that "even the tendency to give individuality by providing special construction may usually be avoided with advantage."[80] According to Steinmetz, the 1912 committee "changed the wording [of this clause] so as to avoid the misunderstanding that we are opposed to progress,"[81] but the AIEE struck the clause out at the last moment. As the technical journals observed, the clause seemed clearly out of place in a code of ethics. This case illustrates how Steinmetz attempted to integrate professional and business interests. In his view the professional engineer had an obligation to promote the well-being of the electrical industry by specifying standard lines of equipment. Defined in this way, standardization was a professional ideal that served commercial demands.

He took a similar position on the question of public dissent by an employed engineer. Steinmetz recognized that when an engineer's professional views differed from those of his employer—on a paper read before the AIEE, for example—the engineer would be faced with an ethical dilemma if he decided to participate in the discussion of the paper. In this instance Steinmetz told his engineers in 1912 that they could not "honorably take the position or defend the official opinion of the Company, against [their] own opinion. In such a case, however, fairness would require [them] to keep out of the discussion altogether."[82]

Steinmetz thus asked his engineers to compromise between professional ideals and business demands—a decision he had made many years earlier in agreeing not to publish design equations. He did not ask his engineers to give up their professional opinions, but he believed that their first obligation was to their employer, as stated in the 1912 code of ethics. This accommodation to business interests was a major difference between the 1907 and 1912 codes. Steinmetz probably supported the change because it codified the current relationships between the engineer and his employer—in line with his philosophy of cultural relativism—and because it would assure the passage of a code of ethics.

Although active outside the institute in the area of social responsibility, Steinmetz did not think the institute was the proper place for an engineer to assume such a role.[83] When a committee suggested in 1919 that the institute support a federation of local engineering bodies, a national engineering council, and an engineering congress, in order to promote the engineer's social role, Steinmetz was opposed. He thought that "many differences of opinion exist, and these differences tend to become more accentuated, and therefore it is very desirable to keep such activities out of the Institute proper, leaving the latter in charge of fields in which there can be universal agreement, that is, science and engineering."[84]

In standardization and ethics Steinmetz thought the AIEE could compromise with business interests and retain its independence. But to bring such social questions as utility regulation and other matters of public policy into the institute invited a takeover by the business community. By assuming social responsibility outside the institute, he attempted to keep it free from commercial interests. This independence, in turn, meant that technically competent engineers would set industrial and professional standards and advance engineering knowledge through AIEE publications.

This philosophy is similar to the idea of autonomy through cooperation that Steinmetz held regarding the relationship between industry and engineering education. The common denominator in his work at Union College and the AIEE was his desire to strengthen the corporation through its cooperation with these and other private associations. His positions on AIEE autonomy, standards, ethics, and social responsibility reflect that nascent corporatism, which helps explain his particular "patchwork of compromises." We will see that the roles he worked out for technical colleges and the AIEE formed an essential step in the development of his theory of corporate socialism.

Standardizing "Steinmetz's Method"

By the first decade of the twentieth century the AIEE's reforms had extended to a subject dear to Steinmetz: the complex-number method he had introduced almost twenty years before. He had responded to one criticism of the method in the interval in regard to power calculations. His classic 1893 paper contained a simple multiplication of complex numbers for voltage and current, which, if done in this manner, did *not* give the correct answer for AC power. He obtained the right answer in this and later papers, but he did not explain how until 1899.[85]

He took a dimmer view of another criticism of his method: the negative sign in the imaginary part of his complex-number expressions, which was due to his use of the Zeuner diagram. In 1903 William Franklin, professor of physics at Lehigh University, debated Steinmetz on this matter before the Schenectady section of the AIEE. Franklin argued from pedagogical reasons that the crank diagram was simpler and better known: "Any junior student is likely to be overwhelmed by the unfamiliar mechanism" of the Zeuner diagram. He could not "imagine that Mr. Steinmetz cares anything about the matter except to the extent that he is concerned with the business of education and to that extent he must, I think, agree with me."[86] Franklin did not persuade Steinmetz to change his method. On the contrary, Steinmetz popularized it more extensively in two revised textbooks and five new ones he wrote while teaching at Union College.[87] Obviously, he had become "concerned with the business of education" and intended to establish his complex-number method as the standard one in electrical engineering curricula.

The question of which method to standardize came to a head when Arthur Kennelly presented an AIEE paper on the topic in June 1910. He noted that the rotating-vector and complex-number methods, "as developed in textbooks, are, at present, in a state of great and unnecessary confusion as to [the] direction of rotation" of the vectors. In a survey of

over sixty textbooks he found three systems: polar, clockwise crank diagram, and counterclockwise crank diagram. The first two yielded a negative sign, while the last one gave a positive sign. Two-thirds of the textbooks used the last method. Since this "confusion has existed for more
than twenty years and is not confined to any one country or language,"
Kennelly recommended that the "direction of rotation and representation should be standardized by mutual international agreement." When a
consensus could not be reached at the meeting, Steinmetz agreed with
his colleagues to submit the matter to the International Electrotechnical
Commission (IEC) for arbitration.[88]

Kennelly was in a better political position than Steinmetz to carry the
matter that far. Both men had served on the AIEE standards committee
and the U.S. National Committee to the IEC since these groups, which
had overlapping memberships, were established. Kennelly, however, was
secretary of the former and president of the latter when he presented the
question of vector rotation to the IEC meeting in Rotterdam in August
1910. Kennelly wrote Dugald Jackson, then AIEE president, that it
"looked as though we could have had a decision at once, in favor of the
'direct' representation [Kennelly's choice], and we were asked which we
recommended. We said that we preferred not to state our opinion at that
time because great writers, like Dr. Steinmetz and Dr. Arnold (of
Karlsruhe), should first be given an opportunity to express their views."
The IEC, whose president was Elihu Thomson, voted to invite briefs
from both sides to be presented by national committees for a decision at
the next annual meeting. Kennelly hoped Steinmetz would write a paper
on the topic and thought his own 1910 paper could serve for the other
side. Kennelly told Jackson that his paper "was intended to be reasonably
impartial, but if it leans to either side, it opposes Dr. Steinmetz."[89]

At a meeting of the U.S. National Committee in November Kennelly
recommended that his paper and Steinmetz's discussion of it be sent to
the IEC. Steinmetz opposed this as "unfair" and suggested, instead, that
two briefs be prepared. Each would present its method as it would be
taught to the student, without mentioning the other side, then compare
both methods. The committee compromised and advised the AIEE
board of directors to send Kennelly's paper, Steinmetz's discussion, and
the briefs by Ernst Berg and Franklin to the IEC. The board thought the
paper and discussion were not sufficient and asked Berg and Franklin to
present their briefs at the AIEE convention to be held in Schenectady in
February 1911.[90]

When that meeting could not reach a consensus,[91] Kennelly had a relatively free hand in presenting the subject to the annual meeting of the IEC in Turin that fall. Berg privately praised him before the Schenectady meeting for taking a "very impartial stand,"[92] but Kennelly was not that impartial, as he had confided to Jackson. He criticized Steinmetz's sign convention in 1900, used the opposite signs in all his own writings, and referred to the crank diagram as the "direct" and the polar diagram as the "indirect" method in 1910. The evidence suggests that Kennelly persuaded the American delegation to the IEC to recommend the following resolution: "In the graphic representation of alternating electric and magnetic quantities, advance in phase shall be represented in the counterclockwise direction." Accompanying this sentence was a note defining the complex-number expression of inductive impedance with a positive sign. An IEC subcommittee, composed of members from Italy, Great Britain, the United States, Germany, Denmark, Belgium, and Holland, voted unanimously for it. The plenary session of the IEC then adopted it on the subcommittee's recommendation.[93] Although Kennelly told the AIEE that the American delegation had not recommended one type of vector diagram, the wording of the resolution made it clear that the IEC ruled against Steinmetz. The AIEE did likewise when its board of directors approved Kennelly's resolution and printed it verbatim in a special addendum to its Standardization Rules just a month after the IEC meeting.[94]

Berg and Steinmetz could hardly believe what had happened. Berg read the decision "with distress" and asked Steinmetz, "What are you going to do about it?"[95] Berg would hold off changing his teaching methods until he heard from him. He was willing to retain the polar diagram and change the direction of its rotation in order to obtain the signs called for by the IEC. The major problem was that teachers could no longer use Steinmetz's books, two of which had been published just before the IEC's decision. Berg told McGraw-Hill in 1912 that he would be using Franklin's books at the University of Illinois until Steinmetz revised his textbooks.[96]

Steinmetz accepted the IEC and AIEE decisions under protest and began revising his books in 1913.[97] Within three years he completed *Waves and Impulses, Engineering Mathematics, Theoretical Elements,* and *Alternating Current Phenomena. Light and Illumination* was finished in 1918, *Transient Phenomena* in 1920. Engineers praised this herculean effort. A British reviewer said, "Since all serious students of electrical engineering will, sooner or later, have to turn to 'Steinmetz,' we are glad that the author decided to use the vector representation that is common to all elementary text books."[98]

Although Steinmetz resisted making these changes, the IEC decision aided the complex-number method. Its main competitor was not Heaviside's vector analysis, proposed by critics of Steinmetz's rather loose mathematics, but the graphical technique of rotating vectors. By revising his books to accommodate the popular crank diagram, Steinmetz promoted the rotating-vector version of his complex-number method, which is still the basis for AC theory in electric power engineering.[99]

His books enjoyed a huge success. They went through several editions and were translated into many languages, including German, Swedish, French, Italian, Japanese, and Portuguese.[100] In 1920 McGraw-Hill brought out the "Steinmetz Electrical Engineering Library," announced by a four-page advertisement, complete with a biography of the "authoritative" Steinmetz. Consisting of nine volumes and over three thousand pages, the collection represented, in the words of McGraw-Hill, the "corner stone and foundation of modern electrical engineering knowledge," the "classics of electrical engineering."[101] For once the advertising hyperbole was fairly close to the mark.

Early Public Image

Teaching, holding office in professional societies, and writing a dozen textbooks brought Steinmetz more and more into the public eye. Newspapers and magazines portrayed him, along with Thomas Edison and Nikola Tesla, as electrical "wizards," sorcerer-inventors wrestling with the forces of nature alone in their laboratories to bring about an electrical utopia for a grateful public. The images of Edison peering into the glow of an alchemical furnace to withdraw a new light for humankind, of the hunchbacked Steinmetz examining the damage done by his lightning generator, of Tesla sitting unharmed under a spectacular discharge of electricity—all of these illustrate the great fascination with electricity at this time. In the public's eye it took a wizard, a Faustian figure, to tame one of nature's most mysterious and destructive forces for human good.[102]

The wizard was the most recent addition to the pantheon of heroic inventors. Earlier Promethean figures like Eli Whitney and Cyrus McCormick brought the gifts of the cotton gin and horse-drawn reaper to Americans in the late eighteenth and mid-nineteenth centuries. Edison, the Wizard of Menlo Park, epitomized the heroic inventor to a later generation when more invention was being done in the research laboratories and engineering departments of large corporations than in the workshops of inventor-entrepreneurs symbolized by Edison. The irony is that Edison, the archetype of the lone inventor, actually ran a large, well-staffed laboratory. Similar contrasts—typical of what anthropologists have seen as the hero's function of resolving cultural contradictions—are evident in the social construction of Steinmetz as the Wizard of Schenectady.[103] The social groups of newspaper and magazine publishers, their readers, General Electric, and Steinmetz himself interpreted the "artifact" of the public Steinmetz from their viewpoints and constructed a cultural myth in the context of prevalent ideas about science and technology and against the commanding figure of the Wizard of Menlo Park.[104]

The creation of Steinmetz's public image began in a modest way when *Success* magazine ran a story on him in 1903. Founded in 1897 by Orison Swett Marden, the "American Samuel Smiles," the magazine was one of many that published the enormously popular success literature at the turn of the century. Steinmetz came to the magazine's attention as one of the prominent industrialists, scientists, and engineers invited to meet Prince Henry of Prussia in New York City in 1902. Edison, Alexander Graham Bell, and Tesla received top billing among the electrical experts in a lengthy article in the *Review of Reviews*, while Steinmetz, invited as president of the AIEE, was briefly described as a professor of electrical engineering and chief engineer at General Electric.[105]

The *Success* interview helped change all that. The interviewer, Herbert Wallace, called Steinmetz an "intellectual giant among craftsmen, yet unknown, in large measure, to the general public . . . the man who probably knows more about the practical application of electricity than any other man in this country." Referring to the contrast between his physical appearance and intellectual accomplishments, Wallace said—in better taste than some later writers—"He is a giant, but he is not a tall man." He then quoted Steinmetz at length about his life story, particularly the dramatic events surrounding Heinrich Lux's trial, the secret communications via invisible ink, and Steinmetz's flight from Breslau to avoid arrest. The article, which was reprinted in Union College's magazine and a technical journal, became a source for later stories and biographies in the popular press.[106]

At this time newspaper reporters had latched onto Tesla as the "New Edison," the new breed of electrical scientist to replace the sooty empiric of Menlo Park. Twenty years after inventing a practical incandescent lamp, Edison still made good copy, particularly during a publicity campaign over the storage battery in 1901. But many reporters had become enthralled with the darkly handsome, flamboyant Tesla, who had dazzled the scientific community with his demonstrations of high-voltage, high-frequency discharges in the early 1890s. He intrigued the press with his predictions of wireless power transmission to be broadcast from a mysterious tower he was building on Long Island in 1902. Yet when his marvelous inventions failed to materialize—while Guglielmo Marconi had been lionized for transmitting the first transatlantic radio signals in 1901—Tesla lost esteem with the scientific community and the newspapers.[107]

The well-respected Steinmetz was ready-made to fill the void left by Tesla. The author of a *New York Times* story published in 1908 wrote that Steinmetz "makes no dazzling prophecies which hold their meteoric course across a double page in a Sunday colored supplement: he announces no startling inventions to be brought forth on a morrow which never comes. He is of that other variety of scientist, the kind that quietly DO THINGS." The allusion to Tesla was evident. But rather than describe what Steinmetz had actually accomplished, because it was too esoteric, the reporter decided to "tell the story of Steinmetz's greatest achievement—the story of what Steinmetz has made of himself." What followed was the Horatio Alger–type tale of how a penniless, hunchbacked immigrant rose to fame and fortune in the Land of Opportunity on the basis of intelligence, hard work, and a bit of luck.[108]

The story followed the *Success* article very closely, to the point of citing portions of it without credit, and added new material. Clearly fascinated by Steinmetz, the reporter described the large head and frail body, the ever-present cigar, life at Camp Mohawk, and the collection of weird pets and exotic plants in his conservatory. His salary supposedly cut an "impressive slice out of $100,000." In the time-honored tradition of describing electrical experts in alchemical terms, the reporter called Steinmetz's home laboratory a "fit workroom for this modern magician. In the daytime, illumed by the sun's rays, it may have a more prosaic appearance, but at night, lighted only by a mercury lamp and a mercury rectifier, with its seemingly blue fire playing about inside the glass tube, it suggests the alchemist's shop of medieval days. Only the dangling skeleton and the stuffed alligator were lacking to make it scenically complete."[109]

This image was canonized in 1909 when Harry McClure, cousin of the famous muckraking publisher, wrote an article on Steinmetz in *Amer-*

ican Magazine. McClure told readers, "His life reads like a romance. It is one of the most inspiring examples for the young man of which I know." His story also used the caption "Electrical Wizard" for a photograph of Steinmetz—the first instance I have found of his being called a wizard. McClure's word picture of the charismatic engineer with the "Liliputian physique [who] when he begins to talk . . . begins to grow" became a standard source.[110]

A writer who interviewed him for *Metropolitan Magazine* in 1914 said,

When once you sight Steinmetz there is no other man in the room for you. In body he is shorter by a foot and a half than the average. The whole man is built to come to consumation in the unforgettable face and head, carrying that victorious brow. He is clad in a blue flannel shirt and he smokes unceasingly a thin, long cylinder of a cigar, which thrusts out from the bearded, energetic face like the feeler of an insect searching the void. The hair on his head bristles, as if electricity were pushing through the skull and lifting it, his beard is angry; he is a high potential. For he is Steinmetz, Charles Proteus Steinmetz, the greatest electrical engineer on earth.[111]

In 1911 the *New York Times* published a sequel to the article it had run on Steinmetz three years before. The new piece, nearly a full page in the Sunday magazine, had five photographs, three of which portrayed the carefree, outdoor life of swimming and canoeing at Camp Mohawk. The story repeated much of the earlier interview and praised Steinmetz for not being "one of the limelight prophets," an allusion to Tesla. In response to questions about the feasibility of the wireless transmission of electric power and firing bolts of electricity at warships—proposals made by Tesla—Steinmetz dismissed the first as impractical and the second as "trash." Steinmetz admired Tesla's work on the induction motor, but he detested the overblown publicity surrounding Tesla's high-voltage research. In 1908, when Tesla was in financial straits, Steinmetz advised GE not to lend him experimental equipment because "his actions and utterances for some years have been such, that I believe he has become seriously unbalanced mentally." Lending him equipment carried the danger that "in some of his irresponsible newspaper articles, Tesla may claim that the General Electric Company is backing his work, and this would hardly increase the Company's Engineering standard."[112]

Steinmetz was hardly immune to this "irresponsible" newspaper sensationalism. Even the conservative *New York Times* in the 1911 article took his prediction that untapped waterpower would one day be harnessed to replace an exhausted coal supply and turned it into the headline "Electricity Will Keep the World from Freezing Up." The headline was hardly a cold, dispassionate claim and helped turn Steinmetz into the very sort of "limelight prophet" criticized by the reporter. By focusing on his life story, physical appearance, and eccentricities—because his technical work was too difficult to popularize—and sensationalizing what technical statements he did make, these articles set the pattern of Steinmetz's public image for years to come.

The role of GE's public relations group in this process was rather minimal at first. As shown by David Nye, the GE Publication Bureau, founded in 1897, came to serve three groups of clients: managers and engineers through the *General Electric Review*, workers through the *Schenectady Works News*, and the public through advertisements and planted newspaper stories. Under the direction of Martin Rice, the brother of Edwin Rice, the bureau gradually assumed this role and built up a large staff and photographic collection to convey the desired image of the company to each target group.[113]

The evidence suggests, however, that Joseph LeRoy Hayden and the Consulting Engineering Department handled most of the publicity on Steinmetz from the GE side before the 1920s. In 1908 and 1909 Hayden prepared biographical sketches for a technical journal and a business college in New York State. In 1912 an editor of *American Magazine* wanted Steinmetz to follow up its earlier article on him with a Horatio Alger–type piece that would tell "how, as an immigrant boy, you overcame stupendous difficulties and rose to the height you have." Steinmetz turned the task over to Hayden rather than to the Publication Bureau. Nothing resulted at the time, because the editor was looking for a serialized book-length autobiography, which Steinmetz—then busy organizing his new department—apparently did not have the time or inclination to pursue. The magazine retained its interest and became a major organ for creating Steinmetz's public image, publishing three interviews and one biographical article on him by 1922.[114]

The Publication Bureau got into the act in 1911. In May Martin Rice asked Hayden to have Steinmetz go to New York City and "have a good portrait taken . . . We have so many calls for photographs of Dr. Steinmetz that I think it would be fair to charge the expense [$30 a sitting] to the company." (Most of the photographs of Steinmetz published before World War I were taken by Hayden, newspapers and magazines, or portrait photographers commissioned by Steinmetz.) The bureau went further when two of its editors of the *General Electric Review* published articles on Steinmetz's human side and his hysteresis research in *Scientific American* in 1911 and 1913.[115]

In one respect the Publication Bureau was responding to an outside demand for more information on Steinmetz, who had been "discovered" by the press by 1909. Later correspondence shows that GE received requests for photographs of him from a wide range of people and organizations.[116] Magazines like *American*, which favored portraits of the "self-made man," also had their own agendas. But the timing of the bureau's more active interest in Steinmetz indicates that it had another motive: to use him to "humanize" the corporation during the government's antitrust suit against GE. In March 1911 the Department of Justice began proceedings against GE, Westinghouse, and the National Electric Lamp Association, a group of over thirty small companies that were supposedly GE's competitors, under the Sherman Antitrust Act. The case received widespread publicity because it came out that GE owned 75 percent of the stock in National and that the combined "Electrical Trust" controlled 97 percent of the market for light bulbs and obtained "extortionate prices" for them. The government charged GE, Westinghouse, and National with price-fixing and market-sharing agreements in restraint of trade. In October—after the U.S. Supreme Court broke up Standard Oil in the biggest antitrust suit of the period, reinstating the "rule of reason" element of the common-law interpretation of the Sherman Act—GE and the other companies admitted the facts of the case but denied any wrongdoing. Under the terms of the consent decree, which dissolved National, the defendants agreed to stop the price and market-sharing practices, and GE, which had purchased the other 25 percent of National in the interval, agreed to manufacture lamps only under its name. Bigness was not outlawed under this common-law interpretation of the Sherman Act, just "unreasonable" restraints of trade by horizontal, cartel-like agreements.[117]

At a time of public outrage against large corporations, it made good business sense to publicize Steinmetz—the colorful and respected chief engineer of the company, who was beginning to make a name for himself in the popular press—in order to combat the soulless image of the "Electrical Trust." The idea was apparently suggested by vice president and corporate counsel Owen Young, who joined GE in 1913. Young turned to Steinmetz because president Charles Coffin shunned publicity.[118]

Hayden and the Consulting Engineering Department were not left out of this process. In May 1915 Hayden complained to the Publication

Bureau about its coverage of his boss. Hayden informed the bureau that Steinmetz "claims that he is not an 'inventor,' and does not care to be featured as as inventor, but that he is an engineer and investigator, and inventions and patents are only incidental" to his main work. Later that year, when Steinmetz's writings on the European War brought him more into public view, the Publication Bureau asked H. C. Senior, Steinmetz's private secretary, for a "short descriptive article" on Steinmetz. Senior sent the bureau the "only thing we have, which is sort of a biography of the Doctor, and which we send out for such purposes."[119] We will see that GE raised its public relations expertise far above this amateur level in the 1920s, when PR was becoming a well-established profession.

A personal quality that helped make Steinmetz "good copy" was his ability to think well on his feet. The author of one of the *Scientific American* articles remarked that Steinmetz knew no peer in the discussion of a paper at an AIEE meeting. "The resonant voice, the quaint accent, the ultimate clearness and force of the phrases, the close-knit reasoning advancing with rapid but definite steps toward an unanswerable conclusion—that is Dr. Steinmetz thinking on his feet." This praise is corroborated by Arthur Kennelly, who participated in many of these debates, and by the AIEE *Transactions,* where conference transcripts often noted that Steinmetz's remarks were followed by applause or laughter, often at the expense of an opponent. He turned his prowess as a debater to good account during newspaper and magazine interviews. Like Edison and Tesla, Steinmetz had a knack for giving reporters a "quotable quote" and controversial statements that made good headlines. Writing popular articles for the *Züricher Post* in the 1880s and editing the Socialist paper in Breslau no doubt helped him understand the demands of the press, just as Edison's stint as a telegrapher served him well with reporters.[120]

A popular, charismatic lecturer, Steinmetz usually gave several speeches each year to engineering societies throughout the country. During the winter of 1907–1908 he spoke at six local AIEE meetings in New England and the Midwest. The stop at Chicago was an annual one, where he lectured at Fullerton Hall at the Art Institute under the auspices of the Western Society of Engineers and the AIEE. According to a local organizer, they had him speak in this large auditorium because he "always commands [an] audience about four times [the] size [of] our regular quarters." His popularity among electrical engineering groups led to the founding of at least one student organization, the Steinmetz Club at the University of Missouri in 1904. The main activity of the engineering club seems to have been "presenting the 'Blarney Stone' for the annual Kow Tow" during the school's St. Patrick's Day festivities.[121]

While his lectures, like the other publicity surrounding Steinmetz, indirectly benefited GE by improving its image, his speeches to engineering societies had a more direct effect because they were delivered to GE's main customers: utility company engineers. A good example of GE's using Steinmetz in this manner occurred when the Detroit Engineering Society invited him to speak at a meeting in 1915. Edwin Rice, then president of GE, asked him to consider the offer because it would "do us a good turn and at the same time help to advance the cause of science." Steinmetz declined because he was too busy. The society persisted, writing six months later to the editor of a technical journal who had some influence at GE that since the society had grown to be the largest of its kind in the country, "we think we are entitled to hear Mr. Steinmetz." A GE district manager passed the letter to Steinmetz and said, "It would be of great commercial advantage to us if you would consent" to make the speech. The matter soon made its way up to Charles Coffin, then chairman of GE's board of directors. Coffin told Steinmetz that it would please him very much, because of GE's business dealings with members of the Detroit Engineering Society, if he would make the speech. The next day H. C. Senior wrote the society that Steinmetz "now finds" he can address its meeting because he would be in Detroit that September to make another speech to a national engineering society.[122] The Electrical Wizard had gained enough independence to turn down GE's president, but not the chairman of the board.

8 Corporate Socialism

The "corporate reformers" at General Electric, the National Association of Corporation Schools, and the American Institute of Electrical Engineers held a political philosophy that recent scholars have called "corporate liberalism." Used in a variety of ways—to denote a category of business thought embracing "welfare capitalism" and regulatory measures, for example—the term generally refers to a new form of liberalism that resulted from an accommodation between corporate capitalism and American liberalism, which replaced the ideal of economic competition with that of cooperation among business, government, labor, and private associations. In the view of Martin Sklar, corporate liberalism was a "cross-class ideology" shared by capitalists, political leaders, professionals, reformers, trade-union leaders, and others in the Progressive Era who had an interest in building a corporate-capitalist order through administered markets and government regulation. By World War I corporate liberals had created a "middle ground" between laissez-faire capitalism and state socialism. Most did not envision a corporate state like British guild socialism on the left or Italian fascism on the right, but later government leaders like Herbert Hoover put their faith in a liberal form of corporatism based mainly on private associations.[1]

Steinmetz entered this turbulent arena when he returned to Socialist politics in 1911 during a municipal reform movement in Schenectady. Busy organizing the Consulting Engineering Department, he cut back his publications and patents during this period, which left more time for politics (see the Appendix). Out of his experience with municipal socialism and engineering reforms, he developed a political theory more along the lines of liberal corporatism than corporate liberalism. That is, he envisioned a "corporatist" society governed by industrial corporations, voluntary associations, and parliamentary bodies, which had "liberal" elements of welfare capitalism and democratic safeguards. Published in 1916, his theory combined the politics of his youth with his enthusiasm for the business corporation to produce what I have called a "corporate socialism." The theory informed his Socialist activism on the municipal, state, and national levels and influenced the corporate liberalism of GE officials.

Socialism in Schenectady

Steinmetz embraced the dangerous passion of his youth at a propitious time, for American socialism was much more of a mass movement in 1911 than when he emigrated. No longer was it primarily associated with the German-dominated Socialist Labor party (SLP), which he had objected to in the 1890s. The movement was now largely in the hands of the Socialist Party of America and covered a wide political spectrum, from "revolutionists" like "Big Bill" Haywood on the left to "constructivists" like the Austrian immigrant Victor Berger, the first Socialist congressman, on the right. The factional lines were not always clearly drawn, but Eugene Debs, the party's charismatic, perennial presidential candidate, usually sided with the left, while the Russian immigrant attorney Morris Hillquit, leader of the New York City group, occupied the center on most issues. The constructive wing, with which Steinmetz identified, emphasized evolutionary, step-at-a-time reforms—a philosophy well suited to municipal socialism. Formed in 1901 from the merger of the Social Democratic party organized by Berger and Debs and a moderate splinter group from the SLP led by Hillquit, the Socialist party managed to hold such diverse elements as Christian Socialists, ex-Populists, and Syndicalists together reasonably well for over a decade. The party gained a substantial following in every region of the country except the South and was particularly strong in Milwaukee, New York City, rural Oklahoma, and Pennsylvania. In 1911, near the peak of its power, the party had over eighty thousand members and elected mayors in seventy-four American cities, including Schenectady.[2]

Schenectady had a proud labor history before then, earning the name Uniontown-on-the-Mohawk in the labor movement. The Wobblies (International Workers of the World) counted Schenectady as their strongest city in early 1906, before a disastrous sit-down strike at GE later that year decimated the three-thousand-member local. The craft unions, which had opposed the Wobblies, grew stronger, and by 1911 Schenectady had the largest percentage of union workers in the state. Eleven per cent of its population of about seventy-two thousand were union members, out of a factory work force of about twenty thousand. After the defeat of the IWW many of its leaders, like metal polisher Charles Noonan and Herbert Merrill, both of whom managed to stay at GE, joined Local Schenectady of the Socialist party. Founded in 1901, the local grew to over four hundred members in 1911.[3]

The party had little success in Schenectady elections until George Lunn ran for mayor that year. Educated at the prestigious Union Theological Seminary in New York City, Lunn came to Schenectady in 1904 to serve as minister for the Dutch First Reformed Church. His fiery sermons on corruption in the local government upset the church elders and led to his resignation two years later. He then formed a United People's Church and established a weekly newspaper, the *Citizen*: two pulpits from which he began to preach a Christian Socialist message. When Lunn joined Local Schenectady in 1910, the *Citizen* became the party's official paper, and the Socialists decided to run the popular muckraking minister as mayor in the fall of 1911.

The results exceeded everyone's expectations. Lunn won with a large plurality, and Socialists captured over twenty other posts: the president of the common council, city comptroller, city treasurer, two assessors, eight of thirteen aldermen, eight of thirteen county supervisors, and New York State's first Socialist legislator. The elections made national news because Schenectady was the largest eastern city and the first in New York to be swept up in the "rising tide" of American socialism. The Socialists won by running on a municipal reform rather than a revolutionary platform. Realizing that a victory would not usher in a Socialist community because of existing capitalist laws, Lunn and his followers campaigned against the graft and corruption of the street-paving "gang" hired by the incumbents, the high price of natural gas and street-railway fares, and other local issues. They promised a businesslike administration that would show what socialism could do on the municipal level—a winning proposal during the prewar "efficiency craze."

The first question concerned Lunn's appointments. Whom would he choose to head the Socialist experiment in efficient government—local organizers, or outside experts? Although the issue eventually led to Lunn's break with Local Schenectady, they managed to patch things over in 1911. Lunn filled many posts with local leaders and selected several outsiders: Charles Mullen, the Socialist street-paving expert from Milwaukee, as commissioner of public works; Morris Hillquit as special legal counsel; and the young Walter Lippmann, then a Socialist, as his personal secretary. He appointed non-Socialists to the posts of city engineer and city attorney, mainly on the basis of their technical expertise.[4]

The Socialists gained another expert in March 1912 when Lunn appointed Steinmetz to the board of education, which elected him president at its first meeting. Socialists viewed the well-known "Electrical Wizard" as a prize catch for the party. A writer for the moderate *New York Call*, the main English-language Socialist newspaper in New York City, said that Steinmetz exemplified the "high order of appointments made by Mayor Lunn." When Steinmetz joined Local Schenectady that September, the event prompted a front-page editorial in the *Citizen*. Lunn proclaimed that his "preeminence in the world of intellect will make the politicians' charge of 'ignorant fools' [for Socialists] sound stupid." The *Call* printed an announcement and brief biography the next day that "one of the foremost electricians of the world" had enrolled in the cause.[5] Steinmetz thus became (like Helen Keller) a celebrity Socialist for the party, one whose unorthodox views were usually tolerated for his publicity value.

Joining Local Schenectady after Lunn had been in office for nearly a year indicates Steinmetz's understandable reluctance to get entangled in Socialist politics again. We have seen that he probably belonged to Local Yonkers of the SLP but had little to do with socialism after moving upstate in 1893. The only clear evidence of further involvement before Lunn's election is that Isador Ladoff, a member of the National Executive Board of the Social Democratic party and a Milwaukee city chemist who took a position in Whitney's laboratory, gave Steinmetz an autographed copy of his book, *The Passing of Capitalism and the Mission of Socialism*, in 1903.[6]

Steinmetz got more involved through the Intercollegiate Socialist Society. Founded in 1905 by the novelist Upton Sinclair, the ISS had 750 members in nearly forty college chapters by 1911. Its leaders included prominent Socialist intellectuals like Hillquit, John Spargo, the muckraking journalist, and William English Walling, a leading theoretician of the left, as well as electrical engineering professors Vladimir Karapetoff at Cornell and Comfort Adams at Harvard. Steinmetz joined the society in the fall of 1911. That December he and Lunn attended its annual convention in New York City, where he probably met Hillquit and Lippmann, then executive secretary of the ISS, and listened to speeches by Spargo and Victor Berger. Because Schenectady was a regular stop on the Socialist campaign trail after Lunn's victory, Steinmetz also had the opportunity to hear other national leaders like Debs, Haywood, and Charles Edward Russell.[7]

Despite this interest on a national level, there is no evidence that Steinmetz joined Local Schenectady before 1912, even though he had subscribed to the *Citizen* since its founding two years before. One factor was that the local consisted mostly of trade unionists and had few members of the professional class, to which Steinmetz belonged. Exceptions were GE engineers E. Otis Hunt and John Dempster, both of whom Lunn appointed to the civil service board of examiners. As Steinmetz's former laboratory assistant, Dempster became his chief aide in municipal affairs as well.[8]

In accepting Lunn's appointment to the school board Steinmetz was not automatically endorsing socialism, since Lunn had appointed non-Socialists to key positions. Lunn had much support from the well-to-do. Nearly 40 percent of the voters in the GE Realty Plot had voted for him. Charles Coffin, then president of GE, told an anti-Socialist magazine editor in early 1912 that "Dr. Lunn was the best man that had ever been Mayor of Schenectady" and that he was "going to work hand in glove with him."[9] The event that seems to have triggered Steinmetz's decision to join Local Schenectady occurred in August of that year, when he was called before the group to explain his position on industrial education in the city's secondary schools. The *Citizen* reported that the workers referred to the "famous educator" as "Comrade Steinmetz," that he treated them as intellectual equals, and that he and the local both opposed the measure. But it was clear that the local was calling the shots in questioning "their" appointee. Steinmetz reportedly said that because Lunn's administration "is an expression of your organization . . . I should consider it your right to summon me before you in any important matter like this; and I consider it not only my privilege but my duty to come." He enrolled in the local less than a month later—an act that ensured him a greater voice in the decision-making process connected with his own office.[10]

By venturing into Schenectady politics Steinmetz joined many "new professionals" in the prewar municipal reform movement. Nearly one-half of the reformers in Pittsburgh, for example, were from this class, and local groups like the Cleveland Engineering Society took up the issues of municipal ownership, smoke abatement, building codes, and other measures in this period. Several city managers were also former municipal engineers. The most prominent progressive engineer was Morris Cooke, a disciple of Frederick Winslow Taylor who became Philadelphia's director of public works on the recommendation of Taylor in 1911. During his four years in office Cooke applied Taylor's principles of scientific management to improve garbage and water services, to hire more engineers in the highway bureau, and to eliminate graft in city contracts. After exposing the overvaluation of the Philadelphia Electric Company's assests by Dugald Jackson, Cooke persuaded the Public Service Commission to reduce the company's rates, saving over $1 million annually.[11] Steinmetz, who knew Cooke through the National Association of Corporation Schools, could have mounted a similar campaign against the electric and streetcar companies in Schenectady, but he chose to focus on education and city planning. These were much safer activities for a prominent manager of Schenectady's largest employer, which until recently had owned both utilities.

Steinmetz took his post seriously, though, and sought to make the board of education a force for "efficient" reform. In a 1916 speech he stated that the "purpose of education is to produce efficient citizens, to bring up the children so that they can make the best of their life."[12] The city's elementary school system was far from efficient when Steinmetz took office; more than two thousand children had to attend part-time for lack of facilities. Steinmetz and the Socialist administration made good their campaign pledge to provide "One Seat for Every Child" by passing two bond issues totaling $800,000 in their term in office—about one-half of the total amount for bonds issued during this period. The money paid to build three new schools and to enlarge three others. In his first speech on public education Steinmetz broached the innovative idea of paying each child a "living wage" to help prevent the evils of child labor—a radical plan soon followed by more conventional efforts to "engineer" improvements to the school system. He studied local schools, approved standard designs for new buildings, pushed through the bond issues, favored free school supplies, campaigned successfully for more playgrounds and improved medical care, and testified against vocational training in elementary schools before the U.S. Commission on Industrial Relations. By 1916 every schoolchild did have a seat, while eleven nurses, one full-time physician, and five part-time doctors tended Schenectady's twenty-two schools.[13]

While heading the school board Steinmetz took on the related task of establishing a city park system. The Socialists had recognized the desperate need for parks in a crowded industrial city, where workers at GE and American Locomotive had swelled the population to seventy-two thousand by 1910. The issue seems in retrospect like a middle-class reform, but Local Schenectady took a great interest in the proposed parks. The middle and upper classes had yards and access to country clubs, while the working class was confined to yardless dwellings, especially in the congested parts of the city. Schenectady Socialists viewed the issue not only as one of public health but as a battle in the "class struggle" that they could fight (and win) in their community.

But they were stymied by the lack of one vote on the common council, which would have given them the two-thirds majority needed to approve bond issues. (They had obtained the support of one non-Socialist for the school board bonds.) In return for the needed vote, Lunn established a five-member nonpartisan board of parks and city planning in January 1913, consisting of one member from each of the major political parties (Socialist, Democratic, Republican, and Progressive) and one "lady member" to represent the suffragists. Lunn appointed Steinmetz to the Socialist position and the board elected him chairman. In line with the prevalent concept of government by experts, the board hired a landscape architect to prepare plans for the park system. After securing a $300,000 bond issue in June, the board recommended purchasing three properties on the Mohawk River, at Cotton Factory Hollow, and near McClellan Street.[14]

The common council agreed to the first two sites, but the third, which was in the more affluent Twelfth Ward, raised a ruckus in Local Schenectady. Steinmetz opened the debate for the McClellan Street park during a meeting of the local in November by saying, "I believe in party rule in a political party but party members must be guided by their platform and must obey the acts of the party." One of these acts had been to create the nonpartisan parks board and to agree to approve only sites it recommended. There was a danger of losing the entire park system, because the next administration (the Socialists had just lost the 1913 election) could issue an injunction if the board's recommendations were not followed. The opposition was led by Alderman Charles Noonan, the former Wobbly who opposed Lunn and his "constructive" wing of the local on most issues. Noonan argued for adding an entrance to Cotton Factory Hollow, which would provide a nearby park for fifty thousand working-class people in the most crowded part of the city. Noonan said, "If you take the McClellan Street park you, in effect, compromise the Socialist movement to benefit capitalistic and real estate interests. You sacrifice the working class interests by dodging the congested districts and buying parks which will benefit mainly people who do not need them."[15]

The common council approved the McClellan Street site after Lunn used some questionable parliamentary maneuvers to push the issue through the local. During the meeting of the local, which Steinmetz did not attend, an alderman chided his colleagues: "It was all right to get Steinmetz into the party and say how nice it was to have such a great man with us. Then some of you suddenly dropped respect for everybody including yourself, and want to show how clean-cut you are for the working class."[16] The remarks reveal Steinmetz's ambiguous status in the local. His reputation made him an asset to the party, but Noonan and other union leaders understandably questioned the commitment to the "class struggle" of a member of the upper class—a highly paid engineering manager who lived in a mansion in the spacious GE Realty Plot. Steinmetz also equivocated on political actions directly affecting GE, like strikes and raising property taxes, as we will see in Chapter 9.

These concerns had not been paramount when Local Schenectady nominated him to run for president of the common council in 1913. The *Citizen* reported that the "name of Steinmetz as a candidate was a surprise to many [at the meeting of the local] and the excitement that followed this announcement was intense. Shouts of 'A winning ticket,' 'A sure victory' could be distinguished from the multitude of exclamations that followed." Steinmetz replied that he would "work for the welfare of Schenectady until the whole public are brought to right thinking—Socialism." The *Citizen* declared that "no better running mate for Dr. Lunn could be found than Dr. Steinmetz." The paper erred in one respect: Steinmetz did not conduct a campaign. While Lunn stumped the city night after night during the hotly fought contest against the "fusion" ticket of combined Republican, Democratic, and Progressive candidates, Steinmetz stayed in the laboratory. His sole effort was to write a letter to the *Citizen*, stating that by holding the posts of president of the school board, chairman of the parks board, president of the common council, and (especially) member of the board of estimate and apportionment, he would be able to work for Schenectady "more efficiently and actively than before."[17]

The Socialists ran on their accomplishments. Walter Lippmann, who had resigned as Lunn's secretary after only four months in office, criticized the administration for concentrating on reforms he thought the Progressives could do better. Yet Socialist leaders like Hillquit, who had served as Lunn's special counsel for a brief period, praised the mayor for the progress he had made in a city with laws favoring a capitalist economy. The Socialist campaign booklet described in over thirty pages the party's numerous reforms, including various city-owned enterprises (coal, ice, and groceries), the municipal collection of garbage, reductions in the cost of street paving, and the new parks and schools engineered by Steinmetz. The *New York Times* thought the campaign newsworthy enough to run a Sunday-magazine story. Part of the headline read, "Steinmetz . . . One of America's Hundred-Thousand-a-Year-Men Runs for a Thousand-Dollar Office . . . and Initiates a Unique Speechless, Bandless, Bannerless, Drinkless, Handshakeless Campaign—He's a Socialist."[18]

Lunn did very well, winning more votes than in 1911, but he lost by a large plurality. Despite his "unique" campaign, Steinmetz polled only two hundred fewer votes than Lunn and ran slightly ahead of the rest of the Socialist ticket. Only five Socialists—all aldermen—were elected, giving control of the city council to the fusionists as well. The Socialists were not the only ones disappointed. The editor of a Republican paper in Scranton, Pennsylvania, criticized Schenectady for not electing Steinmetz: "Think of having a man with international fame, with a brain big enough to bring the whole world to his feet, spending his time in the Common Council of a third-rate American city. Surely there is something inspiring about it." The editor remarked that the fusion ticket had red-baited the Socialists, using the American flag as an emblem against them, and thought this was ironic since Steinmetz, as a "creative genius," was "giving our flag its chief glory in the eyes of other nations."[19]

Steinmetz expected to retain his posts on the school and parks boards, since the latter was nonpartisan by law and he viewed the former in the same spirit. But he had little experience with partisan politics. Mayor J. Teller Schoolcraft did not reappoint him to the school board and appointed instead Albert Rohrer, a Progressive and head of the GE test course. The action prompted such headlines as "Steinmetz's Master Mind Removed from Education Board" in the *Citizen* and drew criticism from the nearby *Knickerbocker Press* and the German *Albany Herald*.[20] Schoolcraft also managed to stop the Socialists from planning the city's parks. Shortly before the Socialists left office Lunn reconstituted the parks board as the city planning commission, which consisted of the original five members of the parks board and four new members. The body was still nonpartisan and represented all political factions: it consisted of three Socialists, two Democrats, two Republicans, one Progressive, and one woman. Steinmetz was selected as chairman for a three-year term. Once Schoolcraft took office he introduced an ordinance to eliminate the commission, claiming that it was controlled by the Socialists and had been set up by them at the eleventh hour. Chairman Steinmetz complained, lamenting that "It will be disgraceful to let politics enter into such matters as parks and city planning." When Schoolcraft ran into opposition from non-Socialist aldermen who favored keeping the commission, he cut its budget in February and then gained enough votes to eliminate it and to create a smaller board with no Socialist members in July.[21]

Steinmetz was incensed. He told the common council, "I have devoted quite a little of my time to city planning, more than, it seems, has been justified. While Schenectady is a large and growing industrial city, yet I feel that it is not so important a city for me to devote much of my time [to]. But I am a resident [and] am, therefore, willing to give my time to the city." Men like G. E. Emmons, head of the GE Schenectady Works, had told him that they opposed frequent changes in city administration. Steinmetz agreed, saying that "we should have appointive boards to serve continuously." This philosophy became ingrained in his thinking and served as the basis for his technocratic theory of corporate socialism. An indication of how seriously he took the elimination of the planning commission is that he refused to transfer its records to the new administration until nearly a year later, maintaining that the new parks board was not a legal successor to the commission. Schoolcraft's "exhibition of political spite and partianship," according to Steinmetz, made "it unsafe for me to turn over these records . . . to some irresponsible city official."[22]

The stonewalling tactic reveals that Steinmetz had stepped out of the laboratory and into the political arena once more. He ran for district delegate to the state constitutional convention in the 1914 election, losing again. Local Schenectady placed him on the platform committee in 1915 and again picked him and Lunn to head their municipal ticket that year. The platform was milder than usual, reflecting the philosophy of the "constructive" wing of the local headed by Lunn, Steinmetz, and Bradley Kirschberg, the city chemist and chairman of the platform committee. The more radical faction, led by Noonan and other union leaders, ran rival Socialists in the primary elections to try to regain control of the party. The result, in the words of the *Citizen*, was a "Sweeping Victory for Constructive Socialism."[23]

Lunn campaigned as hard as ever, and this time Steinmetz made at least one appearance—at a mass meeting in Schenectady with Lunn and Meyer London, the Socialist congressman from New York City. Lunn and Steinmetz had an easier task than in 1913, because the other political parties could not agree on fusionist candidates and did not run a combined ticket. Some voters opposed the Socialist stance against the European War and American "preparedness," but this proved to be much less of an issue than the voters' confidence in Lunn. They returned him to office and elected Steinmetz president of the common council. Steinmetz told the *New York Call*, "My election indicates that the people of Schenectady are desirous to have an efficient and progressive government . . . As a Socialist, since 38 years, I see with great joy the day when the people of America will accept the principles of Socialism as their guidance in government."[24]

That day was to be delayed in Schenectady because only two Socialist aldermen were elected to the common council, far below the number chosen in 1911 and 1913. Steinmetz was president of a predominantly Republican body with an overwhelming eleven to two vote against the Socialists. At the council's first meeting he said that, under his direction, the school and parks boards had taken only unanimous actions (an indication of the type of "efficiency" he liked). He realized that things would be different in the common council. Lunn also appointed him to the school board, which elected him president. As president of the common council and board of education and member of the board of estimate and apportionment, he was now in a position to carry on the work that had been interrupted under the fusionists.[25]

But a fierce battle in Local Schenectady dashed those hopes and seriously damaged the city's Socialist party. Ever since Lunn had joined the local, its more radical members had accused him of violating the constitution of the state Socialist party by appointing non-Socialists to office. The issue even came up in regard to the city planning commission, which the Socialist city council had created as a nonpartisan body. In 1914 the local and an investigating committee from the state party exonerated Lunn from this charge and a related one. (Steinmetz served on the special committee for the local.) Tensions subsided somewhat until the 1915 election campaign, when Lunn resigned as organizer for the local over the question of party control of elected officials, criticizing the "radicals and impossibilists who now control the Local."[26] After the local modified its position on "instructing" elected officials, Lunn's followers gathered enough votes to nominate him and Steinmetz as the top municipal candidates and to win the primary against the radical faction.

Lunn's election brought out the old charges. The issue of appointments became more important than in 1911, because only two Socialist aldermen were returned to the common council. Consequently, Lunn's appointment of non-Socialists to office was especially hard to take when the party faced a Republican-dominated city council. In February 1916 Local Schenectady judged that Lunn had violated the state Socialist constitution—he pleaded guilty—but it did not muster the two-thirds vote needed to expel him from the local. The matter was left up to the Socialist state executive committee, which revoked the local's charter for failing to expel Lunn. State Secretary Usher Solomon came to Schenectady in late March to install a new local, to sign up members who would agree to abide by the party's state constitution. Two hundred and fifty Socialists signed up immediately, including all elected officials except Lunn and Steinmetz. The *Knickerbocker Press* and the *New York Times* said that Steinmetz supported Lunn, while the *New York Call* reported that he "was in sympathy with the new local, but asked for further time to consider the matter."[27]

Steinmetz was caught in a quandary. He agreed with the general aims of the local, but he had been friends with Lunn for several years and was aligned with him in the constructive wing of the party. In early 1915 Steinmetz had helped form the Economic Club of Schenectady, a discussion group from the professional class that met at his Wendell Avenue laboratory. Although several "anti-Socialists and non-Socialists" joined, the leaders were the Lunn faction of Local Schenectady. The club also functioned as an alumni chapter of the Intercollegiate Socialist Society. After the Socialist party ousted Lunn in March 1916, the *Knickerbocker Press* reported that he and Steinmetz had formed a "Social Democratic club, constituting mainly the Lunn factionists," to win control of the Socialist county committee at the primary elections in April. The newspaper quoted Steinmetz as saying that the club consisted of Socialists "who are not willing to bind themselves to obey the Socialist local always," a practice he thought discredited the local. The major difference to Steinmetz was that "Mayor Lunn is a constructionist and the anti-Lunn ring are not constructionists."[28]

After the Lunn forces lost the primaries they tried another tactic. Daniel Sweeny, acting chairman of the Socialist county committee, called a meeting on a few hours' notice at Steinmetz's laboratory—a good distance from the Socialist regular meeting place downtown—on the evening of April 11 to elect officers. Fourteen angry, anti-Lunn committee members showed up, called it an illegal meeting, and walked out. A scuffle broke out when they slammed the door on the Lunnites. The headline in the next day's *Schenectady Gazette* read, "Two Socialist Factions Come to Near-Blows at Steinmetz Home Meeting." The Democratic paper, which had a reporter on the scene, couldn't resist saying that the "'brotherhood of man' theory, as set forth in Socialist doctrine, was handed a horrible wallop in Dr. Charles P. Steinmetz's $500,000 laboratory last night." Steinmetz was elected to fill a vacancy on the committee, but this and other actions of the rump meeting were declared invalid by State Secretary Solomon at the regular meeting of the committee.[29]

The setback did not stop Lunn, who tried to regain control of the *Citizen*. Founded in 1910, the paper had been financially and editorially controlled by Lunn until he sold about one-fifth of its shares to a member of Local Schenectady in 1914. When Lunn resigned as editor in January 1916 to concentrate on running the city, Local Schenectady gained complete control of the *Citizen*. Lunn fought back at the annual stockholders' meeting in June but failed to gather enough votes to overturn the local's control. Steinmetz initially supported Lunn. He voted for the mayor at the stockholders' meeting and gave him power of attorney in late July, after which Lunn foreclosed a $1,000 mortgage Steinmetz held on the *Citizen*. Then Steinmetz seems to have had second thoughts. The paper quickly raised enough money to pay the mortgage, with Steinmetz giving the first $100 to lift his own mortgage. He told the *Citizen* that he had taken out the mortgage when Lunn still headed the paper and had promised him power of attorney over it in the spring.[30]

Although the sudden foreclosure led to his break with Lunn, the root cause was not the events in the local, which Steinmetz dismissed as intraparty squabbling, but their differences over the European War. Steinmetz strongly supported Germany, criticized the "preparedness" campaign, and opposed Woodrow Wilson's reelection as president. While the *Citizen* agreed with him on all these issues, Lunn opposed every one of them publicly—even to the point of running on the Democratic ticket for congressman in the 1916 election. That seems to have been the last straw for Steinmetz, who thought Wilson was a national danger. In late September the *Citizen* ran a front-page story saying that Steinmetz refused to follow Lunn to the Democratic party. When questioned about the fights in Local Schenectady during the last few months, Steinmetz replied, "I have never been out of the Socialist Party," and he "exhibited with much pride his Socialist dues card" paid up to the end of the year. The news prompted an editorial in the *New York Call* entitled "Steinmetz Shows the Way." The editor observed that Steinmetz "is a great man; too great to desert Socialism. He has many things to be proud of, and he is proud of his red card."[31]

The great man proved to be unpredictable. Touted for breaking with Lunn in early October, Steinmetz puzzled Socialists by supporting former New York State governor and Republican reformer Charles Evans Hughes for president and other Republicans later that month. The *Call* and the *Citizen* quickly published articles to counter such headlines as "Steinmetz Out to Elect Hughes" that appeared in the German and upstate papers. Steinmetz explained that a nonpartisan, independent committee of mostly Germans and Irish adamantly opposed to Wilson had met at his laboratory to work toward persuading non-Socialists to vote for Hughes and Henry DeForest, Lunn's opponent in the congressional race. Steinmetz advised Socialists to vote for their candidates, not the Republicans, which satisfied the *Citizen* and the *Call*. In a letter to a supporter Steinmetz, who in newspaper interviews had predicted a German victory, explained that he was voting for Hughes, who advocated a strict neutrality, in order "to defeat a candidate, who challenges my right to citizenship" (Wilson did that with his pro-British stand on neutrality). This opposition overrode his commitment to socialism to the extent that he agreed to serve as an honorary vice president at a mass meeting for Hughes and DeForest in Schenectady.[32]

Supporting DeForest brought newspaper stories of a further break with Lunn, especially when Lunn said he would stay on as mayor if elected to Congress. Newspapers interpreted this decision as an action against Steinmetz, who was in line to be mayor if Lunn resigned. Both men issued statements in October 1916 to counter these stories. They were still friends and had agreed, before Steinmetz came out for Hughes, that Lunn should fill both posts. Steinmetz was too busy at GE to be mayor and Lunn's first congressional session would not start until about a month before his term as mayor ended. Nevertheless, after Lunn was elected to Congress the Socialist press lauded Steinmetz as their newest mayor, and Steinmetz changed his mind about holding that office. Believing that he was acting mayor if Lunn was out of town for more than nine days, he tried to raise police and firemen's salaries in early April 1917, shortly after the United States entered the war, and called a public hearing a week later on a related topic without consulting Lunn, who was in Washington. The city attorney declared the meeting invalid. From then on Lunn returned from Washington often enough to prevent Steinmetz from acting as mayor. During the fall elections, when the *Citizen* was being attacked as unpatriotic for its continued opposition to the war,

Lunn indicated why he had not resigned as mayor: "I didn't want to turn the city over to Dr. Steinmetz . . . I felt his sympathies were too strong for [the] enemies of America."[33]

In that environment of superpatriotism, fueled by the Socialist party's antiwar proclamation in April, Steinmetz understandably withdrew from Schenectady politics. Having presided over the common council regularly in 1916, he retired to his laboratory after the council passed a resolution supporting President Wilson in the war on April 9 and the city attorney refused to let him act as mayor a week later. Thenceforth he attended only four of its biweekly meetings and none at all from early July to the end of his term in December. He also neglected Local Schenectady. When the *Citizen* and the *New York Call* praised his break with Lunn, they incorrectly implied that he had sided with the local. He remained a Socialist but did not rejoin the local. According to one report, the local was willing to waive its rules against supporting nonmembers and to run him for reelection as president of the common council in 1917. Steinmetz refused and backed GE engineer E. Otis Hunt for mayor on an independent ticket. Having been fired as commissioner of public works by Lunn, Hunt was similar to Steinmetz in politics in that he was still a Socialist but had not rejoined the local. Steinmetz did not turn his back on the local, however; he was the largest contributor to its 1917 campaign fund. The Schenectady Socialists lost that election and never won another, mainly because "reversible Lunn," as he was called by the Socialists, had defected to the Democrats. Under their banner the popular ex-Socialist served two more terms as mayor of Schenectady and was lieutenant governor of New York in the 1920s.[34]

Steinmetz's Socialist Writings

Steinmetz remained a Socialist and increasingly turned his attention from municipal politics to national affairs. Though not as active in Socialist groups as he was in engineering associations, he supported the Intercollegiate Socialist Society, the Rand School, and the Socialist party. The ISS proudly listed him as an endorser of the society in 1913, and he established an alumni chapter in Schenectady the next year to supplement the chapter started by Lippmann at Union College in 1912. That year Steinmetz and J. G. Phelps Stokes, the "millionaire Socialist," were the second largest contributors to an ISS appeal, giving $50 each. He gave $100 to the Rand School Scholarship Fund for at least two years prior to 1914, which amounted to one-tenth of the total amount in one year. The ISS asked Steinmetz to write articles for its journal and to serve on its executive committee in 1914 and 1915. He begged off but agreed to provide the Socialist Party Information Bureau with technical information on municipal electric utilities.[35]

William English Walling, who had helped establish the left-wing *New Review*, invited Steinmetz to join a group of "favorably known Socialists" as a contributing editor in May 1914. Walling had probably heard about Steinmetz's socialism from editor H. F. Simpson, who had asked Steinmetz to write an article for the *New Review* on any "scientific or social subjects" of his choosing. Lippmann and John Macy, former members of Lunn's administration, joined Steinmetz on the *New Review*'s advisory council in June. (The suffragist leader Charlotte Perkins Gilman replaced Lippmann in July.) While left-wingers like Walling and Louis Fraina dominated the editorial board, the larger council contained right-wingers like Steinmetz and Phelps Stokes. Through the *New Review* and the ISS Steinmetz enlarged his political circle to include such prominent Socialists as Walling, Stokes, Upton Sinclair, Max Eastman, and Algernon Lee before 1916.[36]

This exposure, along with his reading of political economy, his experience with municipal socialism, and his work at General Electric, Union College, and national engineering societies, formed the basis of his theory of corporate socialism. Published as *America and the New Epoch* in the fall of 1916, the theory envisioned a technocratic state based on the model of the corporation—a Socialist utopia that was evolving from contemporary trends in capitalist society. Because Steinmetz thought the European War was accelerating this evolution, the war prompted him to gather his scattered thoughts into a coherent political theory.

The first inklings of the theory appeared around the time of Lunn's election, when Steinmetz began writing about "co-operation" in industry. (As noted by Samuel Haber, "Efficiency and co-operation were the bywords of business in the progressive era.") In a 1911 speech he predicted that there "will come a time when all our nation will have to co-operate in systematic development," partly because of increased electrification. He made his point clearer at a conference sponsored by the Society for Electrical Development two years later. Held at the rustic setting of Association Island in Lake Ontario—the site of GE management camps satirized in Kurt Vonnegut's novel *Player Piano*—the conference brought together representatives of electrical manufacturers, utility companies, government, finance, and the press for two days of speeches and camaraderie. The society hoped that "Camp Cooperation," as the corporate-liberal outing was appropriately called, would convince these often antagonistic groups of the value of working together to promote the electrical industry. Headed by Henry Doherty, president of a large utilities holding company, the society solicited contributions from all sectors of the industry to finance an advertising campaign based on the slogan "Do It Electrically."[37]

Among the invited guests were such notables as utility magnate Samuel Insull, banker Frank Vanderlip, and writer Elbert Hubbard. Many speakers proclaimed the virtues of private enterprise and questioned such means of "cooperation" as government regulation. Steinmetz, however, took the opportunity to address a subject he had debated with Heinrich Lux nearly thirty years before—electricity's ability to usher in a Socialist society. He did not preach the inevitable triumph of socialism to the gathering of capitalists and friends of capital. Rather, he reasoned that electrification would bring about "industrial cooperation," a phrase he later identified with socialism.

Steinmetz argued that electrification of industry would become universal because electricity is the most efficient means of transmitting and distributing energy. Most electrical production was by steam power, which is more economical in large units. Since electricity cannot be stored efficiently and has to be consumed as it is produced, its production and consumption are made most economical by means of large, interconnected systems with even loads throughout the day and night, that is, with a high load factor. To develop such a system required cooperation on two fronts. On the production side, cooperation among utilities was essential to create a national power network. On the consumption side, industries must cooperate in regard to their energy usage to ensure a high load factor.

Implied in this argument was a planned economy, run by technocrats who would engineer this cooperation by deciding which utilities to interconnect and when industries should consume electricity. Steinmetz also alluded to his belief that electrification would help usher in socialism—a form of technological determinism:

The relation between the steam engine as a source of power and the electric motor is thus about the same as the relation between the individualist [i.e., capitalist] and the socialist, using the terms in their broadest sense; the one is independent of everything else, is self-contained, the other, the electric motor, is dependent on every other user in the system. That means, to get the best economy from the electric power, co-ordination of all the industry is necessary, and the electric power is probably today the most powerful force tending towards co-ordination, that is cooperation [i.e., socialism].[38]

In response to a letter asking him about the national grid implied in the speech, Steinmetz acknowledged that he had dealt only with the engineering side of the matter. He hoped "sometime to express my views also on the political and social side of the problem of the transportation of materials and energy, as necessities of life, and thus entities which society must control." He did so three years later when he wrote in *Colliers* magazine that "industrial socialism is the final outcome of the present development of industrial cooperation."[39]

An important step in this evolution was the growth of corporations. Steinmetz had witnessed firsthand the concentration of capital in the electrical industry. He admired the efficiency of the corporation and worked to improve engineering and research at GE. The giant corporation was his home, his identity. GE provided him with palatial research facilities, the freedom to work at home and at his summer camp, and an opportunity to teach and to publish articles and books. In return the company got patents, quality engineering and research, and a public symbol to "humanize" the corporation. It also received his loyalty. He told a magazine writer in 1922, "I belong to the General Electric."[40]

Steinmetz began to speak about the social role of the corporation in addresses before the National Association of Corporation Schools in 1913. The NACS was an appropriate forum because of its corporatist nature as a voluntary association of companies that encouraged its members to cooperate in industrial education. He based his analysis on an observation that the industrial corporation had evolved to encompass four functions: the administrative (management), financial, technical (engineering and research), and "humane" (welfare capitalism) functions. The corporation owed its efficiency to how well it performed the first three functions, but most companies failed utterly in the fourth area. Few provided education for employees or protection against the vagaries of unemployment, accident, sickness, and old age. Steinmetz thought that much of the public antagonism toward the corporation resulted from its failure to perform the welfare function adequately. Organizations like the NACS helped corporations in this area by encouraging industrial education, just as the AIEE and other engineering societies helped them improve the technical function by publishing journals and setting standards. As noted in Chapter 7, Steinmetz thought voluntary associations could cooperate with large corporations in this manner and retain their autonomy.[41]

In his view the main purpose of these corporatist agencies was to help the corporation evolve to a perfected form as part of the larger evolutionary scheme of industrial society. He outlined the steps in this process in an interview with *Metropolitan Magazine* in 1914, when he said that unbridled, destructive competition had led companies to cooperate by consolidating into corporations and trusts, a view shared by corporate liberals of all political persuasions. The corporation's abuse of this power had led to increased supervision and control by state and federal regulatory agencies, which would give way to governmental management, then ownership, and finally socialism. Two other forces leading to socialism were the Socialist party (mainly through education) and union syndicalism (by the IWW publicizing the downtrodden state of the proletariat through strikes).[42]

Although this scheme seems to place the industrial corporation as a way station on the road to a Cooperative Commonwealth, Steinmetz regarded it as the model for the future Socialist society. The intermediate steps in his plan applied to an economy consisting of big corporations, not small firms. He made this point in a newspaper interview during his campaign for the Schenectady common council in 1913. When a reporter asked whether all production and distribution would be owned by the government under socialism, Steinmetz replied, "Perhaps there will be a Corporation of the United States, owning everything, running everything. Why not?" He then compared a future corporate society with the modern corporation. Cities and states would correspond to the local and regional divisions of a corporation, citizens would be shareholders, and elected officials would form a board of directors, which would fill all other positions by appointment. The Corporation of the United States was "a possibility, I might even say a probability." But he would not predict anything further than eventual government ownership of all the means of production and distribution.

The origin of Steinmetz's idea for a Corporation of the United States appears to have been an extrapolation from the conditions he saw in Schenectady. At the beginning of the interview he drew an analogy between city government, which was technically a corporation, and an industrial corporation. The Schenectady board of estimate and apportionment, for example, was like the executive committee of a corporation's board of directors. Steinmetz claimed that the "Socialists have given Schenectady the most businesslike administration that it has ever had. Their watchword when they got into office was: copy the methods of private corporation management."[43] His logic was that if Socialists could run Schenectady efficiently like a corporation, why couldn't they run the entire country on the same basis? He shared these views with municipal reformers throughout the country who had copied the organization and management style of the corporation since the 1880s.[44] While part of a national movement that looked to corporate methods to bring about "social efficiency" in city government, he differed sharply from those reformers by taking the analogy to extremes and postulating a Corporation of the United States.

This corporatist vision, first presented in 1913, informed Steinmetz's writings leading up to *America and the New Epoch*. He referred to "one big corporation" in an interview in 1914 and told the NACS convention the next year that the Socialist party "cannot be antagonistic to the corporation principle, since its ultimate aim, socialistic society, may be expressed as the formation of the industrial corporation of the United States, owned by all the citizens as stockholders." When a delegate replied that the corporation should be the servant rather than the master of humankind, Steinmetz disagreed and described a startlingly clear, corporatist view: "I look forward to the time when the corporation will be mankind, those times when all mankind will form a co-operative industrial organization which in its initial crude form is represented by the modern corporation. The purpose of our organization is to speed that day by completing the activities of the corporation" in the welfare function of industrial education.[45]

Although not up to par in the welfare and educational areas, the corporation was a good model for municipal and national government because it maintained a high degree of efficiency by keeping able people in office, rather than turning them out at election time. This theme— keeping experts continuously in office, the essence of technocracy—runs throughout Steinmetz's writings on political economy. He saw technocracy's advantages at GE and the disadvantages of "inefficient," democratic politics when Mayor Schoolcraft eliminated the parks board and did not reappoint him to the school board. When asked by a reporter in 1914 whether he thought the U. S. government was efficient, Steinmetz replied, "Efficient? . . . Wasteful is the word. What corporation would remove its president every four years?" The corporation was the "most efficient body society has evolved," partly because its leaders had to stand for reelection each year by the stockholders. The present method of government was wrong because it was based on the spoils system.

Steinmetz voiced a similar critique the next year when he wrote that "American city government is very bad; it is incompetent, inefficient and often corrupt" because of rotation in office and the spoils system. The solution, of course, was a corporatist one: run the city like an industrial corporation and hold elections each year. This would result, paradoxically, in continuity in office, not rotation, as long as city officials performed their jobs satisfactorily. Socialists, he noted, tended to keep experts in office longer than other parties. Like other Socialists, Steinmetz objected to the commission government favored by businessman-reformers, not because it eliminated the ward system—the basis of Socialist power in most cities—but because long terms in office made the commissioners less responsible to voters. He favored a continuity in office that was subject to yearly recall by the voters—an attempt to neutralize the antidemocratic nature of technocracy.[46]

Although Steinmetz favored radical social change, the evolutionary nature of his socialism was never in doubt. He told a reporter in 1914, "Public ownership to-day would be disastrous . . . Socialism grows gradually. Political parties come and go . . . But there is no abnormal growth in Socialism and there is no danger of a collapse." Like other evolutionary Socialists, he repeated the dictum that "Socialism is inevitable," because of evolving industrial conditions, and would come about whether the Socialist party existed or not. When the editor of a Socialist paper in Ohio told him that newspapers had interpreted this to mean that he did not look to socialism to carry out his ideas, Steinmetz replied that his "statement has been repeated—with modifications—so many times, that it finally has been made to convey just the opposite impression from what I said." The *Citizen*, however, accurately quoted Steinmetz as saying, "we can not go too fast [in school reform] or we might get out of breath." He opposed the more revolutionary tactics of the IWW yet thought the Wobblies served a useful purpose in frightening corporations into paying attention to the human welfare side of their work.[47]

Steinmetz's ideas were not unique. John Jordan has placed him among several "technocratic progressives" in this period, including Lippmann, Thorstein Veblen, and Hoover, while James Gilbert has noted that several American collectivists—such as King Camp Gillette, the industrialist and inventor of the safety razor—shared Steinmetz's view of the corporation as a model for an industrial state. Most American Socialists considered corporations and the trusts to be steps on the road to the Cooperative Commonwealth. On the left, the theme is explicit in Debs's early writings, and Walling stated in 1916 that Karl Marx "predicted the coming of a monopolistic period in industry and welcomed it." Walling pointed out that the Socialist international congresses of 1900 and 1904 criticized the trusts for oppressing the workers but agreed not to oppose their formation, which was inevitable under the development of modern industry predicted by Marx.[48] More moderate Socialists stressed the evolutionary nature of industrial consolidation. Hillquit wrote that the "modern trusts, thus transformed into cooperative enterprises on a large scale, will in all likelihood become the starting point of the socialist system of industrial organization." Charles Vail, a Christian Socialist minister, echoed the sentiments of many when he wrote that the present "trust is simply Socialism for the benefit of the few" and that the "trusts must combine into a great trust—the Nation." Reflecting these sentiments was the motto on the masthead of *Wilshire's Magazine*: "Let the Nation Own the Trusts." As noted by Donald Stabile, these ideas probably owed more to Edward Bellamy's popular utopian novel *Looking Backward* (1888), in which society had formed into "one great business corporation," than to Marx. But Socialists could also cite passages from Marx and Engels to support the role of the corporation and trust.[49]

Steinmetz was imbued with the ideas of the evolutionary Socialists, beginning with his exposure to Ladoff's *Passing of Capitalism* and Wilshire's *Socialism Inevitable* (1907). Lunn expressed similar ideas. He wrote three editorials in 1911 on "Industrial Evolution" and opposed the breakup of Standard Oil because "owned by the people, this big machine would be utilized for the common welfare of all, operated in the interests of the people."[50] Although Steinmetz told a reporter that his socialism came from original thinking and discussion rather than from books, he owned an author's copy of Victor Berger's *Broadsides* (1913) and had listened to Berger's proposal to nationalize the trusts at the ISS convention in 1911. He had an opportunity to talk to Hillquit about these matters when Hillquit was legal adviser to Lunn, and he ordered a copy of Spargo's *Applied Socialism* in 1912. An evolutionary Socialist, Spargo wrote that "great organizations like the Steel Trust represent the progress already being made in the direction of Socialism through one channel."[51]

The technocratic utopianism of Steinmetz's writings had many sources. The future society depicted in *America and the New Epoch* resembled Etienne Cabet's centrally controlled production of standardized goods in huge factories run by well-educated workers toiling less than eight hours a day. The main difference, of course, was that Cabet placed much more faith in democracy than did the technocratic Steinmetz. He was familiar with the state-supported producers' cooperatives proposed by Lassalle, and he may have read Bellamy's *Looking Backward*, which shared many of the characteristics of Cabet's book and was very popular when he emigrated. He was also fascinated by the novels of H. G. Wells, who predicted the coming of a Socialist world-state run on technocratic lines.[52]

An important element in the political theory presented in *America and the New Epoch* was his attitude toward the Great War. Steinmetz supported his homeland from the beginning. In July 1914, just one week after the assassination of Archduke Ferdinand, he wrote his sister Clara, "I have rather changed my opinion on the Kaiser & believe he is O.K." He thought it was "quite probable" that Germany would win the war and defeat France and Russia. These private thoughts became public in September when the *New York Times* published an interview with him in the Sunday-magazine section under the headline "Steinmetz, Exile from Germany, Sides with Her in War." As a Socialist, he considered the causes of the war to be economic. England, for example, fought to maintain her commercial superiority. He sympathized with Germany "not because it is Germany, but because Russia threatens European civilization" with the "spread of Slavic rule." He was not surprised that the German Socialists supported the kaiser rather than upholding the principle of international socialism against all wars, because they "realized the danger that threatened from the east." Recalling the eugenic doctrines discussed in his Socialist circle at Zurich, he also analyzed the war in terms of the racial traits of the belligerents. The subject was now fashionable in the United States and became a staple of his writings on the war.[53]

While Germans shared his concern about Russia, Steinmetz felt the issue strongly because he had grown up near the German-Polish border when Russia ruled Poland. In late 1914 he wrote an article for the *New Review* entitled "Russia the Real Menace," which predicted that a Russian victory would undo all that socialism had accomplished in Germany by submerging the country "under an autocracy based on the illiterate masses" of Russia. This would "be disastrous to civilization, and set back the coming of Socialism." The very next month—January 1915—he contradicted himself by agreeing to be the honorary president of the Russian Chamber of Commerce of America. Established to foster trade between the United States and Russia, the organization clearly opposed

Steinmetz's position of neutrality toward Russia in the war. But he told Ivan Narodny, its managing director and a political refugee from the czar, "I gladly accept the office . . . as I have always taken a very great interest in the slav nations and their great future, being myself partly of Slavish descent." Although Steinmetz appears to have done little more than lend his name to the organization—which was ineffective anyway, because the Russian government denounced it owing to Narodny's political stance— after joining the group he no longer publicly named Russia as the "real menace."[54]

Paradoxically, Steinmetz supported Germany in the war while serving as honorary president of the Russian Chamber of Commerce. Two weeks after accepting the post he wrote satirically to his sister Clara that she should not worry about their relatives in Breslau: "You know as all the good papers tell you, that the Cossacks are the defenders of civilization and of morality against German atrocities and barbarism." More seriously, he told a retired rear admiral that the Russian Slavs were a "very promising and capable race" but were "not yet fit for leadership" of the white race, because over 80 percent of the Russian population was illiterate. A German defeat would mean "increased militarism, as it would be merely an armistice [an accurate prediction]. That is the reason why I rather sympathize with the German side." He supported the Fatherland by donating $10 a month to the German-American Literary Defense Committee for about a year and one-half prior to 1917 and by upholding the German Socialists' right to back their government in the war.[55]

He soon found a regular forum for these views. In mid-1914 Waldemar Kaempffert, managing editor of *Scientific American*, wrote a letter of introduction to Steinmetz for Atherton Brownell, an associate editor of the *Philadelphia Public Ledger* who was setting up a National Editorial Service. The service would syndicate editorials from about thirty specialists in different fields and had already signed up David Starr Jordan, chancellor of Stanford University, and other notables. Brownell suggested that Steinmetz write on recent advances in electrical engineering, Tesla's proposal to transmit power by radio waves, or the effect of the war on electrical invention. Brownell got more than he bargained for when Steinmetz sent him more than a dozen editorials by early 1916, most of which were published. Steinmetz avoided the war's technical side and concentrated on political and economic issues. Bearing headlines like "England's Interference with Neutral Trade," the editorials informed readers of nearly twenty newspapers from Boston to Los Angeles that he favored a strict neutrality at a time when anti-German sentiment was growing daily. One even defended Germany's use of submarine warfare shortly after the sinking of the *Lusitania*.[56]

Steinmetz wrote other pieces on socialism and the war, the war and invention, and the racial standing of the major powers. In "Residuum of the 'Melting Pot'" he claimed that the United States had ceased being an Anglo-Saxon nation in the early 1880s when the percentage of Anglo-Saxons fell below 50 percent. Consequently, the United States should be neutral in the war, not pro-British. He repeated this argument in May 1915 at a neutrality meeting in Schenectady sponsored by the local branch of the powerful German-American Alliance and the American Truth Society, an Irish organization active in pro-German propaganda. The main speakers at the mass meeting were Steinmetz and Jeremiah O'Leary, president of the American Truth Society.[57]

Steinmetz's position on the war led to some difficulties with the largely pro-British American press. When Germany's submarine warfare prompted Secretary of the Navy Josephus Daniels to consider forming a Naval Consulting Board in July 1915 to mobilize the nation's inventors, newspapers listed Steinmetz among the prominent men being considered, along with Alexander Graham Bell, Orville Wright, and Henry Ford. The *New York World* interviewed Tesla, who recommended Steinmetz as a "very brilliant expert, one of the very finest living. Of course, he is a German, but that ought not to make any difference." It did not, at first. Steinmetz applauded the selection of Edison—the "greatest living American" in his view—to head the board. He praised Daniels's plan because it brought the United States in line with Germany, where "all engineers and inventors are always at the disposal of the government," and thought that Daniels meant to strengthen national defense rather than prepare the country to enter the war. Many newspapers throughout the country—from the East Coast to Texas—praised the choice of the Electrical Wizard Steinmetz. The *New York World* reported that if the engineering and scientific societies, which were to select the members of the board, did not name Bell, Wright, or Steinmetz, Secretary Daniels would do so because "their services would be indispensable."[58]

Not all newspapers felt this way. The editor of the *New York Press* warned the American public that "Steinmetz is a persistent and virulent German propagandist so sedulously devoted to the Prussian purpose and so seditiously antagonistic to the policy of the Government of the United States and to the spirit of the people of the United States that in spoken and published campaigns he seeks to blacken the American name before our own people, to indict our exercise of right and our observance of neutral obligations, to poison our very national existence." A strongly Republican paper, the *Press* was particularly incensed about a recent Steinmetz editorial in the *New York Evening Mail*, which had been secretly purchased by German agents. The *Press* said it would be a "grave danger to the Nation" if Daniels appointed him to the naval board.[59]

The newspaper sent an interviewer to see Steinmetz the next day. Clyde Wagoner, then a reporter for the *Schenectady Gazette*, warned him that the *Press*'s interviewer said he "would have to be biased in writing the interview by causing you to appear pro-German." Steinmetz then asked the vice president of the Dey Electric Company, for whom he consulted, to intervene with the newspaper's editor. The *Press* dropped the interview in favor of written answers to questions about the war. Steinmetz argued that the United States would use unrestricted submarine warfare if it was invaded by Germany and predicted that a U. S. embargo of all exports to the belligerents would end the war. The *Press* did not print his rather disingenuous reply that he was not partial to Germany but was strictly neutral and had been made to appear partial by statements taken out of context. Instead, the editor rhetorically asked, "As a National Advisory Board member, would we get the services of Dr. Jekyl Steinmetz who knows we will never have a war? Or would we get those of Mr. Hyde Steinmetz who defends a Lusitania murder on the ground that we may be at war and have to commit like murders?" The *Press* was not alone in its criticism. A Massachusetts man objected to Steinmetz's serving on the board, partly because he was a "socialist crank of the first order and is therefore opposed to the present government of the United States."[60]

When Daniels named the members of the Naval Consulting Board in September, Steinmetz was not on the list. The AIEE chose Frank Sprague and Benjamin Lamme. Two of Steinmetz's colleagues at GE— Willis Whitney and William Le Roy Emmet—were selected by the American Chemical Society and the American Society of Mechanical Engineers, respectively. The *New York Press* claimed that the "exclusion of Steinmetz" resulted from its "disclosures of his German leanings." It is unknown what effect the *Press* had in fact had, because Bell, Wright, Ford, and other prominent inventors mentioned as probable members were also not named to the board. The AIEE may have been swayed by the *Press*'s arguments, though, because Steinmetz's file in the AIEE archives contains a copy of the *Press*'s editorial against him. The AIEE president at the time was John J. Carty, the patriotic chief engineer of AT&T who joined the Signal Reserve Corps and served in France during the war.[61]

The *Press*'s criticism did not prevent Schenectady voters from electing Steinmetz president of the city council in November 1915, nor did it stop him from speaking his mind about the war. In declining an invitation in November to sail on Henry Ford's Peace Ship on its ill-fated journey to stop the war in Europe, he commended Ford's idealistic plan and thought that European leaders would "welcome the opportunity to escape from the present hopeless slaughter of millions of their best citizens." Usually he showed his German leanings. In October he wrote a letter to the editor of Hearst's *New York American*, praising its position of "true and honest neutrality" and its support of preparedness. But he did not think that a halfhearted preparedness would do much good: "Germany is invincible, because it has organized the entire nation for national defense." He expressed similar views in a Sunday-magazine interview in the *New York Times* in early 1916. The reporter noted his "very frank and pronounced sympathy for Germany" and quoted him as saying that Germany "is the greatest industrial country in the world. We must all admire her efficiency, war or no war." The interview contained the germ of a central concept in *America and the New Epoch*, written a few months later. Steinmetz predicted that the United States would be invaded by imported goods after the war; the country was "not prepared for such an invasion and to get ready for it she must stop killing the goose that is laying her golden eggs. I mean the big corporation." America was "far behind Germany" in organized production because its attitude toward the corporation was "that of the judge toward the criminal" rather than that of "mutually helpful partners," as in Germany.[62]

The United States could not copy the German method because of its democratic political traditions. In the letter to the *New York American* Steinmetz stated, "We must organize the entire nation for defense, but not in the manner Germany has done: for Germany is a monarchy, and we are a democratic republic, and it is for us to devise a new and democratic way, to show the world, that a democracy can be organized as efficiently for national defense, if not more so, as a monarchy."[63] Steinmetz devised this "new and democratic way" in *America and the New Epoch*. An editor for Harper and Brothers invited him to write a book on the war in the spring of 1916. Steinmetz said it would "not be a war book, but while the war necessarily must be mentioned as one of the outward evidences, and indeed as the culmination of the new epoch, its conduct, actions and incidents have nothing to do with the subject matter of the book." He signed the contract in April and finished the two-hundred-page volume in early August. Harpers published it in October.[64]

America and the New Epoch

Steinmetz was able to write the book so quickly because it was essentially an elaboration of themes he had developed in the articles, interviews, and editorials discussed above. (Writing it during the fight between the Lunn and Noonan factions in Local Schenectady probably also helped him make the decision to break with Lunn and remain a Socialist.) The book's tightly drawn argument was based on the application of socioeconomic "laws." Steinmetz began by saying that the "following does not represent my sentiments, but gives the conclusions drawn from the historical facts which of necessity follow from the preceding causes . . . Sentiment has nothing to do with, can exert no influence on, the phenomena of nature's laws," which included "economic laws." He would proceed deductively as with "any physical or engineering problem," a methodology that was as common to contemporary social scientists, wishing to make their disciplines "scientific," as to scientists and engineers at the time.[65]

Steinmetz's analysis resembled those based on the economic laws of historical materialism. Having read Karl Kautsky's *Das Erfurter Programm*, the German Social Democratic party's official statement of Marxism, while working at Eickemeyer's, he discussed the evolution of society from Neolithic times to the era of corporate capitalism in terms of changes in the mode of production. Following the epochs of slavery and feudalism, the French Revolution and new technologies like the steam engine ushered in the age of individualism, represented by capitalism and economic competition. But unbridled competition and the new forces of mass production led to overproduction, which forced prices down below the cost of production. To fend off the destructive tendencies of competition, capitalists formed large corporations and trusts, thus beginning a new epoch of industrial "cooperation." "Thus, not the 'trusts' are killing competition, but the failure of competition is the cause of industrial consolidation, of the corporations."[66]

As an evolutionary Socialist, Steinmetz differed from orthodox Marxists on several points, especially the importance of the class struggle. He claimed that "'class consciousness' is beginning to become an anachronism," because the diffused ownership of capital in industrial corporations lessened the distinction between classes. The separation of ownership from management, which was coming into the hands of salaried managers, meant that old-style capitalists were in effect committing "race suicide" with the rise of corporations.[67] Such a view helps explain why he ignored the class struggle—an essential part of historical materialism—in describing the transition from one epoch to the next.

Steinmetz's evolutionary scheme did resemble that in Friedrich Engels's *Socialism: Utopian and Scientific*, an influential popular statement of Marxism translated into English in 1892. In Engels's three-stage schema of Medieval Society, Capitalist Revolution, and Proletarian Revolution, taking over the means of "production and communication, first by joint-stock companies [corporations], later on by trusts, then by the State" was the last step in the Capitalist Revolution. This consolidation of industry "socialized" production, thus preparing the way for the third stage—the Proletarian Revolution that would free socialized industry from capitalist control and usher in the Cooperative Commonwealth, in which the "political authority of the State dies out."[68] Steinmetz's theory diverged sharply from Engels's after the second stage. Like the evolutionists, he did not believe in the necessity of revolution, but he placed much more faith than they did in the role of the corporation. As in his speeches before the NACS, Steinmetz said that American corporations needed to improve their human welfare function if they were to avoid public criticism and become the model for the future Socialist society.

Conditions were different in Germany. After the Social Democrats won the anti-Socialist war against Bismarck in 1890, Germany entered the cooperative era. The government solved the human welfare problem—combating the worker's three fears of sickness, accident, and old age—through state-supported insurance programs. It also helped companies form cartels. "The result was that the antagonism of the masses against the corporations, which here in America paralyzes our rapid industrial progress and threatens to destroy our prosperity by interfering with the industries' most effective tool, the corporation, has never appeared in Germany, but consolidation has proceeded unchecked." The resulting prosperity led, of course, to the present war between "individualistic" England and "co-operative" Germany—a battle between the epoch of the past and that of the future. Just as the French Revolution marked the transition from feudalism to individualism, the European War signaled the change from individualism to the new epoch of "co-operative industrial organization."[69]

For understandable reasons, Steinmetz did not mention Bismarck's role in creating this "state socialism" by pushing laws for health and accident insurance and old-age pensions through the Reichstag. Yet he correctly described the pioneering nature of this legislation and the trend toward a corporatist economy based on producer cartels. The cartel movement took off in the 1880s when the government encouraged them following the financial panic of the early 1870s. By 1900 nearly three hundred cartels controlled output, prices, and market shares in major industries. They were particularly strong in the steel, chemical, and electrical industries by the start of the Great War.[70]

America had developed much differently from Germany. Steinmetz believed that the Civil War transformed the United States from the era of slavery to that of individualism, which led to the growth of corporations and trusts, and then the era of cooperation during the "vicious" competition of the 1890s. This cooperation brought about "reactionary" antitrust laws in the United States, because corporations developed in the East while the Populist West clung to the ideals of individualism. The main opposition to the growth of corporations came from the "individualistic temperament" of the Anglo-Saxon race, the labor unions, and the antisocial character of the early corporations, which watered stock and did not take care of its workers.[71]

As in his previous writings, Steinmetz recommended the corporation as the model for municipal government. Such a system would improve efficiency because experts would serve long terms in office. But he also repeated his belief that this system would not work in the United States—at any level—because of America's political traditions. German corporatist methods would not succeed in the United States because "our national temperament is entirely different, is, indeed, the opposite of that of all European nations."[72] While Europeans were monarchial and efficient, Americans were democratic and inefficient, but they made more lasting policies. Thus, each country should use the method best suited to its temperament.

One approach would be for the government to evolve into a technocracy. In such a system production would be controlled for "legitimate demands" rather than "mere profit." If one organization controlled engineering, another sales, another transportation, and so forth, wasteful competition would cease. "All this is not a mere impracticable dream, but it has long been an established fact, has been the operating principle within all the more progressive large industrial corporations, and all that is necessary is to extend methods of economic efficiency from the individual industrial corporation to the national organism as a whole." The "natural" means to this end would be for the federal government, along with its "subordinate" state and municipal governments, to "acquire and operate all means of transportation and communication . . . supervise and control all corporations," regulate working conditions, and provide workers' benefits. Germany had established a similar system, but Steinmetz thought it was unworkable in the United States because the country lacked a "powerful, centralized government of competent men, remaining continuously in office." Professional societies like the AIEE had been helpful in a limited way, such as standardization, but they were not enough.[73]

The answer was the evolution of corporations into an industrial government. The corporation must first perform its human welfare function well and provide continual employment. Steinmetz stated that this "welfare capitalism," to use a modern phrase, "has nothing to do with the broader question of socialism—that is, of the elimination of capital. Socialism has as many followers in the offices of our corporations as it has in the shops, and in no way precludes co-operation within the corporation; indeed, in some respects the corporation may be considered as the first step toward socialism, and the industrial government of the nation by the united corporations as [a] preliminary and crude form of socialistic society."[74] Steinmetz foresaw that industry would evolve into one or a small number of very large corporations, each run by an executive committee. An "*Industrial Senate* as the supreme executive committee coordinates, controls, and directs all the country's industries—that is, governs the country." Companies would not be required to join the system, but economic boycotts would soon drive such "scab" corporations out of business. Officers, technicians, and workers would rise from the ranks in a form of meritocracy. The system—like the modern corporation—would be "essentially democratic in character; there is no autocratic authority, but every member of the organization has a directive power within his field of activity."[75] Such passages indicate how Steinmetz's favored position at GE led to an optimism unlikely to be shared by most corporate employees.

He thought that such a society—an "aristocratic democracy"— would be as "stable as was the classic age or the feudal age of human society, and not self-destructive by its own success, as was the individualistic age." One problem was the possibility of class control, which he proposed to eliminate with the "inhibitory power" of the Tribuniciate. A democratically elected body that would evolve from the present political system, the Tribuniciate would be the equal partner of the Industrial

Senate in a dual government. The Industrial Senate would act as the administrative and executive branch, while the Tribuniciate would have "general supervisory power" to set national policies in the foreign and domestic realm. But it would only have veto power over the Industrial Senate. Steinmetz claimed that the Roman Republic had had such a dual government and was the "most successful and most efficient government the world has ever seen."[76]

In the closing chapters Steinmetz discussed two preconditions necessary for the evolution to this technocratic corporate state. The first was "racial unity." After repeating his racial analysis of the country he noted that each race had its own political characteristics. As "empire builders" and individualists, the Anglo-Saxons were originators but not good organizers or administrators. The Teutons had a "collective or co-operative temperament" and were thus great organizers. The "strong collectivist temperament" of the Celts made them excellent administrators, as seen in the boss rule in New York City. Although these traits complemented each other, Steinmetz did not think they would lead to unity. The Anglo-Saxon characteristics were ingrained in the political system and would be for some time to come.[77]

The second precondition was Steinmetz's often-repeated theme that the corporation needed to improve its human welfare function. This work should spread to all corporations and be given as much importance and as many resources as the technical, administrative, and financial functions received. It would not be safe—even with a Tribuniciate—to entrust the industrial government of the country to the corporations now, because they lacked social responsibility. Such responsibility evolved over time. As corporations grew larger and came to employ thousands, they ceased to be strictly private property, because their actions affected so many people. Steinmetz thought that President Roosevelt's intervention in the great anthracite coal strike in 1902 had established the doctrine of corporate responsibility in practice and in law. More accurately, he criticized existing welfare capitalism as "paternalistic," which he attributed to the individualistic temperament of company leaders. A major problem was how to create independent welfare organizations that were acceptable to both workers and management. A related concern was how to counteract public hostility to the corporation and improve its image. His lengthy criticism of inaccuracies in newspaper stories and interviews and his suggestion that a Jack London or a Kipling should tell the "romance" of the corporation suggest that he was a willing participant in the creation of his public image by GE and the press.[78]

Steinmetz concluded by returning to the issue of the war. No matter who won, the real winner would be collectivism, because all nations—including England—were adopting "cooperative" measures in order to compete with Germany. This was the wave of the future: "Either we enter the coming co-operative era of the world's history and take our place as one of the leading industrial nations organized for the highest efficiency possible under co-operative industrial production, or we fall by the wayside." America could enter the New Epoch by adopting an Industrial Senate. Extending this system to other countries would make war impossible: "International industrial co-operation would be so near socialism, would so imperceptibly merge into it, that nobody would ever be able to see where 'capitalist society' ended and the 'socialist commonwealth' began—though it is obvious that this socialistic commonwealth will be as different from the dreams of us socialists of to-day as every accomplished progress always has been from the first crude ideas of its originators."[79]

What did contemporaries think of Steinmetz's dream, the technocratic corporate utopia that he called socialism? The liberal *New Republic* printed a generally favorable review, which thought that, despite its "dogmatism," the book should provide a "healthy warning" to the "large corporation class." At least two Socialist papers ignored his unorthodox views and praised less controversial parts of the book, most likely because of his celebrity status in the party. The folksy *Appeal to Reason*, the largest Socialist paper with over half a million subscribers in 1917, described his predictions of a shorter workday without mentioning the corporate restructuring needed to bring it about. The *Citizen* followed the same line and used his view of the coming four-hour day as a selling point for workers to support the cause. Some Socialists noted his unorthodox views. The reviewer for the *New York Call*, which had recently praised him for not leaving the Socialist party with George Lunn, thought his independence from Socialist theory made him the right person to open the discussion of the corporation's future role in society. But the reviewer questioned whether labor unions and corporations could cooperate in a manner that would benefit the workers. Florence Kelly, later president of the Intercollegiate Socialist Society, praised Steinmetz for setting his intellectual horizons beyond doctrinaire principles. Other Socialist papers and journals did not review the book, not even the *Masses*, which had recently absorbed the *New Review*, whose advisory council had included Steinmetz.[80]

The Socialists' rather lukewarm response to *America and the New Epoch* dismayed Randolph Bourne, the radical writer and pacifist. Having moved away from Marxism, Bourne wrote in the *Intercollegiate Socialist* that American Socialists might be "genuinely scandalized at the revolutionary ideas" in the book, but they should be the ones most debated by the movement. Bourne described Steinmetz's corporate socialism accurately and in some detail, noting that "instead of 'letting the nation own the trusts' [Wilshire's motto], he would let the trusts own the nation." This prospect did not bother Bourne, because he believed Steinmetz's claim that a reformed corporation would solve the problem of creating a democratic technocracy. Bourne concluded, "If this book does not prove epoch-making for a new epoch, it will imply that American socialism has become intellectually reactionary, given over to party politics and to the decaying philosophy of nineteenth-century liberalism." He repeated this theme in a joint book review in the *Dial*, where he criticized Ida Tarbell's rosy account of welfare capitalism but praised the similar functions of Steinmetz's "industrial state." Bourne's opinions probably owed much to his empathy with Steinmetz's physical handicap and his political views. Both were dwarfed hunchbacks, and Bourne had written in 1915 that whatever the outcome of the war, "all the opposing countries will be forced to adopt German organization, German collectivism"—the main thesis of *America and the New Epoch*.[81]

The book was a bigger success with the "capitalist" press. Steinmetz collected over eighty book reviews in newspapers, most of which had published stories about the Electrical Wizard or had carried his articles for the National Editorial Service. The major papers liked his attitude toward the corporation but did not describe the nature of his corporate state or its socialistic character. A reviewer for the *New York Times* called it a "stimulating volume presenting points of view which ought not to be so novel," while Tansy McNab in the *New York Tribune* spent most of her review telling Steinmetz stories before dispensing gems of wisdom from "one of the sanest and wisest of our citizens." In her view Steinmetz wanted to "legalize big business or monopoly, and give our big industrial units the backing of the political power." The book "ought to be on every legislator's desk." The pro-German *New York Evening Mail* serialized the book's first chapter in mid-1917. Two magazines that had supported President Wilson were more critical. The *Independent* criticized Steinmetz's racial analysis and the *Nation* thought his approach was too Marxist. The *Nation* did admit that "along with much that is bizarre, the book contains illuminating passages."[82]

Although the first printing sold out at the end of 1916 and the Intercollegiate Socialist Society listed it for a course of study in 1919,[83] the book did not prove to be "epoch-making" in American socialism. One reason was that Steinmetz's ideas exacerbated an uncomfortable dilemma for party members regarding their attitude toward the corporation and the trust. During the antitrust proceedings against GE in 1911, for example, Schenectady Socialists had chided President Taft's administration for wanting to break up the "Electrical Trust." The editors of the *Citizen* proudly stated that the Socialists did not want to destroy the trusts, a policy that would better protect jobs at GE. Such a policy must have confused some readers. The *Citizen* and the *New York Call* continually exposed the evils of the trust and the corporation in raising prices and oppressing workers, but they opposed their break-up. One writer for the *Call* informed voters in 1912, "Socialists are not 'trust busters.' We favor trusts. Didn't know that did you? You thought you were 'agin' them, but you're mistaken." Trusts are efficient, economical, and opposed to competition. "We would socialize them, nationalize them."[84]

As Bourne noted, Steinmetz stood this philosophy on its head and would let the trusts own the nation's economy. The corporation was thus a model for a future society, not merely an intermediate step on the road to socialism. This view corresponded to that of collectivists like King Gillette, as noted earlier. Steinmetz's idea of dual political and economic parliaments can also be found in the writings of American Socialists. In 1909 Hillquit mentioned European proposals for a bicameral parliament in a Socialist society, the "political chamber taking the place of the lower house and the economic chamber that of the upper house." And in *Twentieth-Century Socialism* (1910) the civic reformer Edmond Kelly noted the suggestion that in a future Socialist society "all things pertaining to production and distribution might be determined by an industrial parliament . . . subject to the approval of Congress." Spargo criticized such schemes as undemocratic and utopian in his *Applied Socialism,* yet Steinmetz, who owned a copy of Kelly's book as well as Spargo's, looked more favorably on such parliaments.[85]

One reason was his admiration for the German corporative tradition. Although Steinmetz focused on the collectivism of modern cartels, which were promoted by banks and legalized and encouraged by the government, German corporative theories dated to the Middle Ages. Proposals for dual industrial and political parliaments had been common in the early nineteenth century, and the virtues of the "monarchial corporatism" of Bismarck and others were set against laissez-faire liberalism and Marxian socialism when Steinmetz attended the University of Breslau. Bismarck instituted the social welfare measures praised in

America and the New Epoch as part of his broader corporatist plan, while the German war economy admired by Steinmetz emerged from the "cartel corporatism" of Walther Rathenau. President of Allgemeine Elektrizitäts-Gesellschaft (AEG), Rathenau became head of the Strategic War Materiel Office in 1914 at the beginning of the war and organized the economy along collectivist lines. He established war industrial companies for each major industry to procure and distribute war materials. Since the state was the major stockholder in each of these nonprofit groups, he organized virtually all of German industry into self-governing, state-dominated cartels. In a utopian booklet published in 1918 Rathenau predicted that the postwar German economy would be restructured along similar corporatist lines. All companies would join their appropriate trade or vocational groups, which, in conjunction with their industrial associations, would coordinate production and distribution. The state would confer almost sovereign rights on these incorporated syndicates in return for state supervision and a tax on their profits.[86]

Steinmetz had an opportunity to learn about Rathenau's ideas by observing his work at AEG. Founded by Walther's father, Emil Rathenau, as the German Edison Company in 1883, by the start of the war AEG had grown to be one of the two largest electrical groups in Germany through a series of mergers, cartel agreements, and stock purchases of manufacturing and utility companies—all coordinated by large investment banks. Throughout this period AEG did not lose touch with its American cousin GE, which had also descended from the Edison interests. Steinmetz visited the company on his European trip in 1897 and probably knew about its cartel arrangements with domestic and foreign firms. In 1903, for example, AEG and GE agreed to cooperate on such technical matters as the development of the Curtis steam turbine, on which Steinmetz worked, and to divide the world market in electrical machinery between them—AEG to have Europe, GE North America.[87]

Steinmetz's admiration of German cartels also helps resolve the paradox of his idea of autonomy through cooperation, that is, how his political theory would prevent the domination of professional societies and colleges by large corporations. The much different European and American laws regarding trusts and the contrast between the market and political conditions in the United States and Europe help explain the growth of large corporations in the United States at the turn of the century, while Europe had cartels of smaller, independent companies that were horizontally integrated through temporary contracts. These companies preserved their identity and a large degree of autonomy by cooperating according to specified terms in a cartel.[88] Steinmetz probably based his idea of autonomy through cooperation for colleges and professional societies on this model, particularly since private associations played an important role in German corporatist theories.

What did GE officials think of *America and the New Epoch*? Charles Coffin, the chairman of the board, had little sympathy with the new class of corporate liberals and probably ignored it—at best. President Edwin Rice was more receptive to these ideas. In a speech about the recently established National Industrial Conference Board given just after the publication of Steinmetz's book, Rice said that in order to win the economic battle that would surely follow the European War, American manufacturers should recognize that the "day of extreme individualism is past . . . The time has come when co-operation in the broadest sense is essential to the maintenance of our industrial prosperity." Composed of representatives from twelve national associations of industrial employers, the NICB was a powerful "co-operative force" to press for better relations among employers, labor, and the government. The resemblance between the structure of the NICB, organized by corporate reformer Magnus Alexander of GE, and Steinmetz's Industrial Senate is striking—a point probably not lost on Rice or Alexander.[89]

The ideas were not new to either one. Alexander had helped form the NACS, where he heard Steinmetz give his presidential address on corporate socialism. Rice had helped establish such cooperative ventures at GE as the 1896 patent-pooling agreement with Westinghouse, the National Electric Lamp Association combine, and the 1903 cartel agreement with AEG. In effect, *America and the New Epoch* was the corporatism of GE and its national associations writ large across the nation's economy.

9   Building a New Epoch

Steinmetz found applying theory to practice a greater challenge in many respects for corporate socialism than for electrical engineering. The symbiotic relationship between theory and practice that he had mastered at General Electric held few clues to unraveling the mysteries of political economy, a field with poorly understood "laws" and many more "variables" than engineering. Steinmetz based *America and the New Epoch* on his understanding of economic laws and described a corporatist path the United States should take in order not to fall by the wayside. The country seemed to go down that road with President Wilson's "war collectivism," as did revolutionary Russia when it heeded V. I. Lenin's call for state-planned electrification. Steinmetz did not play a large role in these momentous events, but he used his reputation and influence to help build a new epoch within GE, the United States, and Soviet Russia.

War and Postwar Collectivism

The mobilization of the home front during the Great War engendered a large degree of cooperation among business, government, and labor to meet the needs of fighting a highly industrialized conflict. The cooperation was institutionalized by the various public-private agencies formed by President Wilson and the Congress. Some, like the Naval Consulting Board (1915) and the Council of National Defense (1916), were purely advisory, while others, like the Shipping Board (1916), the War Industries Board (1917), the War Labor Board (1918), and the Food, Fuel, and Railroad Administrations (1917), had some executive authority. Their power varied considerably. While the "Great Engineer" Herbert Hoover stressed voluntary compliance in his Food Administration's wheatless and meatless days to conserve food for the war effort, the government ran other industries. The Shipping Board operated and greatly enlarged the merchant marine; Treasury Secretary William McAdoo and a board of industry officials ran the railroads under the Railroad Administration; and the government owned and operated the nation's telegraph and telephone systems, through the Post Office, for about a year at the end of the war. Even before most of these agencies were established, liberals like Walter Lippmann thought the country stood "at the threshold of a collectivism which is greater than any as yet planned by the Socialist Party." The war collectivism, however, came to resemble the corporate liberalism of Wilson's prewar agencies like the Federal Trade Commission much more than socialism. True to Wilson's proclivities, it formed a "middle ground" between laissez-faire capitalism and state socialism.[1]

From Steinmetz's point of view the war agencies represented the country's response to his admonition in *America and the New Epoch* to enter the coming age of industrial cooperation. The editor of a Denver newspaper agreed and thought the government's takeover of the railroads was a "big step toward Mr. Charles P. Steinmetz's plan for the reorganization of the entire government on the lines of a big corporation." The amount of business-government cooperation increased during the war, but it never reached the level of centralized control Steinmetz desired. In the area of industrial mobilization, "dollar-a-year" businessmen formed "cooperative committees" in each trade to help coordinate the purchase of military supplies. Similar committees, called commodity sections, formed the backbone of the War Industries Board under the direction of former Wall Street speculator Bernard Baruch. The WIB thus resembled Steinmetz's Industrial Senate in its organization, but it never came close to having its power. While Baruch and many business leaders favored a strong corporatist body, President Wilson and the Congress preferred a weaker, decentralized agency that would be dismantled after the emergency.[2]

At least one of Wilson's war managers drew much of his inspiration for a corporate state from *America and the New Epoch*. Harry Garfield, president of Williams College, became director of the Fuel Administration in August 1917. He had met Steinmetz in January of that year, when they gave speeches to the Economics Club of Boston, along with William Redfield, secretary of commerce, and Edward Filene, a Boston merchant and proponent of welfare capitalism. Steinmetz summarized *America and the New Epoch*, which had just been published. Five days later—time enough to have read the book—Garfield forwarded a copy to President

Wilson, saying that the "book is full of meat. The only excuse for its initiating crudity & lack of literary merit is that its author is a foreigner & a genius—the Wizard of Schenectady." Apparently unaware of how vehemently Steinmetz had opposed him in the recent election, Wilson said he looked "forward to reading it with a great deal of interest." Robert Cuff has observed that Garfield had "enormous enthusiasm" for the book and that "his speeches are redolent with the book's themes." In early 1919 Garfield recommended that Wilson establish an Industrial Cabinet and a "political cabinet"—which resembled Steinmetz's dual government—to coordinate the postwar economic recovery. The Industrial Cabinet would consist of the secretaries of interior, commerce, labor, and agriculture and the heads of such wartime agencies as the Railroad Administration, the Shipping Board, and the Fuel Administration. Although Wilson did not implement the idea, he did not dismiss it out of hand. In addition to helping shape Garfield's philosophy, Steinmetz's book supported the corporate-liberal ideal of business-government cooperation that was prevalent in the Wilson administration.[3]

The pro-German tone of Steinmetz's editorials kept him from participating in the early war collectivism, as we saw in the furor over his proposed appointment to the Naval Consulting Board. His position at GE also created a dilemma as the company tooled up more and more for the war. Like many large manufacturers, GE began making armaments early in the war, receiving over $30 million in contracts to make artillery shells in 1915. After the United States entered the conflict the company made searchlights, marine turbines, radio tubes, switchboards, and other military hardware. Whitney's laboratory worked on submarine detectors, and Ernst F. W. Alexanderson designed high-powered radio transmitters in Steinmetz's Consulting Engineering Department.[4]

Although national newspapers had reported before the United States entered the war that the Wizard of Schenectady predicted a German victory, the loyalty of the city's most famous citizen, now president of its common council, seems not to have been much of an issue. In fact, shortly after the United States entered the war, Governor Whitman appointed Steinmetz, as acting mayor, and other prominent Germans to a committee to help the state conduct a census of German aliens. Steinmetz "stated that the men of German and Austrian birth employed in the factories are skilled workmen who would be hard to replace . . . they are loyal citizens and the authorities need not entertain any fear of any disturbance or damage to property." The *Schenectady Gazette* helped Steinmetz's image by reporting his views on the draft. In a discussion at his Economics Club Steinmetz disagreed with the speaker, an "extreme Socialist," and declared that a militarism based on a "democratic universal service" was "entirely consistent with Socialism." The Washington-based League of National Unity also appointed him its president in late April to help "eliminate hyphenism."[5]

Nevertheless, Steinmetz was not isolated from anti-German sentiment in Schenectady. A loyalty pledge was circulated in the GE shops and offices in March, and George Lunn, G. E. Emmons, C. A. Richmond, and William Le Roy Emmet—all colleagues of Steinmetz—spoke at a mass conscription meeting in late April. GE conducted a huge Liberty Loan campaign in June, following the example of chairman of the board Charles Coffin, who was active in the Liberty Loan drive. The company subscribed to $5 million in bonds and agreed to match the contributions of its employees on June 6, Liberty Loan Day. Six hundred members of the bond committee marched through the plant amid flying flags and martial music at noon, while representatives from every department spoke to each worker about buying a bond. Nearly fourteen thousand of the plant's twenty-two thousand employees heeded the exhortations of Emmons and Edwin Rice—Steinmetz's friends—and subscribed over $1 million, one-half of the city's quota. At the same time the Schenectady common council voted to buy $40,000 of Liberty Bonds. While the two Socialist aldermen voted against the measure in line with their party's position on the war, president Steinmetz retreated to the laboratory and was absent from the meeting. In July the *Schenectady Union-Star* accused the antiwar *Citizen* of treason, and Emmons fired a Socialist worker because he would not buy a Liberty Bond. Many of Steinmetz's GE colleagues participated in subsequent Liberty Loan drives, including Albert Rohrer, C. W. Stone, and even his adopted son Joseph LeRoy Hayden.[6]

Steinmetz was criticized as the war continued. The *New York World* reported in October that he was one of a handful of prominent Germans who did not respond to a letter asking them to affirm their allegiance to the United States. During the election campaign in November Lunn, as noted earlier, thought Steinmetz sympathized with the "enemies of America." The war put him in a difficult position, because he had kept in touch with his German relatives over the years. He sent $100 to $200 a year, via the Hamburg-American steamship line and by wireless, to his half-sister Margarethe Mache in Breslau from 1914 to the end of 1916.[7] Doing war work at GE would endanger his German family, but staying neutral was becoming increasingly difficult.

Steinmetz made what must have been an agonizing decision in early 1918, when he agreed to join the war effort. The Naval Consulting Board—with which he had been embroiled in controversy three years before—wrote his Consulting Engineering Department in March for help in solving a problem with high-explosive shells. The shells exploded only after penetrating the ground to a depth of one to twenty feet, thus creating craters rather than killing enemy soldiers. D. W. Brunton, chairman of the naval board, asked Steinmetz's group to develop shells that would explode a few feet above the ground, in what he called "probably the biggest single problem of the war." Steinmetz told his staff, "We should by all means endeavor to solve [the problem] or assist in its solution. All our laboratory facilities, those of the Consulting Engineering Department as well as those of my private laboratory, obviously are at its disposition." Edwin Rice, then president of GE, learned of Steinmetz's decision to work on this and related war research in June. Rice told his colleague of a quarter of a century, "We are delighted that you feel disposed to put your services at the disposal of our country in its effort to win the war."[8]

Why did Steinmetz turn against the Fatherland when he had defended it so vehemently as late as January 1917? The main reason seems to have been America's entry into the war. A naturalized citizen since 1895, Steinmetz wrote in an editorial that all Americans had the right to oppose the war: "Only when final action is taken and is irrevocable, then all citizens must rally around the government." When the United States entered the conflict, many "pro–war Socialists" like John Spargo, William English Walling, and Upton Sinclair left the party because of its antiwar stand. Steinmetz remained a member, but he called for a complete mobilization under a "cooperative industrial system of production" in May 1917 and was doing war work nearly a year later.[9] One factor in the delay may have been the Russian revolution. Lenin took power in November and signed the treaty of Brest-Litovsk in March 1918, which officially took Russia out of the war. In deciding to do war research later that month Steinmetz was thus not helping his old enemy, imperial Russia, defeat Germany. But he still had to face the dilemma that working for his adopted homeland meant working against the government of his German family.

Once he had made the decision he turned to war problems with some enthusiasm. Hayden sent a copy of Steinmetz's ideas on the shell problem to Lunn, then a Democratic congressman on the Committee on Military Affairs, and asked Lunn to stop by when he was in Schenectady.

Lunn agreed, thus helping mend the break with Steinmetz. The effort apparently succeeded, because Steinmetz was appointed to the Inventions Section of the War Department in May 1918. By June he was working in five areas: ballistic trajectories, a tungsten-alloy gun, the premature explosion of shells, frictional retardation of shells, and methods to conserve rare materials like tantalum. Rice greeted this effort warmly, but he reminded Steinmetz that GE was not generally involved in these fields and could not spend large sums on them without a government contract. Later Steinmetz advised the Army on processes to produce nitrogen. Little is known of the success of Steinmetz's work, which he continued until at least a month before the armistice. While Brunton found some of his data useful, an ordnance major informed him that others had analyzed the mathematics of ballistics with far greater sophistication. Outside his field of expertise the Wizard of Schenectady was apparently not very helpful in creating the instruments of modern warfare.[10]

He was of more help in other ways. Following the severe fuel shortages of the unusually cold winter of 1917–1918, Fuel Administrator Garfield unexpectedly announced that he would close all factories east of the Mississippi for four days in mid-January and then for nine consecutive Mondays beginning at the end of the month. The announcement unleashed a storm of protest that almost wrecked the Wilson administration's war-planning measures, indicating that a more "scientific" approach was needed. Steinmetz wrote a lengthy article on the U.S. energy supply for the "war convention" of the American Institute of Electrical Engineers that June, called by Edwin Rice, then president of both GE and the AIEE. Steinmetz described past coal production in terms of an empirical equation—similar to his law of hysteresis—and plotted it as a curve, which he called the "result of economic laws, which are laws of nature." An extrapolation from this curve—which gave an indication of coal consumption—convinced him that waterpower would not be able to meet the country's energy demands when coal eventually ran out. He proposed that the United States increase its efficiency of using coal, develop techniques to collect solar energy, and adopt a special type of induction generator to harness the scattered waterpower of rivers and streams, as well as the energy lost in burning coal in furnaces. The generators would feed power into large electrical grids from many decentralized hydro and steam plants. The proposals—many of which, like solar energy and co-generation, were revived during the energy crisis of the 1970s—became the basis for his popular articles on the conservation of energy in the 1920s.[11]

The survey of America's energy supply foreshadowed similar proposals made by advocates of technocracy shortly after the war. The movement came to the fore in the early years of the Great Depression when the radical engineer Howard Scott and his followers formed Technocracy groups across the country, groups whose goal was a government by "technicians" (engineers and other professionals). The public first learned about Technocracy in the fall of 1932 when Scott released the results of its ten-year survey of American industry to eastern newspapers. Scott claimed that engineers had been working on the project at Columbia University since 1920 and that Steinmetz and Thorstein Veblen had been original members of the group.[12]

Veblen greatly influenced Scott's views—through informal discussions and his book *Engineers and the Price System* (1919)—but Steinmetz played a lesser role than that claimed by Scott, who was known for his hyperbole. Steinmetz became involved in December 1916 when H. B. Brougham, an editor of the *Philadelphia Public Ledger,* sent him clippings about the founding of the New Machine that month. Usually viewed as a prototype of Veblen's Soviet of Technicians and Scott's Technocracy, the New Machine was a group of engineers and reformers organized by the scientific management expert Henry Gantt and Charles Ferguson, a writer and newspaperman who later advocated a veblenian technocracy. While with the Department of Commerce in 1914 and 1915 Ferguson served as President Wilson's missionary to American business, preaching self-government in industry along European lines rather than strict federal regulation. Brougham told Steinmetz that Ferguson—the theoretician and general manager of the New Machine—agreed with the technocratic conclusions of *America and the New Epoch* and wanted to get Steinmetz's opinion of the group.[13]

Steinmetz's response is unknown. When the New Machine folded after a few months, its program was picked up in 1919 by the Technical Alliance. Formed by Scott that fall to accomplish the goals set out in Veblen's *Engineers and the Price System,* the alliance included an educator (Veblen), scientists, physicians, engineers, and other "technicians." Scott listed Steinmetz as a member—the source of his later claim that Steinmetz and Veblen were founders of Technocracy—but there is no evidence that Steinmetz did more than lend his name to the group, as he did with several other organizations. The alliance, which did some consulting work to pay expenses, was apparently not well received by mainstream engineers. Herbert Hoover, a leader of the Federated Association of Engineering Societies, told a colleague that the alliance was essentially a "money-making organization" that was "headed by extreme radicals."[14]

Before the alliance went under in 1921 it did some consulting work for the International Workers of the World during the Wobblies' brief flirtation with scientific management. Scott was named the first and only research director for the IWW in 1920 and published studies of waste in the oil, coal, and milk industries for the union the next year. The one on coal made two recommendations that Steinmetz had suggested in his survey of the nation's energy supply: harnessing the waterpower of streams and rivers and utilizing the by-products of burning coal for heat to generate electricity. *America and the New Epoch* may also have influenced Scott, since he proposed a vague corporatist reorganization of industry in 1920, consisting of groups based on occupational functions. Scott wrote that "a Steinmetz may join any one of a number of labor organizations as an individual, but there is no organization of workers which he could join in his industrial capacity of research technologist of the equipment division of the power industry."[15]

Labor and Capital

An important task in building the New Epoch was to improve the corporation's welfare function. Only with a well-developed welfare system would corporations be qualified to form an Industrial Senate, the economic half of the corporatist dual government presented in *America and the New Epoch*. Steinmetz reformed industrial education at GE, but he ran into difficulties when he turned to the company's labor policies.

Part of the problem was his attitude toward labor unions. He told a correspondent in 1913 that "undoubtedly it is an advantage to belong to a union; but at the same time, unionism is a side issue only, has never succeeded in permanently improving social conditions, and never will."[16] He had an opportunity to test this belief in late November, when nearly all of the fifteen thousand workers at the GE Schenectady plant went out on strike to protest the inclusion of two union leaders in a layoff by works manager G. E. Emmons. Thousands marched peacefully to the center of the city with dinner pails in hand and then went home. Mayor Lunn had a good labor-union record at the time—having been arrested for supporting the free-speech rights of textile strikers at Little Falls, New York, the previous year—and came to the aid of the GE workers. He appointed more than thirty of them as special deputies to patrol the picket lines, an action applauded by Socialist papers, and said no strikers would go hungry as long as he was mayor. Lunn negotiated an early end to the strike as a mediator between the strike committee and company officials. Workers returned to the plant after five days, jubilant that laid-off employees would be put on part-time work, that the two union leaders would be rehired, and that Emmons had promised not to discriminate against labor leaders.[17]

Although a member of Lunn's administration and a Socialist candidate in the citywide elections held just before the strike, Steinmetz did not take a public stand on the strike, the largest in GE history to date. His private papers do not reveal his attitude toward it, nor how he felt about the dilemma of being an engineering manager and a member of the Socialist city government that supported the strikers. The dilemma would not resolve itself during these years of high labor unrest, as workers struck three times in 1915 while the Schenectady Works was busy filling munitions orders for the war. In the fall union leaders demanded that the workday be shortened from ten to eight hours—when Steinmetz was running for president of the common council and Lunn for mayor. Emmons proposed a nine-and-one-half-hour day with a 5 percent increase in pay and promised to reduce the workday to nine hours and to give another 5 percent raise a year later. A committee of the Metal Trades Alliance, which included most of the GE unions, accepted Emmons's offer, but the machinist unions (numbering between twenty-five hundred and three thousand workers) objected strongly. The action led to a walkout on October 4 of nearly all of GE's Schenectady workers, thirteen thousand of whom were members of unions belonging to the alliance. This time Emmons refused to budge and waited out the strike. The Metal Trades Alliance voted ten to one on October 19 to reject Emmons's offer, then began superintending the picket lines so that the company would have no excuse to call out the militia. The strike was relatively peaceful but long, and the alliance accepted Emmons's offer on October 25. The machinists, whose refusal of the compromise had led to the strike, stayed out for another two weeks, making the strike the largest and longest at the Schenectady Works until World War II.[18]

Out of office but running for mayor, Lunn said little about the strike publicly until he urged a compromise of a fifty-hour week on October 22. The idea languished until the machinists presented it as their counteroffer a week after the other workers had returned to the plant. The strike ended with a whimper; the machinists returned to work without a victory on November 5, just three days after Lunn had been elected mayor and Steinmetz president of the common council. Lunn did not made the strike a campaign issue, nor did he offer to mediate it, as he had other labor disputes since becoming a Socialist. The reasons are not clear, but it did not hurt his campaign that he had helped settle the 1913 strike in five days as mayor, while the 1915 strike had gone on for nearly five weeks under the fusionists. Recalling this contrast, Emmons, who did not support Lunn in either campaign, congratulated him on his victory and said GE was ready to cooperate with his administration. Steinmetz made one campaign appearance with Lunn and others at a "monster mass meeting" in Schenectady in late October, after all workers except for the machinists had returned to the plant. The timing of the meeting and Lunn's reluctance to make the strike an issue made it easier for him to follow Lunn's lead. Nor did he have to cross a picket line, since he could conveniently work at home in his laboratory.[19]

Shortly after the strike Steinmetz became more involved in labor matters outside Schenectady—a much safer activity than commenting on a GE strike—when the Labor Center Association of New York City appointed him to its advisory board. Then he thought unions were more than a "side issue" and told the *New York Times* that fighting among labor organizations "seems so unnecessary in a movement so progressive and humanitarian, with its underlying principle of fraternity." Warring groups like the Wobblies, the Socialist Labor party, and the American Federation of Labor should adopt the "engineering method" and join the Labor Center Association to cooperate with each other, just as capital was cooperating by combining into corporations and trusts. He also supported labor indirectly by taking a stand against scientific management, which he had had the opportunity to observe in GE's shops. Although his watchword was *efficiency,* Steinmetz, like many Socialists, criticized Taylorism for viewing the worker as a cog in the industrial machine.[20]

In general, however, he disliked labor unions. They did not figure either in the evolution of society described in *America and the New Epoch* or in the running of the future corporate state. Corporate executives, not labor leaders, would form the Industrial Senate. Steinmetz thought that unions were not a political factor in America because of the workers' upward mobility. Instead, unions were a reactionary force—like the misguided trustbusters—because of their "unjustified and illogical" opposition to the growth of corporations. Steinmetz said this opposition had been warranted during the industrial warfare of the late nineteenth century, but he left little room for unions in a future society where corporations and the government would protect workers against accident, sickness, and old age; would regulate working conditions and hours; and would develop improved technology to reduce the working day to four hours. Technocrats, not union leaders, would solve the labor problem.[21]

Steinmetz had proposed elements of this corporatist ideal to GE management in March 1914, just four months after the 1913 strike. In a "confidential" report he criticized the high age requirement of the company's pension plan, established in 1913, and thought the federal government could provide a more equitable system. In the area of profit sharing, he recommended that GE establish a "wage dividend," similar to that given by some Edison utility companies. His corporatist philosophy was apparent when he said that the dividend would make all employees "practically stock holders and thereby interested in the welfare of the company, and less inclined towards anything which might interfere with the prosperity of the company, [such] as a strike." The dividend would increase productivity, would allow GE to raise or lower wages indirectly in good times and bad without incurring the wrath of labor unions, and would help erase class lines between workers and the "ruling classes."[22]

When the company introduced a 5 percent dividend in early 1916, shortly after the 1915 strike, the *Citizen* printed an argument for it that was pure Steinmetz. The bonus resulted from the "recognition of the coming of a new era, that of co-operation; and that the final and ultimate outcome of this is Socialism, however great the step may appear from a wage dividend to a Socialist society . . . [The] acceptance of this principle of co-operation by such a big company shows the recognition of the far higher industrial and productive efficiency of co-operation; and illustrates how much greater efficiency would result from complete co-operation, or Socialism." Union leaders opposed the dividend and probably doubted whether GE management was working toward socialism. They suggested instead that the bonus bribed workers not to go on strike, because it was based on continuity of service—a fairly accurate reading of "Comrade" Steinmetz's confidential proposal.[23]

The semiannual wage dividend became part of GE's evolving system of welfare capitalism, one of the most extensive in industry. The company paid two 5 percent bonuses in 1917 of over $1.3 million to nearly twenty-two thousand employees who had over five years of continuous service. By 1919, seven years after Unitarian minister Albert Clark had become the first director of the company's Welfare Department at Schenectady, GE had a wide range of such programs. The pension plan had not changed much since 1913, but workers who had ten years with the company got a one-week paid vacation. The General Electric Mutual Benefit Association, established by Clark in 1913, replaced traditional fraternal orders organized by workers in paying some accident, sickness, and death benefits. GE created company hospitals, bowling teams, and educational courses before the war and a works council (shop committees plus a joint committee of workers and management) at Lynn on the demand of the War Labor Board. Ronald Schatz has pointed to these programs as one factor in the decline of unionism in the 1920s, while David Montgomery has emphasized the role of government repression and widespread unemployment in creating an environment conducive to the success of corporate welfarism against the unions.[24] Ironically, Steinmetz proposed many of these welfare measures—as he had the reforms in education, technical standards, and ethical codes—because he thought they would lead to corporate socialism, not stablize corporate capitalism.

His socialism was evident after the war, even during the government's attacks on radicals. In 1919—during the hysteria of the Red Scare brought on by the Seattle general strike and a wave of strikes involving about one-fifth of the nation's work force—Steinmetz remained a Socialist, albeit one who criticized labor unions. *American Magazine* published an interview with him in April entitled "The Bolshevists Won't Get You, but You've Got to Watch Out!" As an evolutionary Socialist, he disliked Lenin's revolutionary tactics and similar methods used by the Wobblies. He thought bolshevism was a symptom of an industrial disease, which American corporations could cure by improving their welfare function. GE approved of these sentiments and reprinted portions of the interview in the *Schenectady Works News*.[25] Steinmetz extended this analysis to the labor movement in a speech, also delivered in April, in which he told a gathering of factory managers that the "most formidable obstacles in bringing about 'industrial cooperation' often are imposed by the labor unions." By abandoning their traditional fraternal roles and becoming "mere fighting organizations," they came under the control of the "most radical men," who were "less able and less willing to establish relations of a cooperative nature." The "best that can be expected is neutrality. To bring about cooperation against the open hostility of the labor unions, or by fighting them, appears almost hopeless." Steinmetz favored a "true" open shop and a "well-kept" industrial city, like Schenectady, where strikes, "more than were desirable," were orderly and peaceful. Paradoxically, the Schenectady machinist unions, which had precipitated the 1915 strike, sponsored a Socialist forum talk by Steinmetz on public education at their headquarters two weeks after this speech. Their action is another example of how Socialists and labor leaders tended to ignore the unorthodoxy in the philosophy of the Wizard of Schenectady because of his publicity value for the movement.[26]

Steinmetz leaned more toward labor as the Red Scare intensified. In the fall of 1919 Governor Al Smith of New York, who had earlier named Steinmetz along with Bernard Baruch and others to a postwar reconstruction commission, invited him to participate in an industrial conference regarding labor unrest in the state. Steinmetz told the conference, "I am a Socialist because I am convinced that in some future time capital will cease to be necessary, but to-day and for a very long time capital and labor both are necessary . . . If . . . these two necessary enemies—capital and labor—cannot get along peaceably with each other . . . one has to go, and that one could only be capital." He realized that "society is very far from socialism, and therefore I [would] like to see exist in society capital and labor, and find a means of co-operation."[27]

Steinmetz thus extended his earlier idea of cooperation between corporations and the government to include cooperation between management and labor (which became an element of corporate liberalism). He favored a form of "industrial democracy" based on his prewar proposal of a wage dividend. In his lecture to factory managers in April 1919 he argued that since both capital and labor were necessary at the present time, "it appears just and reasonable that both should share in the profits and in the management of industry." Previous means to that end had not succeeded. Most welfare work was paternalistic, the bonus system favored the employer, and workers saw shop committees as tools of management. He thought that nonpaternalistic welfare schemes would help, but the best measure was a wage dividend. Instead of receiving an annual bonus, as in his earlier plan, workers would receive "labor stock" in a company, based on their yearly pay. Workers would earn dividends on this stock, just like those who held "capital stock," and would vote for a labor representative on the board of directors. Although he noted that some utility companies had some form of wage dividend, his proposal would have been a large step toward establishing the type of industrial democracy demanded by many Socialists and other reformers at the time, particularly by the Guild Socialists in Britain. The proposal, however, was not especially radical. Steinmetz thought that "whatever is done, must be done by gradual development, by slow evolution and not by revolution." Only workers with ten years of service would receive the dividend. "Thus the experiment would be perfectly safe."[28]

He also became involved with the League for Industrial Democracy. Established in 1921 as the successor to the ailing Intercollegiate Socialist Society, which had suffered from anti-Socialist sentiment during the war, the league asked Steinmetz to be its first president. He had not been very active in the ISS since joining it in 1912, but he had established an alumni chapter in Schenectady and had written an article for the society's journal on socialism and invention. He declined the presidency of the league, probably because of the press of work, and was elected instead as its first vice president. The league counted a number of prominent reformers and Socialists among its officers, some of whom became well known in the 1930s. Joining the reformer Robert Morss Lovett as president were Evans Clark of the Labor Bureau, a research group for organized labor, Socialist Florence Kelly, and Arthur Gleason of the Bureau of Industrial Research as vice presidents. Economist Stuart Chase, who organized the Labor Bureau after the collapse of Veblen's Technical Alliance in 1921, was treasurer. Norman Thomas, the Socialist party's perennial presidential candidate in the 1930s, chaired the executive committee, and Harry Laidler, the unflagging organizer of the ISS since its founding, was executive director. Steinmetz knew at least three of the officers. Gleason had interviewed him for *Metropolitan Magazine* in 1914, he had served with Chase on the Technical Alliance, and he had known Laidler through the ISS since 1911. The league reelected Steinmetz as vice president in 1923, but he had even less to do with it than with its predecessor.[29]

He became more involved with industrial democracy in Schenectady. In early 1922 GE joined the widespread movement among American corporations to establish "company unions" by proposing an Industrial Representation Plan at the Schenectady Works. Six delegates chosen from over three hundred elected workers met with six management representatives to formulate a plan to give employees a "voice in matters pertaining to their employment and working conditions." Workers from voting districts throughout the plant would be elected to shop committees and a joint committee with management, as was the practice in the works council set up by the War Labor Board at Lynn during strikes at several GE plants in 1918. Emmons, now a vice president, promised not to discriminate against labor organizations, but most postwar employee representation plans did not look as favorably on labor unions as did the wartime works councils. The worker committees approved the Schenectady plan, but the entire work force, many of whom were veterans of the prewar labor struggles, voted down the proposal overwhelmingly in March 1922. In June John Broderick, secretary of GE's Manufacturing

Committee, published a booklet entitled *Pulling Together*, which described a Bellamy-like utopia achieved by a similar employee representation plan instituted by a midwestern manufacturer who saw himself as a "steward" for both capital and labor. In his view this approach was much better than negotiating with "autocratic" labor unions under the collective bargaining procedures introduced during the war. The book clearly had management's blessing, since the company-sponsored employee magazine, *Schenectady Works News*, published it as a supplement at the end of the year. Along with other reformers, Steinmetz saw employee representation as a forerunner of more radical types of industrial democracy—with workers owning "labor stock" and having a vote on the board of directors in his case. Consequently, he wrote in an introduction for Broderick's book that industrial democracy "in one form or another is rapidly growing in favor and in many instances where it has been honestly tried has led to increased cooperation." But rather than trust the outcome of another election, GE announced the formation of a works council at Schenectady in the spring of 1924 without putting the question to a vote. The council was, in effect, a company union with little power to bargain for worker demands.[30]

By looking to company unions and other welfare reforms, rather than to traditional labor unions, to resolve the conflict between capital and labor, Steinmetz showed in theory, as well as in practice, that he sided with management on most issues. He may have sympathized with GE strikers, but there appears to be no record that he came forward and took a stand. The conflict between political ideals and commercial demands in this case was similar to the dilemma he faced in regard to the professional issues of autonomy, standards, and ethics, which he resolved with a "patchwork of compromises." His compromise in regard to labor troubles was to retreat to the laboratory and let others, usually Lunn, handle the situation. He was farther removed from the workers than Lunn, who had made his Socialist reputation standing up for strikers at Little Falls and on GE picket lines. At Wendell Avenue and Camp Mohawk—far removed from the battle between capital and labor on the shop floor—Steinmetz worked out theories of an industrial utopia, where contented workers toiled only four hours a day in a technocratic, corporate state.

He acted in a similar manner as president of the Schenectady common council when it took up the question of GE's tax assessment. In the spring of 1917 he agreed with the proponents of a higher appraisal, said the workers' homes were assessed far more than those of the upper classes, and thought all property should be taxed at full value. GE took the case to the state legislature and the state supreme court, seeking a reduction in assessed value from $11 million to $6 million. The issue came up during the fall election, won by the Republicans. A headline in the *Citizen* declared that the city would now be "Dominated by Corporation Lackeys and Labor Exploiters," since John Miller, GE's attorney in the assessment case, would be the new city attorney! The lame-duck common council resolved to do something about GE's assessment and voted in December to hire a tax expert to aid their case against the company.[31] Steinmetz was absent from the meetings, as he had been since the summer. In this case he probably took the advice he had given engineers in his Consulting Engineering Department and remained silent when his opinion conflicted with that of his employer.

Electricity and Socialism

The most sustained area in which Steinmetz attempted to apply the theory in *America and the New Epoch* to practice was in electrical engineering. He breached the subject of electricity's role in creating a Socialist society at Camp Co-operation, but he did not bring it up again until after the war. The impetus came from the Russian revolution. In 1920, during the throes of the economic crisis facing Soviet Russia, Lenin proclaimed, "Communism is socialist power plus the electrification of the whole country . . . Only when the country has been electrified, and industry, agriculture, and transport have been placed on the technical basis of modern large-scale industry, only then shall we be fully victorious" over capitalism within and outside Russia. Electrification had many virtues. It would "accelerate the transformation of dirty, repulsive workshops into clean bright laboratories worthy of human beings." Electrifying the home would ease the burdens of millions of "domestic slaves." And electricity would "make it possible to raise the level of culture in the countryside and to overcome, even in the most remote corners of the land, backwardness, ignorance, poverty, diseases and barbarism."[32]

To achieve these results Soviet engineers drew up a ten-year plan to build regional power networks covering the entire country. Such systems were not without precedent in 1920. Regional networks, larger and more advanced than those planned by Lenin, had "evolved" in the United States, Britain, and Germany prior to the war. Still larger systems had been proposed in all three countries shortly after the war, including the socialization of a unified electric supply system in Germany in 1919. Although Lenin's project owed much to these precedents, both in technology and in its projected social impact, it stood out for its comprehensiveness, degree of state control, and ideological imperative. All of Russia must be electrified to bring about the Communist state.[33]

Lenin's program fired the imagination of Steinmetz. He probably first read about it in the *Citizen*, which published lengthy excerpts from Lenin's 1920 speech, including the famous dictum equating electricity and communism.[34] Most American Socialists admired the Russian revolution, but Steinmetz had not at first, as shown by his interview on the Bolshevists during the Red Scare. He became an ardent supporter after Lenin announced his technocratic plan to electrify Russia. In early 1921 he gave a speech at the Schenectady Unitarian Church in which he praised Albert Einstein, who had received the Nobel Prize in physics that year, and Lenin as the "two greatest minds of our time." Lenin was great because he was bringing order out of chaos in Russia. Steinmetz also changed his mind about Lenin's political methods. He later wrote that one of the "great achievements of the Russian Revolution" was the "elimination of the will-o-the-wisp of so-called 'democracy' and [the] realization that the social revolution can be accomplished only by the dictatorship of the minority which knows what it wants."[35] Steinmetz saw that Lenin intended to use technocratic methods to accomplish for Russia the dream he himself held for the United States—state-planned electrification—and he wanted to take part.

In February 1922, two months after the Russian electrification project was formally approved, Steinmetz sent a letter to Lenin via a courier, and expressed his

admiration of the wonderful work of social and industrial regeneration which Russia is accomplishing under such terrible difficulties.

I wish you the fullest success and have every confidence that you will succeed. Indeed, you must succeed, for the great work which Russia has started must not be allowed to fail. If in technical and more particularly in electrical engineering matters I can assist Russia in any manner with advice, suggestions or consultation, I shall always be very pleased to do so as far as I am able.

Lenin wrote back in April that he was grateful for the offer, but the absence of diplomatic relations between the United States and Russia made it difficult for him to accept it. Instead, he would "take the liberty of publishing your letter and my reply to it in the hope that by this means many persons living in America or in countries bound by trade agreements, both with [the] United States and with Russia, may help you (by information, translations from Russian into English, etc.) to fulfill your desire to help the Soviet Republic."[36]

Lenin's suggestion that Steinmetz publicize, rather than help engineer, the electrification of Russia did not dampen his enthusiasm for the project. Harold Ware, an American Communist and agricultural engineer who was in Russia to introduce tractors onto Soviet farms, hand-delivered a copy of Lenin's reply and an inscribed photograph of Lenin to Steinmetz at his GE office in the spring of 1923. Ware vividly recalled Steinmetz's pleasure in receiving the photograph, which he reportedly hung on the wall in his office, and his excitement when discussing Lenin's electrification plan: "Young man, do you realize what Russia has been doing? In this short time they have developed a standardized, planned electrification scheme for the whole country. There's nothing like it anywhere. It's wonderful what they have done. I would give anything to go over there myself and work with them."[37]

By then Steinmetz had already helped by publicizing the project in the United States. As promised in Lenin's letter, he received information from Russian engineers and officials, one of whom called him "our beloved comrade and leader in electrification as the liberator of human labor."[38] Steinmetz then wrote two articles for *Electrical World* in the fall of 1922, in which he informed American engineers of the Soviet plan for electrifying the country through eight regional power networks and described the opening of one plant near Moscow. Although plants were to be interconnected only within these regions, Russian engineers designed the separate networks as a system, bearing in mind the different types of natural resources in the country and the transportation of fuel between regions. At the end of the first article Steinmetz stated that the cost of the first part of the project would be $570 million and that its completion in the planned time of ten years as part of Lenin's New Economic Policy would "depend on the assistance which can be enlisted from America."[39]

Electrical World's reaction to pouring money into Soviet Russia was predictable. The journal's editor doubted "whether this or any material amount of American capital can be coaxed into Russia until both its ideals and its practice of government are radically changed." Despite Steinmetz's technical reputation, the editor questioned the second article's accuracy because it was prepared at the "instance of the putative Russian government." The journal grudgingly admired the comprehensiveness of the project, though. A month prior to publishing Steinmetz's articles, it characterized Lenin's plan as "sane and far-reaching in its effect." The editor concluded with the provocative thought that "good and lasting things grew out of the destruction wrought by the French Revolution, and who knows but that an electrified Russia may arise in the future due in part to the visions of the Moscow dictator."[40]

The large orders needed to electrify Russia could, of course, aid the American electrical industry, a point not lost on GE. The company had a great interest in Russia in any event because, during the war, the Russian branch of Allgemeine Elektrizitäts-Gesellschaft (AEG) had been "Russianized" as the Russian General Electric Company, with American GE and the Russian government as partners. Lenin nationalized the company when he came to power, which prompted GE officials to search for means to recover their large investment. In 1920 Owen Young, corporate counsel and vice president of GE, suggested to Gerard Swope, president of International General Electric, that a three-member delegation negotiate directly with Lenin, which was apparently not done. Chairman of the board Charles Coffin also resisted the suggestion that GE trade with Russia to help relieve unemployment during the business depression of 1921. Coffin thought it would be unprofitable because of the unstable political conditions in Soviet Russia.[41]

GE sang a different tune when its leadership changed the next year. Swope and Young, who was also chairman of the board of directors of the Radio Corporation of America, then owned by GE, AT&T, and Westinghouse, were being groomed by Coffin in early 1922 to head GE. In May Coffin and Edwin Rice—the team that had run the company since its founding forty years before—stepped down in favor of Swope as president and Young as chairman of the board. In the middle of the transition to these leaders, who became well known for their corporate-liberal view of business, GE reevaluated the Russian situation, particularly since Walther Rathenau, president of AEG and minister of reconstruction in the Weimar government, was involved in opening up Russia's electrical business to Germany and the United States. In late March 1922—about six weeks after Steinmetz wrote to Lenin—Maurice Oudin, vice president of International GE, informed the U.S. State Department that he believed the time was ripe to resume business with Russia and that GE and AEG were discussing an agreement to this effect. Lenin released his correspondence with Steinmetz to the press a few weeks later (about a week after Rathenau, then foreign minister, signed the Rapallo Treaty reestablishing diplomatic and commercial ties between Germany and Russia on April 16). The front page of the *New York Times* carried the headline "Steinmetz Offers Russia Help in Electric Projects." Two weeks later, in early May, the Soviet Electro-technical Trust invited

International GE to participate in a joint mixed-capital company to help develop the electrical industry in Russia. But all these negotiations fell through, and GE did not become a large supplier of electrical equipment to Russia until the late 1920s.[42]

The timing of these events suggests that Steinmetz assisted his employer in its dealings with Russia, but he apparently made his Soviet contacts independently of GE. The courier of his letter to Lenin was the Russian engineer B. W. Losev (also spelled Losseff and Lassoff), a graduate of MIT who might have worked at Schenectady. George Galvin, a Socialist physician in Boston, asked Steinmetz to find Losev a position with GE in 1915. Losev was secretary of the New York branch of the Russian Technical Aid Society in 1922 and knew Abraham Heller, U.S. representative of the Soviet Union's Supreme Council of National Economy. A "millionaire Communist" who held economic concessions in Russia, Heller wrote Steinmetz that upon seeing the story in the *New York Times*, he gathered that "comrade Losseff presented your letter to Lenin." Steinmetz sent copies of the correspondence to Heller, whom he had probably met when they served on the administrative board of the Rand School. After writing an introduction for Heller's book on Soviet industrialization, Steinmetz sat on the advisory committee of the Kuzbas mining colony in Siberia and on the board of the *Soviet Russia Pictorial*.[43] GE's role in these matters is unclear, but it had good reasons not to discourage these well-publicized overtures to Russia. The company could easily have thwarted them since the GE Publication Bureau cleared all employee publications, including Steinmetz's technical articles.[44]

Besides supporting Lenin's program in the technical press, Steinmetz carried the issue of state-planned electrification into the political arena by agreeing to run for New York state engineer and surveyor in 1922. The "Oracle of Schenectady" was an excellent choice as a candidate. He had reached new heights of publicity that year when newspapers touted him as "Modern Jove" for hurling thunderbolts about the laboratory and Lenin released their correspondence to the press. Such a figure was bound to draw attention when the New York branches of the Socialist and Farmer-Labor parties nominated him for state engineer in a joint campaign effort. Vladimir Karapetoff, who had run for the office as a Socialist several times, was the first to be nominated. Then "there was a wave of enthusiasm for the little wizard of Schenectady that put him over," according to the *New York Call*. The *Call*'s photo caption read, "Socialists Name Electrical Wizard." Non-Socialist papers also took much interest in Modern Jove, Socialist Chief Engineer of General Electric, and Friend of Lenin running for the top engineering office in the state.[45]

About a week after the convention, a delegation appointed by the state Socialist party informed Steinmetz in person of the nomination. Headed by Schenectady leaders Herbert Merrill, now state secretary of the Socialist party, and Charles Noonan, erstwhile foe of Lunn and Steinmetz in Local Schenectady, the committee persuaded him to accept the nomination. As in his first Schenectady campaign in 1913, Steinmetz did not make any public appearances. He was slated to address two rallies in Manhattan and Brooklyn in mid-October, and Westinghouse was to broadcast his speeches by its radio stations "all over the country," according to the *Call*. Both times, however, Steinmetz sent his regrets at the last moment. One suspects that GE officials had a hand in the matter once they learned that the Electrical Wizard would address these rallies via the radio network of its foremost competitor. (He might also have canceled in order to prepare for Edison's much-publicized visit to Schenectady later that month.)[46]

In any event, reporters had earlier gone north to sit at the feet of the Oracle of Schenectady, to hear what he would do if elected. What they heard was his dream of a technocratic Socialist state. First he would investigate the personnel in the state engineer's department, remove political appointees, and bring in "socially minded" engineers like his Socialist colleague Karapetoff. Next he would conduct a survey of the state's natural resources, much like the survey planned by the Technical Alliance, and then submit a plan for the "co-ordinated development" of these resources. Transportation would be gradually socialized, beginning with the state owning and operating a fleet of barges, which would force down rail and canal shipping rates. The centerpiece of his program was the development of the state's waterpower. One proposal was to harness more fully the hydroelectric power of Niagara Falls by diverting the Niagara River, which would dry up the falls. To overcome resistance from the tourist industry, he would turn the falls back on for Sundays and holidays. Steinmetz was serious about this; he had proposed a similar scheme during the fuel crisis in 1918, a plan ignored by the government and the utilities. With this experience in mind, he now told voters that such technocratic plans were impossible under capitalism and required a "Socialist world" to implement them.[47]

Much more practical was his plan to harness the state's rivers and streams by means of the small, automatic generating plants he had proposed in his 1918 article on America's energy supply. Steinmetz wrote more on this topic the next year and made it the major theme of his campaign in 1922. Although the scheme for turning Niagara Falls on and off like a faucet made better newspaper copy, he succeeded in drawing more publicity to the development of the state's scattered waterpower, an issue on which Al Smith, the Democratic gubernatorial candidate, held a view similar to Steinmetz's.[48]

It was an uncomfortable issue for the electrical industry, however. The editor of *Electrical World*, who had criticized Steinmetz earlier that year for promoting Russian electrification, commented on the election by noting that he was well qualified technically, but pointed out that he was also a Socialist. The editor feared that the "general public will not differentiate between Steinmetz the engineer and Steinmetz the politician, and hence unwittingly Steinmetz will bring hurt to his industry" by advocating a program whose feasibility required the public ownership of utilities.[49]

The election was not much of a contest, especially since the Socialists had split into the Socialist Party of America, the Communist party, and the Communist Labor party in 1919, which sapped much of the movement's strength. The united Socialist and Farmer-Labor candidates ran about a million votes behind their Republican and Democratic counterparts. But Steinmetz ran far ahead of his ticket, polling nearly three hundred thousand votes, about two and one-half times that of his fellow Socialists and over 12 percent of the total vote. Predictably, he received a large vote in Schenectady County, about three times that of the regular Socialist vote. He also did well in New York City, whose newspapers had carried many stories about the Wizard of Schenectady in the past year. He polled over two hundred thousand votes in the greater New York area and ran ahead of the Republican candidate in the Bronx by ten thousand votes.[50] The *New York Call* was elated by the record vote, while many newspapers ran a syndicated editorial about the "cowboy sage" Will Rogers challenging readers to recall who defeated Steinmetz. Rogers confessed his ignorance, but said he didn't "know the difference between a short circuit and a long shot." The editorial thought the election was a "rather sad commentary on the state of mind of voters who do their political shopping on the strength of party labels" rather than on expertise.[51] The *New York Times*, however, heaved a sigh of relief. After noting that the Socialists had put him on the ballot to give them "much-needed prestige," its editor said that a Steinmetz victory "would have been an appreciable loss to science" because his "proper place" was the GE laboratory.[52]

Steinmetz had other ideas about his proper place and continued to speak out for his view of a just society. Following the election he made a speech at the Unitarian Church in Schenectady, where he spelled out his political beliefs in some detail. His thinking had undergone some changes since his recent political defeat. As in 1916, he described three economic systems: capitalism, socialism, and cooperation. The ideology of laissez-faire capitalism was "increasingly being abandoned, as unsafe, by our leaders of industry and finance, especially since the experience during the last war." But it was "still largely held by labor and represented by the conservative union leaders," a continuation of his criticism of labor unions. He had identified socialism as the final outcome of industrial cooperation in 1916 and as recently as 1919, but he now observed that socialism was out of consideration, "regardless of whether it is justified or not," because "only a small percentage of Americans accept this viewpoint today," a reference to the recent election results. The view "increasingly gaining ground among the industrial leaders" was that of "cooperation," in which "capital and labor are necessary for industrial production, and both, therefore, have rights and duties in the industry." Capital had a right to a "fair return" on its investment and continuous dividends, while labor had a right to a "living wage" and continuous employment. To ensure these rights required nonpaternalistic cooperation between capital and labor. Steinmetz then proposed a plan similar to his industrial democracy scheme of 1919, whereby labor could "participate in the profits of the company through dividends, and in the management of the company through the Board of Directors." Since "any rash change or radical modification is liable to be disastrous," he recommended that the "transition to industrial co-operation therefore must be gradual, by evolution."[53]

He delivered a similar message in a speech before the Babson Institute in February 1923, in which he declared that both capitalists and Socialists were working toward a corporate technocracy. Lenin had already made great progress toward this end:

You may not admire Lenine [sic] and the soviet organization of Russia. His motive may be right or wrong, but this much is at least clear. He has organized and maintained a government in Russia through one of the most trying periods that any nation has ever experienced and he has been able to do it because he is using the same system of control that is being used by the best managed corporations in this country. He considers all his followers as stockholders and appoints and promotes the officials of his government on the basis of their efficiency and fitness to perform the task required of the office.[54]

Steinmetz concluded the speech by describing a future society similar to that outlined in *America and the New Epoch*, with production and distribution planned along corporatist lines and a popularly elected tribunal having only inhibitory powers. Lenin had achieved his goal through political revolution, but industrial evolution would be the path of the United States toward a technocratic corporate state. Congress and the executive branch, for example, would evolve into the tribunal.

Steinmetz's philosophy, then, had not changed much since 1916. He called his new system "cooperation" and said socialism was unlikely, but he still believed in basing the economic branch of the government on the corporation and saw the wage dividend as one means to achieve a corporate state. As noted earlier, he had, however, extended his concept of "cooperation" to cover that between labor and management, as well as that among corporations and between them and the government. The new emphasis meshed well with the corporatist ideologies prevalent during the early 1920s, when groups ranging from Guild Socialists on the left to advocates of company unions on the right to corporativists on the far right in Mussolini's Italy called for various types of industrial cooperation to replace cut-throat competition. Even Herbert Hoover wrote in 1922, "Today business organization is moving strongly toward cooperation." A vehement anti-Socialist, Hoover advocated an "American individualism" based on corporatist trade and professional associations, rather than a corporate state ruled by the technocratic, industrial government of large companies proposed by his fellow engineer Steinmetz.[55]

The reaction to Steinmetz's 1922 speech was mixed. Many newspapers praised his preference for "cooperation" over socialism, but the *New York World* was more skeptical, saying that the prospect of capital and labor running industry was as unlikely to occur as was "doctrinaire socialism." The severest criticism came from an orthodox Socialist. Henry Kuhn, former national secretary of the Socialist Labor party and successor to Daniel de Leon as head of the SLP, wrote that Steinmetz's speech did "not betray the slightest intimation" of the goals of socialism. It was, instead, a "distinct attempt at revamping capitalism . . . The 'Socialism' of Dr. Steinmetz seems to consist of a strong desire to prolong, if possible, the existence of the capitalist system, by making it less obnoxious and more palatable. But even that is a utopian undertaking. It can't be done."[56]

Steinmetz, Young, and Swope

In a general sense corporate liberals did try to revamp capitalism à la Steinmetz in the 1920s and 1930s by improving the cooperation among capital, labor, and government through public and private corporatist agencies. Most did not, however, recommend the centralized corporate state favored by Steinmetz. Two of the most influential and articulate proponents of this corporate-liberal ideology were Owen Young and Gerard Swope, respectively the chairman of the board and president of GE. Outside the company Young concentrated on the international economy through his work on the Dawes and Young plans for German war reparations in the 1920s. Swope made his mark in 1931 with the Swope Plan, a proposed corporatist restructuring of the U.S. economy that influenced the industrial policies of the early New Deal.[57] There is no record that Young and Swope simply borrowed the ideas of Steinmetz—who had gained a national reputation as a political economist—but they knew his positions on economic matters, and the similarity of their views, especially Swope's and Steinmetz's, is remarkable.

Young was indirectly exposed to Steinmetz's philosophy when he commissioned a report on the volatile labor situation at GE in 1919, following companywide strikes at four plants the year before. Atherton Brownell, Steinmetz's editor at the National Editorial Service during the war, wrote the report after interviewing workers at GE plants in Schenectady, Erie, Pittsfield, Lynn, and Fort Wayne, while posing as a writer for *McClure's Magazine.* His first conference in Schenectady was with his friend Steinmetz—a meeting that apparently shaped and reinforced Brownell's views. A former editor of the magazine of the National Association of Manufacturers, Brownell confessed in the report that he had a "very strong prejudice against labor unions" and held a "somewhat radical view of the relations that may some day exist between capital and labor." He had studied GE a year earlier for another project and had concluded that the company "was developing along the lines of the highest and most idealistic Socialism." This socialism was likely the corporate variety favored by Steinmetz because Brownell warned Young against the radical, IWW element in the plants. He also suggested that GE establish a department of industrial relations, staffed by professionals, and a com-

panywide employee representation plan. Steinmetz, whom Brownell recommended as a member of a Joint Board of Men and Management to
study the latter issue, had made similar proposals.[58] As we have seen, GE
did not act on Brownell's suggestion until early 1922—shortly before
Young and Swope took office—when it created a Joint Committee on
Industrial Representation (without Steinmetz) to plan a company union.

The connection between Steinmetz and Swope was more direct. In
September 1922, after the Schenectady workers had voted down the
company union, Swope went to see Steinmetz—first at Wendell Avenue
and the next morning at Camp Mohawk—rather than summoning him to
GE headquarters at New York City or to an office at the plant. It was an
unusual step for the busy president of GE, who was infamous for sched-
uling two-, three-, and five-minute office appointments and keeping to
this strict timetable.[59] One result of his conferences with Steinmetz may
have been the distribution of *Pulling Together* to every employee in
December.

Three months later, in March 1923, Steinmetz sent Swope a "confidential" eighteen-page report entitled "Industrial Cooperation." The
first part repeated his Babson Institute speech on capitalism, socialism,
and cooperation, in which he argued for the superiority of the corporatist
system. In regard to employee representation plans, he repeated his
assertion, made during the war, that democracy was inefficient in industry as well as in politics. He recommended that "if real cooperation could
be established, why should labor not be led by the same leaders and in the
same manner as capital." Ensuring this cooperation—which implied
replacing labor unions with a nonelected company union—should be
one of the corporation's major functions and the responsibility of a professional industrial relations department. Engineers were the "least
qualified" for this work because their training stressed finding the one
best answer to a problem, rather than looking at all sides of an issue.
Workers were too poorly educated and ministers preached too much.
Most suitable would be a "liberal broad-minded lawyer."[60] As noted
earlier, GE imposed a company union at the Schenectady Works a year
later, in 1924, without putting it to a vote.

Swope made his major contribution to American corporatism with
the Swope Plan, proposed in a speech to the National Electrical Manufacturers' Association in September 1931 as a means to recover from
the depression. Swope's biographer relates how his staff sifted through
hundreds of economic-recovery programs prior to drafting the proposal,
yet the Swope Plan bore a striking resemblance to the corporatist state

outlined in *America and the New Epoch*. The plan called for the carteliza-tion of the economy under the direction of trade associations in each industry. The associations—made up of all companies in an industry having more than fifty employees—would coordinate production and distribution and ensure that their member firms took care of the workers' welfare by providing death, accident, and old-age benefits. The Federal Trade Commission would supervise the companies and trade associa-tions. The associations would thus perform the functions of Steinmetz's Industrial Senate—running the economy and providing for the workers' welfare—while the federal government, through the FTC, had the in-hibitory power of his Tribuniciate. Both Swope and Steinmetz preferred to rely on the self-government of industry rather than a democratically elected body to make economic decisions.

Many other people besides Steinmetz had favored such a corporatist system, including the National Chamber of Commerce in 1931. Swope acknowledged, "There is nothing new or original in what I am proposing. I am merely bringing together well-chosen propositions that have found support, including some that have been put into practice."[61] Herbert Hoover, while commerce secretary, had tried to establish the rudiments of a corporatist state based on professional and trade associations. But as president, he denounced the Swope Plan in no uncertain terms, calling it Fascist and the "most gigantic proposal of monopoly ever made in histo-ry." The main difference, of course, was that Hoover preferred voluntary cooperation, rather than the centralized, technocratic control inherent in the schemes of Swope and Steinmetz. Swope probably first encountered corporatism during the war when he served as a dollar-a-year man as assistant to General Hugh Johnson in the Army's supply department and the War Industries Board, of which Johnson was an Army representative. After becoming president of International GE in 1919, Swope resumed his prewar friendship with AEG's Walther Rathenau, who had set up the German Army's supply organization, and discussed their common expe-riences in the war. Their companies engaged in a bit of postwar collectiv-ism by renewing their international cartel agreement in 1920 and signing it in early 1922. Swope visited Rathenau, then foreign minister, in Berlin in June 1922, just a few days before Rathenau was assassinated by right-wing anti-Semitic officers. Rathenau's ideas about a corporate state organized around industry-supported cartels meshed well with the cor-poratism Swope heard about a few months later during his talks with Steinmetz.[62]

The Swope Plan fell on deaf ears with Hoover, but President Roosevelt and the New Dealers resurrected its main elements in the National Recovery Administration in 1933. NRA director Hugh Johnson asked his former assistant Swope to help draft the legislation, and Swope served on the NRA advisory board and helped draft the planning code for the electrical industry. The principles behind the short-lived NRA resembled the Swope Plan, minus the welfare provisions, many of which later came under the government's auspices as "social security." As Larry Gerber has written, the failure of the NRA in 1935 led to a shift from the corporatism of the 1920s to the pluralist political economy of the later New Deal and beyond. Resolution of the conflicts between the competing interests of business, government, and labor replaced the elusive goal of attaining the cooperation among these groups envisioned by Steinmetz and other corporatists.[63]

10 Modern Jove

Already well known as an Electrical Wizard, Steinmetz reached the height of his fame in 1922 when the popular press enthroned him as Modern Jove for creating lightning in the laboratory. The image of a hunchbacked German Socialist taming one of nature's deadliest forces inside a large corporation proved irresistible to the national media, which made him a bigger celebrity than he had been during the war. Newspapers and magazines quoted his views on religion, politics, science, and future technological wonders. In the public's eye Steinmetz was a Wizard of Science, a symbol of the new breed of scientifically trained engineers who were daily surpassing the feats of Thomas Edison, the Wizard of Menlo Park. Newspapers had viewed Edison as a scientist in the 1880s, much to the consternation of "pure" physicists like Henry Rowland, but later writers usually pictured him as an anti-intellectual, homespun Great American who berated "impractical" scientists. This attitude changed as the American public came to recognize the importance of industrial research during the war and again called Edison—and other inventors and engineers—scientists, a practice common until World War II. In the case of Steinmetz—called one of the most "visible scientists" in a recent study of the public image of American scientists—writers began to grapple with the meaning of science for invention and engineering when they compared him with Edison during the war.[1]

Wizard of Science

Newspapers started to place Steinmetz alongside Edison in the pantheon of great American inventors when he was being considered for Edison's Naval Consulting Board in 1915. In that year a Cincinnati paper was proud that Steinmetz, the inventor "second only to Edison," was visiting its city. The *New York Times* put them on a par in 1916 as the "two

Steinmetz in his "office" at Camp Mohawk, about 1920. Source: Hall of History Foundation, Schenectady, N.Y.

men who have done more than any others to advance electric lighting."[2] Steinmetz's German accent and university education contradicted the anti-intellectualism and apple-pie image of the midwestern man-boy Edison, and newspapers and magazines had difficulty popularizing his highly mathematical engineering theories. Yet they found the new wizard's physical appearance and career to be "good copy," as we have seen. The incongruity between his deformed body and intellectual achievements fascinated the public and reporters, bringing forth such sobriquets as the "Little Wizard with the Big Brain." The meteoric climb of this penniless, misshapen immigrant to the top of his profession invoked the mythology of the "self-made man" used so well for Edison, and Sunday supplements had a field day describing the weird pets, plants, and political views of the mastermind of General Electric.[3]

Steinmetz helped create this image. He discovered at a young age that his physical appearance made him a public figure, whether he wished it or not. A dwarfed hunchback drew attention—in public and private. By the time he came to Schenectady he seems to have accepted his deformity and turned it into an asset, cultivating a public persona rather than becoming an embittered recluse. In keeping the menagerie at Liberty Hall, riding a bicycle on city streets, and running with the Socialists, Steinmetz became a local celebrity, the "funny, rare little animal," the protected "pet oddity" of Schenectady and GE—in the words of novelist John Dos Passos.[4] At work he gained an international reputation by publishing more than a dozen books and giving numerous papers at engineering societies over a period of thirty years. Newspapers, magazines, and GE followed his lead and made him a national celebrity, the pet Electrical Wizard.

The Steinmetz myth centered on the story of a benevolent corporation giving an eccentric genius freedom from bureaucratic constraints in exchange for engineering miracles. GE supposedly paid him $100,000 per year and gave him unlimited freedom to work in his home laboratory, at Camp Mohawk, or in a canoe at the camp. One day GE placed a no-smoking sign in Steinmetz's building. Steinmetz, the incessant cigar smoker, packed up for home and answered his callers, "No smoking, no Steinmetz!" Elements of the myth are a variant of the machine-in-the-garden pastoralism identified by Leo Marx. Perhaps in an attempt to resolve the conflict between the ideal of rugged individualism, epitomized by the civil engineering heroes of contemporary fiction, and the bureaucratic demands of the corporation, reporters and GE public relations writers put a Wizard, not a Machine, in the Garden. This type of pastoralism is similar to that of George Innes's 1855 painting *The Lackawanna Valley*, in which a steam railroad is set in a bucolic valley to harmonize technology with nature, or of R. F. Outcault's 1881 painting of Edison's laboratory in winter at rural Menlo Park.[5] The Garden of Camp Mohawk, on a peaceful creek midway between the Adirondack wilderness and the city of Schenectady, and the Garden of his home laboratory next to a well-cultivated conservatory were the pastoral settings for Steinmetz's Faustian labors, just as Menlo Park was for Edison's. GE obviously could not give all employees the freedom Steinmetz enjoyed, but its products, according to the myth, were creating more leisure time for consumers. Steinmetz in fact had predicted that electrification would shorten the working day to four hours.[6] The Wizard in the Garden, whether Edison at Menlo Park or Steinmetz at Camp Mohawk, thus represented Paradise regained through technology.

The myth was a product—a social construction—of Steinmetz's actions, newspaper hyperbole, and GE public relations. We saw that the salary story began in 1908 when the *New York Times* reported that his salary "cuts an impressive slice out of $100,000." The *Times* repeated the qualified statement three years later, but called him a "Hundred-Thousand-a-Year" man in 1913. The myth became canonical the next year when an Albany paper ran a story with the headline "To Earn $100,000 a Year 'Do Things That Others Don't Do,' Says Steinmetz." Newspapers and magazines picked up the story, often with the angle that a Socialist made such a large salary. Steinmetz, who did make the considerable sum of $18,000 per year in 1911, tried to correct the legend in a magazine interview a decade later. Disappointed that his subject did not in fact receive $100,000 a year, the interviewer quoted him as saying that he was not paid a fixed salary but could spend whatever he needed on his laboratory. Company records revealed after his death that he was paid a regular salary and laboratory expenses.[7] The "no smoking" story began in 1893, the year Steinmetz joined GE, when he introduced "smoking to the sacred halls of my office in the factory where no one had ever smoked previously." But the "no Steinmetz" part was apparently apocryphal—and a good punch line.[8] He spent most of the summer at Camp Mohawk, and he actually worked some in a canoe—an image GE helped publicize in the early 1920s.

Publicity for Steinmetz fell to its lowest level after the United States entered the world war—most likely because of his pro-German stance—then rebounded to new heights when the war ended. During the reign of science and the expert, newspapers cited his views on current scientific wonders, described his technological prophecies, and reported his opinions on the social and political questions of the day. The Wizard of Schenectady surpassed the Wizard of Menlo Park in publicity in 1922—in the number of popular articles, but not in newspaper coverage—the year he offered to help Lenin electrify Russia, ran for New York state engineer, and was heralded as Modern Jove (see the Appendix).

The Jovian legend began with a front-page Associated Press story on March 3, 1922. The *New York Times* headline and leading paragraph read, "Modern Jove Hurls Lightning at Will—Million-Horse-Power Forked Tongues Crackle and Flash in Laboratory . . . Schenectady has a modern Jove who sits on his throne in a laboratory of the General Electric Company and hurls thunderbolts at will. He is Dr. Charles P. Steinmetz, electrical wizard, who announced today he has succeeded in producing and controlling an indoor thunderstorm with all the characteristics of its natural brother except the thunder clouds." Major newspapers in New York City, Washington, D.C., Chicago, and San Francisco

The Wizard of Menlo Park and the Wizard of Schenectady, 1922. Source: Hall of History Foundation, Schenectady, N.Y.

published the AP story—all used the lead paragraph just quoted—and Steinmetz became known as the "Thunderer," the "Jove of Schenectady."[9] An integral part of the legend was the supposed inspiration for the lightning generator: a dresser mirror struck by lightning at Camp Mohawk. The story, briefly mentioned in some newspapers, was told in contemporary magazines and in later biographies. One day in August 1920, the story goes, Steinmetz arrived at his camp and found that lightning had struck a tree outside his cabin and then arced inside and shattered a mirror. Undaunted by the damage, he pieced the mirror together in order to study a preserved picture of the lightning flash. Two years later, after much research, he built the lightning generator and astonished the nation by hurling thunderbolts about the laboratory— thunderbolts forged by calculations done in the Garden of Camp Mohawk.[10]

In October 1922 GE invited the seventy-five-year-old Edison to visit the plant he had founded in Schenectady nearly forty years before, and company photographers took a famous picture of the two men hunched over a table, inspecting the damage done to line insulators by Steinmetz's lightning generator. The story of Edison's visit made the front page of major newspapers, while many smaller ones published the photo of Steinmetz and Edison. There are over thirty copies of the photo in the Steinmetz archives—he subscribed to a clipping service—many with captions like "Wizard Meets Wizard." The photo appeared in newspapers large and small throughout the country from October to March of the next year. In December 1922 *Current Opinion* ran an article on Edison's visit with the title "Wizards of Science Astonish Edison." Although the story dealt mostly with Edison's visit to Whitney's laboratory, the editors chose as the lead picture that of Edison, the Wizard of Menlo Park, being instructed by Steinmetz, the new type of Wizard representing science and scientific engineering. A Philadelphia paper called Steinmetz a "Modern Franklin," the "Wizard to Whom Edison Pays Homage."[11]

It was not all publicity. Steinmetz and Edison had a mutual respect more genuine than that orchestrated by public relations men for the camera. Edison, not known for his mathematical prowess, told reporters that he got along so well with Steinmetz because Steinmetz did not talk mathematics at him. We have seen that Steinmetz admired Edison as one of the founders of his profession. One of his first articles on electricity had been written on the subject of Edison's improved phonograph in 1888, and he had applied for a job at the Edison Machine Works when he emigrated. Once he made his mark in electrical engineering, he talked with Edison at a meeting of the Association of Edison Illuminating Companies in 1910 and asked for a signed photograph. They did joint research on condensite the next year—in which Steinmetz thought more of Edison's work than Whitney's—and Steinmetz knew Edison well enough to ask him to support a candidate for president of the American Institute of Electrical Engineers. During the utility company meeting, a rather candid shot was taken of the two wizards, in which Steinmetz supposedly communicated with the hard-of-hearing Edison by tapping Morse code on his knee. Newspapers first published the photograph in 1915, when Steinmetz was being considered for Edison's Naval Consulting Board.[12]

Steinmetz paid high tributes to Edison in interviews and speeches. In a 1914 article on Edison the *New York Times* quoted Steinmetz as saying that Edison was the "man best informed in all fields of human knowledge." When asked about the Naval Consulting Board the next year, he told reporters that Edison was the "greatest living American." He thought that the "public also has another illusion that Edison is only a practical and not a theoretical man—he is both. I have talked the theory of electricity with him and Edison's great mind knows theory too." As president of the Illuminating Engineering Society in 1916, Steinmetz spoke at a banquet where Edison was made an honorary member. He outdid did his earlier accolades, calling Edison the "greatest of these giants who have made the modern world." GE could not have written a better script for its "gnarled genius of alternating current" to pay tribute to Edison, the battle-scarred enemy of AC, whom GE now revered as its founder, even while it supplanted his direct-current system.[13]

Following a war in which the public came to recognize the importance of chemistry in winning the conflict, Steinmetz grew in stature over Edison. Newspapers looked more to the younger scientific wizard than to Edison for expert advice on such topics as the radio signals Marconi had supposedly received from Mars, the light-emitting substance in fireflies, and the theory of relativity. After interviewing Steinmetz about Edison's reputed latest invention, a telephone to call the dead, a newspaper ran the headline "Steinmetz Laughs at Edison's Spirit Phone—Some One Spoofing the Wizard, He Believes." (Edison later said he had made the story up as a joke.)[14] The promotion of Steinmetz to an authority who could evaluate Edison's work from the throne of science is evident in a juvenile biography of Edison, first published in newspapers as a series of cartoons in 1927. In one panel the author used the image of the "late C. P. Steinmetz, electrical genius," to represent scientists like Lord Kelvin and Pasteur who admired Edison.[15]

The booklet illustrates the difficulty imagemakers had in creating a consistent myth for Edison in an age when invention was increasingly being done by corporate laboratories. The booklet's drawings show the usual Horatio Alger scenes (a railroad stationmaster teaching the young Edison telegraphy as a reward for saving his son from being run over by a train) and the heroic efforts of the Menlo Park days (Edison bringing forth the incandescent lamp from an alchemical furnace). But the booklet also depicted Edison as part of a modern research and development team (shown in a white lab apron admiring the improved storage battery with two co-workers) and as a researcher revered by scientists. The writers of the booklet, which was reprinted as late as 1962, attempted to combine into one myth the contradictory images of Promethean inventor, R&D manager, and cohort of scientists. They tried to create, in the view of some students of modern myths, a paradoxical reality that would resolve the cultural contradiction between the ideal of the lone inventor and the existence of the R&D laboratory. More successful, in my view, were the Steinmetz mythmakers. The photograph of him in comfortable clothes doing calculations in a canoe on a pastoral creek presented a more consistent paradoxical reality—an individual genius granted freedom by a "progressive" corporation to help bring about an electrical utopia. The image was greatly distorted—few besides Steinmetz had this freedom, and even he was not completely independent)—but it tapped a powerful myth symbolized by such mass-media heroes as the cowboy, movie star, and sports figure. All, like Steinmetz, were rugged individuals who worked for large organizations.[16]

Manufacturing a Legend

It is highly unlikely that Steinmetz's mythmakers understood the structural elements of myths or the functions of cultural heroes pointed out by anthropologists. Rather, they responded to a subject who was inherently good copy, and the GE public relations writers consciously used the pastoral and other images of Steinmetz to humanize the corporation. In the case of the Jovian myth GE had two main motives in fostering the legend. The company felt compelled to answer technical criticisms of its lightning arresters, and it wanted to combat public sentiment against the electrical industry in the early 1920s.

Let us examine the technical motive first. We have seen that Steinmetz's theories of lightning and protective devices led to the development of the aluminum-cell lightning arrester in 1906 by GE engineer Elmer Creighton. In 1918 Steinmetz announced that his Consulting Engineering Department had invented the oxide-film arrester, which worked on the same chemical principles as the aluminum-cell arrester yet overcame the disadvantages of a wet electrolyte. It was not a fire hazard and did not require daily charging.[17] His engineers also developed apparatus to test lightning arresters and insulation. Creighton tested the aluminum-cell arrester at 2,300 volts, N. A. Lougee subjected the oxide-film arrester to 25,000 volts, and F. W. Peek tested sphere gaps at 200,000 volts.[18]

Before 1921 Steinmetz and other engineers considered these relatively low-powered devices adequate for testing lightning arresters. But in June of that year the AIEE Protective Devices Subcommittee on Lightning Arresters issued a report critical of GE's arresters. It was this report, in addition to and perhaps more than the famous lightning stroke the year before, that stimulated Steinmetz to build the lightning generator upon which so much of his popular fame rests. Newspapers and hagiographers state that he built the generator to "perfect" arresters, but they have obscured this point and hidden its significance. The significance is that the design, construction, and popularization of the lightning generator counteracted severe criticism of the operation of GE's aluminum-cell and oxide-film arresters.

The AIEE subcommittee's report was a summary of answers to a questionnaire sent to utility companies. Although the aluminum-cell and oxide-film arresters were the overwhelming first choice of operating engineers, the report quoted strong criticisms of these devices by prominent engineers. They complained that the aluminum-cell arrester was a fire hazard, required excessive maintenance, and had poorly designed mechanical features. Although the oxide-film arrester alleviated these problems, it was lumped with the aluminum-cell model by one critic, who said that both were "hair-trigger types of arresters based on fine hair theories and obscure chemical reactions."[19]

In November 1921, four months after the subcommittee's report, Lougee and Joseph LeRoy Hayden announced the development of the lightning generator in a GE journal article that went unnoticed by the technical and popular press. The lightning generator probably did not seem as spectacular as the 1 million volts produced by GE engineers at the high-voltage laboratory in Pittsfield, Massachusetts, that September.[20] The main difference between Steinmetz's generator and these experiments was that he produced much more power at a lower voltage. It was the first impulse generator with power approaching the estimated energy of a lightning discharge—it delivered 120,000 volts at 10,000 amperes—but its technology and design were not revolutionary. The main innovation was the use of direct current rather than alternating current as the charging source. The device that made this possible was the kenotron, a vacuum-tube rectifier developed by the GE Research Laboratory.[21]

After reporting in a GE journal in December 1921 that the oxide-film arrester had passed strenuous testing using the lightning generator, Lougee brought the matter up again at the AIEE convention in mid-February. At this meeting Creighton defended his aluminum-cell arrester against the subcommittee's report. During the discussion of the paper Lougee upheld GE's film-type arresters against the criticism that they operated on "fine hair theories and obscure chemical reactions." "The best answer to all this," said Lougee, "is to state briefly the arrester's behavior when connected to the lightning generator," a reference to his tests in December.[22]

If this use of the lightning generator was not occasion enough for GE to publicize Steinmetz as Modern Jove, Westinghouse's announcement of a new arrester at that very meeting was. During the discussion of Creighton's paper Joseph Slepian revealed that Westinghouse would soon introduce an "autovalve" arrester that had the good features of the chemical arresters without their disadvantages. Slepian, who had filed for a patent on the autovalve arrester the week before, explained that the Westinghouse spark-gap device did not work on "obscure chemical phenomena."[23] GE's response was swift. On March 2, two weeks after the meeting, reporters announced to the nation Steinmetz's creation of artificial lightning. *Electrical World* ran a picture of the generator a week later, and national magazines followed suit in a much more elaborate fashion in July. After Edison's visit to Schenectady in October the legend of Steinmetz, the Wizard of Science, astonishing Edison, the Wizard of Menlo Park, was in full flower.[24]

In November the AIEE subcommittee praised electrical manufacturers for their great strides toward developing equipment for testing arresters and hoped that improved devices would follow. The subcommittee did not repeat its criticism of the film arresters, especially after they observed experiments with Steinmetz's lightning generator and had their picture taken in front of the now-famous machine. In February 1923 Westinghouse announced the perfection of the autovalve arrester, which had passed tests on that company's own Steinmetz-type impulse generator. And in 1924 the Duquesne Light Company of Pittsburgh ran field tests on arresters using a similar generator Duquesne engineers had built.[25]

The national press did not herald the Westinghouse or Duquesne engineers as Modern Joves. Lightning generators were becoming commonplace by 1924, especially since engineers had produced 2 million volts in GE's Pittsfield laboratory the year before. That event rated an AP story, and engineers Giuseppe Faccioli and F. W. Peek were touted in one paper as "Modern Joves." A reporter also noted that "Like Prof. Steinmetz, [Faccioli] is a cripple." But Peek and Faccioli did not receive the heroic-scientist type of publicity awarded Steinmetz, whom newspapers had cited as an expert on the work at Pittsfield.[26] The main reason, of course, was that Steinmetz's physical appearance and eccentricities had made him much better copy than any engineer at the time, or any scientist except perhaps Einstein. GE tried to publicize Peek and Faccioli, but it found that newspapers preferred to tell the public about the gnomelike wizard Steinmetz, just the type of figure likely to create lightning in the laboratory.

GE's second motive for publicizing Steinmetz as Modern Jove met a less immediate objective than selling lightning arresters, but one that promised to be more beneficial in the long run. The company undertook a massive advertising campaign in the early 1920s to sell electrical consumer goods in its diversification into this field—a period in which advertising had changed from featuring the magical power of electricity to emphasizing its utility, safety, and modernity. After Gerard Swope and Owen Young took the reins of the company in early 1922, GE hired Bruce Barton's advertising firm to supplement the activities of its rather stodgy Publication Bureau. Barton, who became a legendary figure in the annals of American advertising, flooded popular magazines with theme ads on GE products and the progressive virtues of electricity. The latter "institutional" ads supported the concurrent campaign by the National Electric Light Association against the public ownership movement, which was gaining ground at the municipal, state, and national levels in the 1920s. GE had a strong interest in the NELA. Young served on its powerful Public Policy Committee, and GE owned one of its principal members, Electric Bond and Share, a giant holding company of electric utilities.[27]

The publicity for Steinmetz—which can be considered as a form of free institutional advertising—reached its peak in 1922–1923, in the midst of these public relations efforts and at the beginning of another campaign: the fight against new antitrust charges. In January 1922 Samuel Untermyer, a prominent reformer and attorney for the New York State Joint Legislative Committee on Housing (the Lockwood Committee), said GE was an "intolerable monopoly and should be suppressed." Untermyer claimed that the company's lamp monopoly drove up the price of light bulbs and thus housing costs. GE responded by inviting state and federal authorities to look into the matter. U.S. Attorney General Harry M. Daugherty ordered an investigation to see if GE was complying with the 1911 consent decree, the basis of Untermyer's charges. In the meantime, GE vice president Anson Burchard testified before the Lockwood Committee that the 95 percent market share of electric lamps held by GE and Westinghouse, its patent licensee, did not constitute a restraint of trade, because the position had been obtained under the patent laws and, besides, other companies had had the opportunity to enter the market without the patents.[28]

Burchard's remarks raised a few eyebrows among newspaper editors, but Young had confidence in Daugherty. The attorney general spoke against government ownership at the 1922 NELA convention at the invitation of Young's Public Policy Committee, and Young told GE officials that he preferred Daugherty in office to an unknown politician. Daugherty sat on the investigation for two years. But in March 1924, shortly before President Coolidge forced him to resign over the Harding administration scandals, he sued GE and Westinghouse under the antitrust laws for controlling lamp prices. GE won the case in 1926 on the basis of the monopoly protection provided by patents.[29]

The timing of the Jovian story was probably dictated by the AIEE's criticism, but the publicity also served to combat the antitrust charges. The story appeared in early March 1922, two months after Untermyer's charges. Bringing Edison to Schenectady in October created an even bigger media event, with Steinmetz's lightning generator astonishing the older Wizard. Edison's first stop was the Research Laboratory, where still and motion-picture cameras recorded Whitney, Swope, and Burchard— who had testified at the Lockwood Committee hearings—greeting him at the door. Newspapers and magazines followed GE's lead and praised the value of the laboratory's contributions to society. The theme anticipated the argument used by the firm in the 1924 anti-trust case: GE's patent monopoly (won by research in its laboratories) benefited the public because it produced a more efficient product (the gas-filled tungsten lamp) at a lower cost. To back up the claim, GE pointed out that it had reduced its lamp prices four times since the antitrust charges were made in 1922.[30]

The exact part played by the GE Publication Bureau and outside advertising and public relations agencies in publicizing Steinmetz is difficult to assess because of the relative scarcity of archival material. In an oral-history interview Clyde Wagoner, a Schenectady newspaper reporter who joined GE and started its News Bureau in 1920, recalled, "I always felt that if we had a good story . . . fourteen copies and a dozen photos would do the trick. If the story was worthwhile the wires . . . would use it and if the pictures were good the photo syndicates would do the job. This not only saved considerable money for the Company but worked with splendid results. I made it a strong point to cultivate the acquaintance of editors by serving them with news." A private letter backs up this account. In 1923 Wagoner asked Steinmetz for a copy of a speech he was going to make in Boston a "couple of days in advance for publicity purposes." Wagoner's efforts paid off; at least four medium-size newspapers carried the same lengthy article on the speech, which Wagoner probably sent out as a news release.[31]

The Jovian legend had a similar pedigree. The AP story was published verbatim and in its entirety from a news release written by Wagoner. A longer, more technical version appeared in the *Schenectady Works News* on the same day. The pastoral images of Steinmetz working in a canoe at Camp Mohawk were first published in the *New York Times* Sunday picture section in the fall of 1923. The *Times* staff probably came up with the caption "Jupiter Turns Neptune," while the photographs were courtesy of GE. A story with similar photos on life at Camp Mohawk, probably written by Wagoner, appeared two weeks later in the *Schenectady Works News* and filled out the pastoral scene used by later writers.[32]

Like Edison, Steinmetz did not play a passive role in this mythmaking. He regularly gave speeches at national engineering conventions, where he delivered "quotable quotes" to newspaper interviewers on the future of electricity. In addition to writing for Atherton Brownell's editorial service during the war, he wrote articles and was interviewed for popular magazines like *Ladies' Home Journal, American Magazine, Collier's, Harper's, Good Housekeeping,* and *Scientific American.* In 1922 the Hearst newspaper chain agreed to pay him $500 per article to write a series of popular articles on electricity and civilization.[33]

Steinmetz enjoyed playing the part of the Electrical Wizard, but he thought the role had a wider purpose. In *America and the New Epoch* he blamed the corporation for the public's hostility toward it. Since the corporation had been "too self-centered, self-satisfied" to appreciate the effect of secrecy on public opinion, its enemies—the muckrakers and trustbusters—explained the corporation to the public with disastrous results. But when the corporation itself tried to publicize its "social and industrial activity," the result was "so obviously untrue that it carries no conviction." Giving only positive accounts invited disbelief, giving a balanced account would lead to distortion by the press, and saying nothing—as in the past—left the corporation helpless against attacks. "The only remedy apparent seems to be to entirely throw open the discussion, give information to the fullest extent, and count on the public gradually realizing what is unfair representation and what is reasonable." Steinmetz recommended that writers who were not connected with the corporation or the "muck-raking crowd" take up the task of depicting the "wonderful field of romance and interest" of the "huge modern industrial corporation."[34]

Such passages leave little doubt that Steinmetz favored the type of publicity campaign conducted by Swope and Young. His friend Atherton Brownell had made similar suggestions to Young in 1919. Brownell complained about GE's public silence on labor issues and called for a more professional public relations department, which GE established in the 1920s. Steinmetz assisted these efforts by writing introductions for Charles Ripley's *Romance of a Great Factory* (1919) and John Broderick's *Pulling Together* (1922). Both were essentially public relations pieces, although only Ripley worked for the Publication Bureau. Steinmetz's articles on the social value of electricity also helped the company's publicity campaign in the 1920s and often appeared in the same magazines as Barton's GE ads.[35] The irony—which we have noted before—is that Steinmetz wrote these articles to speed the coming of corporate socialism, while Young, Swope, and Barton supported similar efforts to promote corporate capitalism.

The two ideologies occasionally came into conflict. As Modern Jove, Steinmetz became more valuable to GE but also more autonomous. When Young decided to publicize Steinmetz after the first antitrust suit in 1911, the hunchbacked genius was a great asset, even though he became known as a Socialist. The aims of his corporate socialism meshed well with GE's objectives, and as president of the common council he stayed out of the Socialists' attempt to raise the company's assessment. The only evidence I have found of possible company interference in his political views before 1920 was that he asked Brownell to remove the GE affiliation from his war editorials, which were becoming stridently pro-German.[36] When GE began publicizing him during the second antitrust crisis in the 1920s, it dealt with a more independent and unpredictable Wizard.

The conflict between Steinmetz's politics and those of his employer came to a head over the question of the public ownership of utilities during his campaign for New York state engineer in 1922. GE supported the NELA's publicity drive against public ownership because it owned Electric Bond and Share. Steinmetz's attacks on trustbusting supported GE's battle against the antitrust charges, yet he could also be interpreted as favoring public ownership. In a prewar article he listed government supervision and "national ownership" as a step toward corporate socialism. *America and the New Epoch* saw federal "operation and practical ownership of industries" as a last resort if companies could not establish a corporate state on their own. After the war he said that the spread of electrical networks across state lines called for federal supervision and control, and he praised Lenin's state-owned electrification plans.[37] GE may have encouraged the overture to Russia, yet the much-publicized correspondence with Lenin in the summer of 1922 helped create the impression that Steinmetz favored public power.

That was the interpretation during the 1922 election. Both the Socialist *New York Call* and the *New York Times* said that Steinmetz would advocate "public ownership and control of the hydroelectric power of the State" in the campaign speech to be broadcast by radio. The telegram that first canceled the speech was not that explicit, however. He promised to "push without regard for any selfish interests the immediate development of the great water power resources of New York State." He was even more ambiguous when he canceled the speech the second time, saying he would "accomplish something in developing for all the people the abundant natural resources of our state and more particularly the great water powers which are now running to waste."[38]

As noted earlier, *Electrical World* thought Steinmetz was calling for public ownership—and at a dangerous time. The League of California Municipalities placed a public power bill on the referendum for the November election, which *Electrical World* and the electric utilities fought. The NELA denounced it in a pamphlet entitled "Shall California Be Sovietized?" By then over 40 percent of city power plants were municipally owned, and Senator George Norris of Nebraska was calling for federal control of the unfinished hydroplant at Muscle Shoals, Alabama—a government project started at the end of World War I to provide cheap electricity for nitrate plants. In New York Democratic ex-governor Al Smith was campaigning for the governorship on a public power platform not much different from Steinmetz's, and Steinmetz published the technical articles on Soviet electrification shortly before the campaign began.[39] In the end Steinmetz did not "bring hurt to his industry," as *Electrical World* had feared. Instead, he canceled his speeches and prepared the lightning generator for Edison's visit to GE. The resulting publicity helped GE improve its image during the antitrust investigation, thus indirectly helping the utilities fight the public ownership movement by showing the virtues of private control.

Smith was reelected and came out strongly for public ownership of waterpower. In 1923 he appointed a Water Power Committee of the State Conference of Mayors to suggest appropriate legislation. The chairman of the committee was Lieutenant Governor George Lunn, who promptly announced that he would ask Steinmetz for advice. Newspapers in the Niagara Falls area were outraged. A Buffalo paper complained that Lunn, a former Socialist, "has given the job of framing the bill to Charles P. Steinmetz, another Socialist." The paper remarked that the Republican-controlled legislature would not stand for state-owned waterpower—an accurate prediction, since that body repeatedly defeated proposals for a New York State Power Authority until Franklin Roosevelt became governor in 1930.[40]

The Buffalo papers, *Electrical World*, and GE need not have been so concerned. A close reading of Steinmetz's writings would have convinced them that he favored corporate ownership and control of industry. He made the point clear in a speech at the Babson Institute after the election: "Businessmen make a great mistake in thinking that socialism means government operation. Only the labor leaders and the radical politicians are talking government operation [and they want it not] for the benefit of the community but rather for their own selfish purposes . . . Successful government operation of industries, railroads, or utilities is inherently impossible under a democratic form of government."[41] Many newspapers printed a short extract from the speech proclaiming that the Socialist Steinmetz had declared public ownership to be "impossible."[42] They failed to give the rest of his argument: efficient public operation was impossible under a "democratic form of government" and required the type of technocratic state Lenin was building in Russia.

The episode was typical of how groups at opposite ends of the political spectrum used Steinmetz for their own purposes by quoting him out of context. The NELA no doubt liked the parts of the speech where he identified Senator Robert La Follette, a vocal proponent of public power, as a dangerous "radical politician" for wanting to break up the trusts. Consequently, the Rocky Mountain division of the NELA co-sponsored a speech by Steinmetz in Denver in 1923. According to a later federal investigation, the division's public relations committee was one of the most active in the NELA's campaign against public ownership. Inviting Steinmetz, the foe of La Follette and friend of Lenin, to speak was risky, but he did not disappoint his hosts. He thought that Congress should have approved Henry Ford's offer to buy Muscle Shoals, and a Denver paper carried the headline "Revolution Like Russia Impossible in America, Declares Dr. Steinmetz."[43]

On Jupiter's Throne

The controversy over public power indicates that while GE could help enthrone Steinmetz as Modern Jove, it could not control what he said in his speeches. The company was thus only one element—although an important one—in the social construction of the Steinmetz legend. Much of the publicity for Steinmetz in the 1920s came from newspapers and magazines that adopted the pet Electrical Wizard as their expert on the role of science in modern society. Newspapers called on him to give opinions on three main subjects: the theory of relativity, science and religion, and the future of science and technology. GE provided photographs and other resources to help keep his name before the public as a scientific authority.

The publicity surrounding Einstein's theory of relativity was unprecedented for the work of a "pure" scientist in utilitarian America, and it illustrates the growing popularity of scientifically trained experts like Steinmetz at the expense of the Edisonian inventor after the war. When Einstein visited the United States in the spring of 1921, his unfathomable theory of relativity, picturesque pipe and violin, informal persona, and quotable quotes made a great hit with the New York newspapers. Fascinated that such a modest man had created a revolutionary theory of time, space, and the universe that supposedly only twelve people in the world could understand, reporters covered his two-month stay as if he were visiting royalty. At the same time, Edison's employment questionnaire, which tested general knowledge on a wide range of topics, was causing a controversy in the press because college graduates regularly flunked it. Looking for a good story, reporters gave the test to Einstein, who promptly put it down upon failing to know the velocity of sound! Einstein said he did not need to memorize such knowledge, which he could look up in a handbook—a criticism of the philosophy behind the questionnaire. He then criticized Edison directly for saying that a college education was of little value.[44]

Although Einstein's visit reopened the controversy over the undemocratic nature of a new physics that only experts could understand, his personal magnetism helped improve the public image of the "pure" scientist in the 1920s. One indication is that in 1929 the promoters of the fiftieth anniversary of Edison's invention of a practical lamp wanted so badly to place Edison in the pantheon of great scientists that they persuaded the popular Marie Curie to attend the ceremonies and Einstein to broadcast a radio message to Edison from Berlin. Einstein delivered a carefully worded statement that praised Edison's technical accomplishments without using the word *science*.[45]

The Wizard of Schenectady had much more in common with Einstein than did the Wizard of Menlo Park. During Einstein's visit Steinmetz accompanied him on a tour of a transatlantic radio station and shared center stage with him in a group photograph of leading scientists, engineers, and executives from GE, AT&T, Western Electric, and RCA.[46] The visit prompted Steinmetz to start reading Einstein and to begin popularizing the theory of relativity. Ernest Caldecott, minister of the Schenectady Unitarian Church, recalled that he asked Steinmetz if he was one of the twelve who understood Einstein and, if so, would he give a talk on relativity at the church. Steinmetz answered yes to both questions. He delivered three more lectures to local AIEE meetings and also included a discussion of relativity in a magazine article on science and religion. An editor for the *New York Times* remarked that, as "one of the favored dozen," Steinmetz in a "few admirably written and beautifully simple pages really does what so many other writers vainly have tried to do" toward making Einstein's work comprehensible to the layperson.[47]

Steinmetz continued to popularize relativity in 1922. In April he gave a speech, via GE's radio station WGY, on the existence of the ether that stirred the public debate about Einstein's theories in the United States. The *New York Times* thought that Steinmetz's denial that the ether existed was worthy of a story, and at least one reader responded that not all scientists had abandoned the ether. Edgar Larkin, director of Mount Lowe Observatory and science editor for the Hearst newspapers, quoted Steinmetz as an authority on the question in his science column. Larkin had read a copy of Steinmetz's radio address and thought it was "one of the most comprehensive scientific productions since Newton."[48]

Larkin's hyperbole aside, Steinmetz was a successful popularizer of Einstein. His *Four Lectures on Relativity and Space* (1923) gave a simplified mathematical description of Einstein's work. Returning to his graduate-level training in mathematics, Steinmetz used analogies, simple equations, geometric diagrams, and stereo-optic images to explain the general consequences of the theory, the relativity of length, mass, and time, the famous equation relating energy and mass, and the relationship between the curvature of space and a gravitational field. Although not reviewed by the *New York Times* or general magazines, the book was praised by *Electrical World* as useful to engineers.[49] It went through at least five printings and was republished in the mid-1960s. In the case of relativity, newspapers elevated Steinmetz to the status of a pure scientist—which they had also done for his work on artificial lightning—a further indication of the growing stature of scientists at the expense of inventors in the 1920s.

Einstein's theory was also involved in the renewed debate over science and religion. The lengthy cease-fire in the warfare between science and religion, negotiated by liberal theologians in the accommodation over Darwinism in the 1880s, came to an end in the 1920s. The Scopes "monkey trial" (1925) gained most of the publicity, but the question of how the new physics affected religion cropped up shortly before then. In 1923 the Nobel Prize–winning physicist Robert Millikan prepared a Joint Statement upon the Relations of Science and Religion. Signed by thirty-five eminent churchmen, politicians, educators, and scientists—including Michael Pupin and Herbert Hoover—the statement declared that there was no inevitable antagonism between the separate spheres of science and religion.[50]

Steinmetz had made similar comments the year before. An agnostic who attended the Schenectady Unitarian Church only a few blocks from his house, he was often thought to be Jewish, probably because of his surname. In 1922 readers of the *Jewish Tribune* listed him eleventh in a poll of the "twelve outstanding Jews in America," just ahead of *New York Times* publisher Adolph Ochs, and *American Hebrew Magazine* included him in its annual issue on "Who's Who in American Jewry." Steinmetz stated his views on the subject of science and religion in an invited magazine article earlier that year and in a speech in November before the Schenectady Unitarian Church. After reviewing the history of the subject, he repeated the time-honored argument that science and religion were "not necessarily incompatible," because science dealt with "immutable laws of nature" discovered by applying human logic to sense perceptions, while religion considered matters of belief beyond the senses. Science and religion were "different and unrelated activities of the human mind," and neither could speak authoritatively in the domain of the other. The theory of relativity supported these conclusions by showing the limitations of science. Sense perceptions—and thus nature's laws—were based on the concept of absolute space and time, but Einstein had shown them to be relative.[51]

Steinmetz's speech, a much shorter version of the article, gained a good deal of notice, especially since it came near the peak of his popularity during the 1922 election. *Literary Digest* published nearly the entire lecture and noted that coming "as they do, from a preeminent scientist and investigator, these words are widely quoted in the daily press, with the assurance that they will bring comfort to those laymen who fear that a study of science as it is taught today will invariably lead to atheism." Although opposed by some ministers, Steinmetz's views received support from many newspapers. A Philadelphia paper cited Steinmetz's opinion, along with those of Newton and Pasteur, to back up the joint statement issued by Millikan in 1923.[52]

Steinmetz's technological prophecies—the third area in which he spoke out about science and society—received more press than his remarks on relativity or science and religion. He had a longstanding interest in utopian speculations, dating to his college debates about the failure of the Icarian colonies. He enjoyed reading futuristic novels as an adult and exchanged several of them with Westinghouse engineer Benjamin Lamme, especially the works of H. G. Wells. Steinmetz owned numerous books by Wells and recommended several of the English Socialist's novels to Henry Ford. This reading carried over into *America and the New Epoch*, as we have seen.[53]

Steinmetz based most of his technological prophecies on an article he published in *Ladies' Home Journal* in 1915, entitled "You Will Think This a Dream." Actually written by an interviewer, the article envisioned an all-electric world. Electricity would become so universal that all fires and combustion would be outlawed in the city. Electricity would cook all meals, heat and cool temperature-controlled houses, run industry, and propel all forms of transportation. Electric autos would be garaged and recharged in the basement where the coal furnace used to be, while residents upstairs would listen to entertainment piped in over "wireless telephones," which upgraded Edward Bellamy's prediction that music would be piped in by telephone. The cost of living would drop because electrically produced nitrates would be used as fertilizer, and electric machinery would shorten the working day to six hours. Electricity could do so much because locating generating stations at the source of energy supplies and transmitting it into urban areas would make it too cheap to meter. Instead, electricity would be taxed like water, so much per outlet rather than so much per faucet (the practice at the time). Like earlier writers of technological utopias, many of whom portrayed an electrical age, Steinmetz claimed that his vision was a simple extrapolation of present-day trends: "The means for all these things are here now."[54]

The title of the article indicates how far most American homes actually were from this electrical utopia in 1915, even those among the nearly 25 percent of urban and rural households that had electricity at the time. Electric sewing machines, fans, stoves, saucepans, chafing dishes, teakettles, and other items had been available since the 1890s, but high electricity rates made them too costly for most families. Added to this was the task of standardizing a system of electrical outlets, plugs, and interchangeable appliances from a plethora of competing schemes, a process not settled by manufacturers until 1917. The electrical industry started promoting heating devices like flatirons and toasters at the turn of the century to provide a daytime load for central stations to balance the nighttime load of electric lighting. But sales did not rise markedly until the NELA began to encourage the standardization of outlets and plugs in 1915 and utilites saw the domestic market as a profitable one. The manufacturing value of electrical household appliances and supplies rose steadily from about $2 million in 1899 to about $24 million in 1915, then almost doubled in 1916, the year after Steinmetz's article. He probably based part of it on the "house without a chimney," the highly publicized all-electric home in the GE Realty Plot designed by engineer H. W. Hillman in 1905. Ironically, Steinmetz's utopian vision—reminiscent of Etienne Cabet's Icaria and Lenin's plan to electrify Russia—stands in sharp contrast to recent scholarship on household technology. Ruth Schwartz Cowan has argued persuasively that "labor-saving" devices like electric lighting and vacuum cleaners eliminated drudgery but created "more work for mother," not less, because it saved the work of her helpers (men, children, and servants) while giving her more to do.[55]

Steinmetz elaborated on the themes of this article in later writings. He reduced the workday from six to four hours in *America and the New Epoch* and in a newspaper interview in 1918. He wrote about the technical and economic advantages of generating electricity at the source of energy supplies and told a newspaper in 1917 that electricity would be too cheap to meter. In 1922, when radio broadcasting had become a reality and electrical appliances were more common, he added new prophecies to his repertoire with two magazine articles. The first recognized that since electric heating was more expensive than coal, homes in the future would be well-insulated and would use a regenerative system of ventilation to make electric heating economically feasible. The second article added the prediction that scientists would develop "energy crops" that could be burned for fuel in place of coal.[56]

These predictions received wide publicity. One newspaper said that Steinmetz "in the role of prophet is even more interesting than in his well-known part of electrical wizard. For he is no dreamer like Edward Bellamy, nor is he a fictionist like H. G. Wells. On the contrary, he is a hardboiled scientist and electrical engineer, whose motto is 'efficiency,' whose name is joined in the scientific world with those of Edison and Marconi."[57] The writer probably did not realize that the "hard-boiled" Steinmetz was a fan of Wells. Nor did other newspapers know about his private criticism of Tesla's mental state when they reprinted an interview with him on the possibilities of wireless power in early 1922. Steinmetz was careful to state that transmitting radio beams of power would be impractical because of dispersion, but he did say that Tesla's idea of tapping stationary electromagnetic waves sent around the globe might be feasible if the right wavelength was chosen. Newspapers generally ignored these qualifications and printed headlines like "Steinmetz Forecasts Radio Power Transmission."[58] He caused more of a furor in a speech on future inventions later that year. To his idea of energy crops he added one to crossbreed bacteria as a food source to supply protein directly and bypass the food chain from bacteria to plants to animals to humans. Papers carried headlines that he "Would Feed World on Microbes" and other stories of his "biological engineering."[59] These schemes met his goal of social efficiency in his attempt to engineer society, but unwittingly or not Steinmetz had joined Tesla in making fantastic predictions in the newspapers.

The widest coverage came from an invited article he wrote on the next hundred years for *American Legion Magazine* in 1923. Steinmetz reviewed the inventions of the past century and had "no doubt that the transformation to be visualized a hundred years from now will be more complete than the most fanciful among us might conjure." He repeated his predictions for an all-electric future and added a paragraph stating that it would not be of any real importance unless the lot of workers was improved. He then introduced the subject of electricity and socialism to the conservative Legionnaires—and clothed it in nineteenth-century Utilitarian philosophy—by stating, "I have no name for this world of the future. It could be called collectivistic, socialistic, or whatever term might be given a civilization governed by the principle that the end of all effort is the greatest possible health and happiness for the greatest possible number of human beings." One way to ensure this was to use technology to reduce the workday to four hours and the working year to seven months. Steinmetz predicted that the United States would remain the bulwark of individualism. But the "collectivistic tendencies of the Slavic peoples . . . will make of them the dominant race of the future."[60]

Coming at the height of his popularity, the story was picked up by newspapers across the country in several forms. The *New York Times* published a front-page account, several newspapers printed a much shorter version, the Associated Press carried a one-paragraph story, and the United Press listed six of Steinmetz's prophecies. The predictions even made their way into the *Wall Street Journal*.[61] At least two papers carried cartoons of the shorter workday and five-month vacation of the future. A Pittsburgh editor responded, "If we ever do get the four-hour workday predicted by Doc Steinmetz, the chances are that soon after it has been established somebody will become dissatisfied and start a movement for a three-hour workday."[62]

Most papers focused on two predictions: the four-hour workday and the rise of the Slavic race. The syndicated columnist Herbert Kaufman marveled that "not many soothsayers could share" the front page with the world's headlines and thought that Steinmetz's predictions were "not only probable, but an underestimate."[63] Others disagreed. The editors of at least three New York City newspapers thought a four-hour workday was dangerous, because workers would not know how to use their leisure time wisely. (Steinmetz had predicted that people would live in the suburbs and grow their own food or engage in other "productive diversions.") Liberal columnist Heywood Broun replied that he would gladly try to get along with more leisure time. While the *New York Tribune* had declared that a "moral equivalent for hard work has never been discovered," Broun said he would try to find a "moral equivalent for labor."[64] A Florida critic of the prophecy that the collectivist Slavs would rise above the Anglo-Saxons said, "It is easy to call Steinmetz a Socialist, and let it go at that; but the question he raises cannot be disposed of so easily."[65]

Steinmetz lived to see increased electrification in industry and the introduction of radios and electrical appliances into the home, and we now have temperature-controlled houses. But many of his forecasts were wide of the mark, as shown by George Wise's study of technological predictions. Like Bellamy, Tesla, and other frequent predictors, Steinmetz was wrong slightly more often than he was right. (Edison fared somewhat better in Wise's study, but Ford did not make a single accurate prediction.)[66] Steinmetz would have been disappointed to see that the universal electrification of railroads and the widespread use of electric vehicles did not come to pass in the United States. Having helped Eickemeyer develop trolley-car motors, he believed in electric traction so fully that he spent a considerable amount of time outside GE promoting electric vehicles. His ten-year effort to introduce a practical "electric" illustrates the difficulty of realizing part of his utopian dream.

Steinmetz predicted that electrics would supersede gasoline cars in a speech to the Electric Vehicle Association in 1914. The electric car had competed effectively with gasoline and steam automobiles at the turn of the century. But the introduction of the Model T in 1907 and the invention of the electric self-starter in 1912 spelled the doom of the steamer and the electric. Promoters of the latter saw it as a vehicle to complement the speedier and longer-range gasoline automobile and marketed it as a luxury item with limited range—due to its heavy lead-acid storage battery— for the urban upper class. Edison developed a lighter nickel-iron battery in 1909, but it was not well suited to vehicular use and the sale of electric cars declined sharply thereafter, especially when thousands of inexpensive Model T's began rolling off Ford's new assembly lines in 1913. But when Steinmetz gave his speech the next year, the prospect for electric vehicles—particularly electric trucks—seemed bright, paradoxically because the growing number of gasoline cars raised gas prices severalfold when oil production could not keep up with demand.[67]

Reflecting that optimism, and going considerably beyond it, Steinmetz declared that the electric would take off once the automobile was no longer the object of a popular touring sport, requiring high speed and long range, but became a "useful commodity, a business and pleasure vehicle." The gasoline auto craze would thus go the way of the bicycle craze of the 1890s, especially if utilities cooperated with electric automakers to provide inexpensive, flat-rate recharging and garage services. That was feasible because the electric's battery would be charged during off-peak hours, thus helping to even out the central station's load curve. Steinmetz, who purchased a Detroit Electric in 1914, made a similar speech to the NELA convention that year and calculated that electric vehicles would be a $75 million business for utilities in ten years.[68]

He then helped form the Dey Electric Automobile syndicate to manufacture and sell an electric invented by Harry Dey, a New York City electrical engineer. Steinmetz acted as consulting engineer, while Horace Parshall and Hayden were two of the company's directors. Although the extent of his involvement is unknown, he made suggestions about finances, personnel, marketing, and the schedule for completing ten trial cars. The Dey syndicate announced the venture in early 1915, calling a news conference at Steinmetz's home. The *New York Times* devoted part of its Sunday magazine to the enterprise with an interview with the Wizard of Schenectady. Steinmetz claimed that Edison had solved the battery problem and described in some detail the technical merits of the patented Dey electric—how the design of its motor eliminated the differential and thus reduced the weight of the car, and how it automatically recharged the battery when going downhill. Many newspapers picked up the story of the Wizard's new car, and *Current Opinion* reprinted the interview in a story entitled "Forecast of a Revolution in the Automobile in the Near Future."[69]

The public heard nothing further until after the war, when newspapers reported in January 1920 that an electric vehicle designed by Steinmetz would be built in Baltimore. Baltimore and Washington investors had formed the Steinmetz Electric Motor Car Corporation to introduce an "industrial truck" and lightweight "delivery car" that were similar in design to the Dey Electric. The company's prospectus took liberal advantage of Steinmetz's reputation, calling the car the "crowning triumph of Dr. Steinmetz's career" and reprinting a letter from him about the technical merits of the vehicles. The company expected that one thousand trucks and three hundred cars would be produced within eighteen months. Interested persons could buy stock from the Steinmetz Syndicate in New York City. The endeavor caused a furor at GE when the syndicate solicited company stockholders to invest in the enterprise. GE issued a statement to its stockholders in March that it was not connected in any way with the Steinmetz Syndicate and had not approved or known about the solicitation. The notice prompted the *New York Times* to wonder if there had been a break between GE and Steinmetz.[70]

The Steinmetz vehicle made its debut two years later when an electric truck climbed a steep hill in Brooklyn during a stunt staged for the press in early 1922. The presence of the Electrical Wizard at the event and at a press luncheon afterwards guaranteed free advertising for the venture in New York City newspapers. The Steinmetz Syndicate also ran ads that appeared on the same day as the story. One in the *New York Call* asked readers to send in a postcard to "learn how you may still share in the profits of Dr. Steinmetz's company." The incongruity of appealing to the profit motive in the Socialist *Call* apparently did not bother the promoters. But one wonders why Steinmetz, as chairman of the board and chief consulting engineer of the company, did not intervene.[71]

In October the Steinmetz company announced the completion of a five-passenger coupe at its experimental shop in Syracuse, New York. Steinmetz was slated to be the first person to ride in the lightweight vehicle, which had a speed of 40 miles per hour and an advertised range of 150 to 200 miles, but he could not come to Syracuse because of an illness. Newspapers reported that he later made several trips to Syracuse in preparation for sending test vehicles to the Baltimore factory.[72] The coupe apparently did not pan out, because the next ads for a Steinmetz vehicle, in March 1923, featured a good-sized electric truck. *Printer's Ink* ran a story on how the ads were "built around the personality of the famous inventor." Newspapers then carried an interview with Steinmetz that electric trucks would lower distribution costs and thus food prices, and the Steinmetz company exhibited its wares at the National Electric Truck Show in June.[73]

The venture soon folded. Shareholders were offered stock in an oil company for their unlisted, preferred stock in the Steinmetz company, which had neither paid a dividend nor issued financial statements. More information about the firm came out in December, shortly after Steinmetz's unexpected death, when a shareholder sued Edward Powers, a stockbroker for the Steinmetz Syndicate, for grand larceny in making false representations. Although Powers said that the company was making 135 cars per month, the investor discovered that it had made only 48 vehicles altogether and had sold only 30. The syndicate had sold $1.4 million worth of stock yet could account for only a fourth of that and had only $840 in its treasury. The Baltimore directors of the company— including Herbert Wagner, president of Consolidated Gas and Electric in Baltimore—quickly distanced themselves from Powers and the Steinmetz Syndicate. Powers was a former vice president of the Steinmetz company who had resigned some time before to sell shares through the syndicate. The *New York Times* observed that the sorry episode unfortunately dragged Steinmetz's good name through the courts and showed that he was out of place in business.[74]

The proper place of Modern Jove, of course, was on Jupiter's Throne. One sign of his growing stature as a cultural hero comes from a widely published 1923 newspaper story entitled "Don't Worry Lowbrows the Stigma's Gone!" A column of cropped photographs, showing only the brows of six celebrities, supposedly refuted the adage that people with high brows were smarter and more successful than those with low brows. The faces of the "lowbrows" were so familiar that even small papers like the *Des Moines* (Iowa) *News* did not identify the owners of the famous brows.[75] The celebrities were Babe Ruth, Steinmetz, President Warren Harding, Henry Ford, Edison, and Jack Dempsey. All are still household names today—except Harding and Steinmetz.

Steinmetz's popularity in 1923 was clearly evident during a railroad journey to California that fall. The trip was part business and part vacation with the Haydens, whose three children were now teenagers. Their father arranged sightseeing visits to the Grand Canyon, Yosemite, and Hollywood, scheduled Steinmetz to speak at engineering meetings and civic functions in several cities, and set up conferences with GE engineers and customers on the six-week tour. Local newspapers heralded the visit of the Electrical Wizard to their cities, particularly in Colorado— where the NELA helped with publicity—and also in California. Photographers snapped shots of the actor Douglas Fairbanks giving Steinmetz a tour of Hollywood studios, and a San Francisco paper published three days' worth of stories on his visit, including a picture of Steinmetz and the Hayden clan (which was good copy because the kids were taller than

Steinmetz). A chance conversation about science and religion with William Jennings Bryan on the train home seemed a fitting end to the triumphal tour. Upon returning to Schenectady in mid-October, Hayden wrote a GE vice president that the trip had been "most gratifying." Steinmetz spoke to "record-breaking audiences in Denver, Los Angeles, San Francisco and Salt Lake City." These were luncheon talks to local chambers of commerce or Rotarians and evening speeches to technical groups and the general public, where the attendance ranged from one thousand to four thousand at each city.[76]

But the strain of the trip was too much for the fifty-eight-year-old Steinmetz, who had suffered from a heart condition for some time. The *New York Times* comforted its readers on October 22 that reports of a turn for the worse in his health were unfounded. He was just resting in bed and was "anxious to get up and get about perfecting his present pet project—development of generators of 'man made lightning.'" His nurse talked to him on the morning of Friday, October 26, just after he woke up, and asked one of the children to bring him breakfast. Some minutes later Billy Hayden carried a breakfast tray to his grandfather's room and found him lying motionless on the bed with a book on the physics of the air on his bedside table. His physician reported that he had died from "acute dilation of the heart, following a chronic myocarditis of many years standing, which is a weakening of the heart muscles."[77]

Fate of the Steinmetz Legend

Newspaper coverage of Steinmetz's sudden death at the height of his popularity was extensive. In an article entitled "A Hunchback Who Played with Thunderbolts," *Literary Digest* published extracts from some of the "thousands of newspaper obituaries" that had appeared by mid-November.[78] The *New York Times*, an admirer of Steinmetz despite criticizing his socialism, and the *San Francisco Chronicle* ran lengthy front-page stories on his death and career. The *New York Tribune* and the *Washington Post* printed more modest accounts on their obituary pages. Many papers covered the unfolding story of tributes to Steinmetz from scientists and other notables on Friday and Saturday, the hundreds of mourners who filed past his coffin lying in state at the Wendell Avenue house on Sunday, the private funeral service at home and burial on Monday, while fifteen hundred looked on, and the reading of his will on Tuesday, October 30.[79]

GE and the city of Schenectady paid their respects. The corporation flew its flags at the Schenectady Works at half-mast for a week and gave orders to all its plants and offices at home and abroad to stop work for five minutes at the time of the funeral. Martin Rice, manager of the Publication Department, gave an address on GE radio station WGY the evening Steinmetz died and then closed the station for the night. Schenectady schools and Union College closed for the day of the funeral, city offices closed for the afternoon, and the city maintained the five minutes of silence observed by GE. Owen Young, Gerard Swope, Ernst Berg, and George Lunn were among the honorary pallbearers, and Edwin Rice and Ernest Caldecott spoke at a memorial service on Wednesday at the Unitarian Church.[80]

The reading of Steinmetz's will gained as much publicity as these events. Newspapers printed the will, made out during the war, in full as the centerpiece of a story that Steinmetz, the $100,000-a-year Socialist, left an estate that would probably not meet his relatively modest bequests: a $24,000 trust fund for his sister Clara, $1,500 to two half-sisters in Germany, and the remainder to Joseph LeRoy Hayden, the executor of his estate. The story that the "man who 'made lightning' probably could have made millions" but left little more than the $1,500 GE life insurance policy given to employees made its way onto the front page of newspapers and into magazine articles. Instead of feeling duped by the $100,000 legend, many writers fostered the myth that Steinmetz had refused a salary and asked only for laboratory and living expenses. They made his supposed lack of interest in money a virtue of the idealistic scientist—or Socialist, depending on the paper—who wanted only to do research that would better the lot of humankind.[81]

When Hayden settled the estate in 1924 he found that Steinmetz was not poor—a story that did not make the papers. Probate records show that Steinmetz had assets of slightly over $57,000. After all debts, taxes, and funeral expenses were paid and Clara's trust fund was set up, the Haydens were left with a little over $5,000 and the Wendell Avenue house, which was not included in the evaluation because Steinmetz had deeded it to Hayden in 1922. Stocks and bonds worth over $44,000 made up the bulk of his assets, most of which were in Edison utility companies and GE. A listing of worthless stocks included his ill-fated business ventures and idealistic investments in Dey Electric, the Steinmetz Syndicate, the *Milwaukee Social Democrat,* and the Russian-American Industrial Corporation. Nearly 30,000 marks of Allgemeine Elektrizitäts-Gesellschaft bonds were also deemed to be without value because of the inflation rampant in Germany at the time.[82]

The main theme of the newspaper and magazine obituaries was how a penniless, hunchbacked political refugee became famous in the Land of Opportunity, yet the media tried to come to grips with the meaning of his technical accomplishments as well. Writers had not been concerned with this problem before, because they could always refer to his position as chief consulting engineer of GE for his scientific credentials and then get on with the more exciting task of describing the Little Cripple with the Giant Mind who threw thunderbolts about the laboratory or dreamed up an all-electric Socialist utopia. After his death writers listed many tributes from scientists, engineers, and inventors, most of whom praised him in a general way.

Those by Cornell professor Vladimir Karapetoff and Edwin Rice provided some assistance to the popularizers. Karapetoff, who had worked summers at GE and assigned many patents to the company, saw himself as Steinmetz's successor at GE and as head of the electrical engineering profession. The editor of the *New York Times* liked Karapetoff's "happy phrase" that GE gave Steinmetz the freedom to try to generate electricity out of the square root of minus one. "That, doubtless, was what the man often seemed to be doing to those to whom mathematics as he knew it was equally incomprehensible and useless," continued the *Times*. "Fortunately his employers—no genius ever had better and few as good—took a different view."[83] The same editor commented on a eulogy by Rice: "To most of us it means nothing" when Rice described Steinmetz's research in magnetic hysteresis, but this work seemed to have improved the quality of electric motors and generators. In its obituary the *Times* had pointed to his lightning-generator experiments as an example of "pure science." But "Mr. Rice gives no support to the theory that Steinmetz lacked practicality—that he lived in the abstract world of 'pure' science. He both denies and conflates that theory by crediting the man with a great number of useful inventions and describing several of them." The editorial reveals how poorly the newspapers understood Steinmetz's work, even the commercial significance of the lightning generator.[84]

Michael Pupin, Steinmetz's longstanding rival, tried to put the Jovian legend into better perspective. Writing in a New York City paper, Pupin complained that the press had latched onto the lightning generator and ignored Steinmetz's more important contributions to electrical engineering. There "were quite a number of men in the country who led him in mathematics, and quite a number of other men who led him in the mastery of the electrical science, and yet few people have heard about these men, but they have all heard about Steinmetz." He thought the reason lay in the fact that Steinmetz worked in applied science, closer to the "problems of the people." (Pupin himself gained some of this recognition when his autobiography won the Pulitzer Prize in 1924.) The *New York Tribune* came closer to the mark when it said that the "utter dissimilarity of this tiny genius to any man of his times was what captivated the popular fancy."[85] To these observations I would add that newspapers, magazines, GE, the public, and the fluent Wizard himself interacted in a complex manner to construct the Steinmetz legend. The image they constructed made good copy and tapped the powerful, paradoxical myth of a cultural hero succeeding as a rugged individual in a bureaucratic organization.

The mythmakers did not retire with Steinmetz's death, especially GE, which no longer had to worry about an unpredictable Socialist Wizard making embarrassing statements about a Cooperative Commonwealth based on the corporation.[86] The company had other worries, however. It faced the antitrust suit over the electric lamp business brought by Daugherty in early 1924, and George Norris introduced a Senate resolution in December calling for an investigation of GE's part in the "Power Trust." The very next day Young and Swope announced that GE was giving up control of its utility holding company Electric Bond and Share and distributing its stock as a dividend to GE shareholders. GE thus hoped to avoid monopoly charges by ceasing to own its main customers. Yet Norris and Senator Thomas Walsh of Montana persisted and persuaded Congress to request the Federal Trade Commission to investigate the Power Trust, especially GE's role in it. In 1927 the FTC concluded that GE had controlled only one-eighth of the country's electric generating capacity when it owned Electric Bond and Share. Senator Norris and his colleagues were not satisfied and initiated a more thorough FTC investigation, which led to the passage of the Public Utility Holding Company Act in 1935 and the exposure of the NELA's unethical public relations and political campaigns. (The industry dissolved the discredited NELA and reconstituted it as the Edison Electric Institute in 1933.) At the same time Samuel Insull—Edison's protégé who built Commonwealth Edison on political deals, business acumen, and state-of-the-art GE technology—became a scapegoat for the depression's financial misdealings when his multitiered holding-company pyramid of electric utilities crashed in 1932. Concurrent antitrust charges against RCA, the cartel formed by GE and other companies, meant that GE and the rest of the electrical industry were under fire politically for a decade or more after Steinmetz's death.[87]

One response, as in the past, was to publicize Steinmetz in order to humanize the corporation. GE had better resources for the task after his death. It merged the Publication and Advertising Departments into the Publicity Department in November 1923, with Martin Rice as manager, and created an advertising council in December with Bruce Barton as its chief outside adviser. The new department lost no time in capitalizing on Steinmetz's image as Modern Jove. In 1924 it released a film of Steinmetz making lightning for Edison during his visit to GE and sent the famous photograph of the two Wizards examining the damage done by the lightning generator to three audiences—investors, workers, and the public—by publishing it in a stockholder pamphlet on his high-voltage work, an employee magazine, and a nationwide magazine advertisement entitled "Steinmetz." The pamphlet reprinted several tributes, including one by Commerce Secretary Herbert Hoover, and ended by quoting the *New York Times*'s comment that "no genius ever had [a] better" employer than GE.[88]

The Publicity Department expanded this work considerably through the work of John Winthrop Hammond. A former newspaperman, Hammond joined GE in 1920, the same year Clyde Wagoner started the News Bureau. He was interested in history and wrote news releases on Edison's visit to Schenectady and the fortieth anniversary of the Pearl Street central station in 1922. At this time he began to collect primary source material for a history of GE and interviewed several employees, including Steinmetz. He published the first book-length biography of Steinmetz in 1924 and a history of GE in 1941.[89]

We can better understand the underlying ideology of Hammond's biography if we look at notes he made in the mid-1920s for his study of GE. He thought the "legend of Prometheus has some place in the picture," as did the "earliest known, timid efforts of man, groping in the dimness, to penetrate the veiled realm of electricity. Behind the screen that hid this realm lay the making of a whole new world for mankind; lay the modern magic that has eclipsed the ancient lore of story tellers and fairy tales."[90] Hammond (or an editor) excised the purple prose from his books, but some of the magic seeped through in his biography of Steinmetz. Like many newspapermen, Hammond was awed by the Wizard of Schenectady and earnestly believed that famous men courted his advice. Consequently, he published unretouched photographs of the visits of

Steinmetz, Einstein, and company inspecting the RCA radio station at New Brunswick, New Jersey, 1921. Source: Sarnoff Library, Princeton, N.J.

Retouched photograph of Steinmetz and Einstein, 1924. Source: Hall of History Foundation, Schenectady, N.Y.

Edison and Marconi to the Schenectady Works, but also a doctored group photograph showing only Einstein and Steinmetz. Brushed out of the original photograph, taken during Einstein's 1921 visit, were some fairly prominent people, including David Sarnoff, Irving Langmuir, and Ernst F. W. Alexanderson, chief engineer of RCA and Steinmetz's protégé.[91]

GE continued to publicize Steinmetz regularly in the 1920s and 1930s. The company gave $25,000 to establish four scholarships at Union College as a memorial to Steinmetz in 1924. Two years later Hammond wrote a biographical magazine article and a book-length "boy's life" of Steinmetz (emulating the 1911 boy's life of Edison).[92] The staff artist of the Publicity Department painted a portrait of Steinmetz from a photograph in 1930, which GE gave to the newly established Charles P. Steinmetz High School in Chicago in 1936.[93] In Schenectady Wagoner tried to keep Steinmetz's "memory alive with stories with some new angle in the local press every April and October, dates of his birth and anniversary of his death." GE orchestrated a three-day celebration of his birthday in 1934—the fiftieth anniversary of the AIEE—marked by public tours of the Wendell Avenue house and speeches by Whitney and Dugald Jackson. Jackson wrote a GE engineer after the meeting, "Since the Publicity Department of the General Electric Company are anxious to make as much of Steinmetz's reputation as possible[,] why would it not be desirable for them to issue a Steinmetz number of the General Electric Review" reprinting the speeches. The *Review* was booked that year and published a special photograph section on Stein-metz the next year on the seventieth anniversary of his birth.[94]

GE took advantage of the 1939 New York World's Fair to place his name before a much larger audience. Crowds of people from the United States and abroad walked past the GE building, outside of which a twelve-ton stainless steel lightning bolt, 129 feet tall, "struck" a pool of water below. The thunderbolt drew visitors to the main attraction of the exhibit, Steinmetz Hall, in which two lightning generators discharged 10 million volts of electricity across a thirty-foot gap at regular intervals. The "thunderclap" accompanying the spectacular fireworks in the "caver-nous, weirdly lit chamber full of mysterious machinery resembling some-thing out of a horror movie" (newspaper writers now had a modern referent for their alchemical images) came from generators whose basic design was the same as that pioneered by Steinmetz nearly twenty years before. The GE exhibit was one of the most sensational and popular ones at the fair, attracting over 7 million people out of about 26 million fairgoers. Steinmetz Hall drew as many as twenty-two thousand people a day and its engineers, "mortal Joves" in the words of the *New York Times*, put on over four thousand demonstrations, the climax of which was the "creation of crackling, zig-zagging arcs of electricity, colored red, green and yellow by vaporized chemicals." Steinmetz Hall overshadowed the other parts of GE's exhibit: electrical appliances, including a television studio, and wonders from Whitney's laboratory, highlighted by a levitating aluminum dish. The "House of Magic," as the GE Publicity Department dubbed the laboratory, had been well received at the 1933 Century of Progress Exhibition in Chicago and drew many more visitors in 1939. But the huge lightning generator eclipsed the House of Magic in New York, just as Steinmetz proved to be better copy than Whitney in real life.[95]

Steinmetz Hall marked the peak of his posthumous fame with the general public, when public relations efforts could count on at least two generations of Americans who had read about the Jove of Schenectady. The decline began with an ill-fated project to turn the Wendell Avenue house into a museum. Prominent Schenectady citizens suggested the measure in 1934, when Hayden announced his intention to sell the house. Two years later the state legislature appropriated $2,000 for the state's Conservation Department to maintain the house for the first year as a museum if it was purchased from Hayden and deeded to the state. The one-year appropriation lapsed before the Schenectady Junior Chamber of Commerce could obtain the money, but it raised the necessary $25,000—including donations from GE and $100 from students at the Steinmetz High School in Chicago—and bought the house in 1938, as Steinmetz Hall was being planned for the World's Fair. Apparently because the maintenance cost was projected to be too high, the state did not follow through and vandals ransacked the house to such an extent that the state declared it beyond preservation in 1941. The legislature appropriated money that year to transfer fifteen hundred of Steinmetz's books to the state library in Albany and to demolish the house and landscape the grounds as Steinmetz Memorial Park. Neither GE nor the citizens of Schenectady rallied enough support to the cause, and the house was razed in the fall of 1944. A park was created on the site, and GE placed a bronze plaque commemorating Steinmetz on a stone erected by the state in 1945.[96]

This rather pathetic affair notwithstanding, GE did not relinquish Steinmetz as a symbol. The company printed at least two Steinmetz booklets in the 1940s and 1950s and established a committee to plan celebrations for the centennial of his birth in 1965. The committee did not succeed in petitioning Congress to have a postage stamp issued in his honor, but it supported the publication of several magazine articles. One in *Life* featured the brushed-out Einstein photograph, which had become a standard image in the iconography of Steinmetz. Poetic license was also evident in a 1950s GE pamphlet entitled "Steinmetz: Latter-Day Vulcan." The cover showed a drawing of Steinmetz dressed like a 1950s engineer in white shirt and tie, patiently piecing together the celebrated mirror struck by lightning at Camp Mohawk.[97] Ironically, the Vulcan title contradicted the myth of the Promethean inventor. Vulcan forged Jove's thunderbolts, but he was also the Roman counterpart of Hephaestus, the lame (hunchbacked?) Greek god of fire who bound Prometheus for stealing fire from his forge and giving it to humankind. The confusion in mythology is matched by the incongruity of a shirt-and-tie technocratic Wizard playing with a mirror in the Garden of Camp Mohawk—an unsuccessful attempt, in my view, to resolve the conflict between individualism and bureaucracy.

GE skirted the issue of Steinmetz's socialism in these publications. Invariably, company writers retold the tale of Steinmetz fleeing Bismarck for democratic America to escape prosecution for his Socialist beliefs. Hammond had described his Socialist political campaigns in Schenectady and New York State, but later writers in the Publicity Department usually did not refer to this aspect of his life. W. W. Trench, secretary of GE, addressed the issue squarely in a radio broadcast in 1940: "Dr. Steinmetz was as you probably all know a life time Socialist." But "when he came to America where men knew freedom, and did not have to hold secret meetings and flinch with fear when a door opened, he dropped all this militant activity. He became an employee of a great corporation" and wrote about the "need for such large scale private enterprise."[98] In the hands of GE Steinmetz's socialism was either in the past, a corporate philosophy (which was closer to the mark), or an idealistic aberration that could be easily overlooked.

A good example of how easily it could be forgotten occurred in 1926. When Governor Gifford Pinchot proposed his quasi-public Giant Power scheme for Pennsylvania—which was similar to Steinmetz's plans in that electricity would be transmitted by high-voltage lines from mine-mouth power plants—critics charged that it would lead to socialism. One opponent, former congressman James Burke, went so far as to bring up the specter of Lenin: "Is Pennsylvania to lead America in following Russia into the dismal swamp of commercial chaos and financial disaster? Are we to forget our Franklins, our Westinghouses, our Edisons, our Stein-metzes, and all the geniuses whose names light up the horizon of industrial progress?"[99] Either Burke was unaware that Steinmetz had offered his help to Lenin four years before, or he dismissed it as the forgivable foible of an eccentric genius.

Socialists, of course, took a different view and continued to honor Steinmetz as a hero of the left, an interpretation that did not make it into mainstream publications because of the decline of socialism in the 1920s. The *New York Leader,* successor to the *New York Call,* ran a two-inch headline announcing his death across the top of its front page, the *Soviet Russia Pictorial* honored Steinmetz as a "true friend of Soviet Russia," and the League for Industrial Democracy established a college-essay prize in honor of its first vice president. The league, the Socialist party, and the Rand School held a memorial service in New York City, at which Norman Thomas, Algernon Lee, Harry Laidler, and other prominent Socialists were scheduled to speak. Eugene Debs was in the city and wrote a glowing tribute to Steinmetz, the "genius . . . planted in the deformed, weak and almost unsightly body of a dwarf." Although Debs knew him only by reputation, he felt in "close and intimate touch with him because of the spirit he breathed of sympathy and good will to all mankind." His Socialist achievements would have been greater, continued Debs, if he had not been "exploited by private interests for the accumulation of private fortunes." When Debs visited Schenectady in late 1925 he placed a wreath at the grave of Steinmetz.[100]

Whatever Socialists may have privately thought about Steinmetz's corporate socialism, most writers followed the example of Debs and claimed him for the movement. An early example is the book *Pioneers of Freedom,* written in 1929 to "meet a definite need—the need for new heroes for the young people of America." The heroes were an eclectic group of "radicals," including Thomas Paine, Henry George, Samuel Gompers, Debs, John Mitchell, and Steinmetz. The chapter on Steinmetz, "A Visitor from Olympus," began with the standard fare of the Jovian legend, described how a penniless immigrant with a "tragically distorted body" rose to be a great scientist, then veered from other accounts to observe that "Comrade Steinmetz" was a Socialist who believed that electricity would usher in the Cooperative Commonwealth. In 1931 the Socialist party listed Steinmetz, along with Karapetoff and Einstein, in a large group of prominent Socialists. Reminiscences by Ella Reeve Bloor and August Claessens glossed over their past differences with Steinmetz and generally ignored the contradiction of a Socialist being the chief consulting engineer for a large capitalist corporation, in order to keep the "celebrity" within the fold.[101]

Some authors were more critical of the legend. Science journalist Jonathan Leonard wrote in 1929 that Steinmetz's "reputation finally outgrew the actual man to such an extent that it lost even a family resemblance to him. Such publicity-built characters are not human; they have printers' ink in their arteries instead of blood."[102] John Dos Passos was more sympathetic in one of the biographical sketches, "Proteus," in his *USA* trilogy:

General Electric humored him, let him be a Socialist, let him keep a greenhouseful of cactuses lit up by mercury lights, let him have alligators, talking crows, and a gila monster for pets . . .

and the publicity department poured oily stories into the ears of the American public every Sunday and Steinmetz became a little parlor magician,

who made a toy thunderstorm in his laboratory and made all the toy trains run on time . . .

Steinmetz was a famous magician and he talked to Edison tapping with the Morse code on Edison's knee . . .

and he talked to Bryan about God on a railroad train

and all the reporters stood round while he and Einstein

met face to face,

but they couldn't catch what they said

and Steinmetz was the most valuable piece of apparatus General Electric had

until he wore out and died.[103]

The criticism of Leonard and Dos Passos did not alter the public image of Steinmetz as Modern Jove. In 1930 Henry Ford turned Steinmetz's cabin, where the mirror had been struck by lightning, into an authentic piece of Americana by moving it to his outdoor museum, Greenfield Village, and placing it not far from Edison's Menlo Park laboratory, which he had painstakingly reconstructed the year before. Both Wizards were now rooted in the Garden of Greenfield Village. Several hagiographers joined Ford in perpetuating the image of the hunchback who played with lightning. Even Leonard's 1929 biography was entitled *Loki*, after the wise, mischievous, half-human Norse god of fire, who understood Thor's thunderbolts better than Thor himself. A radio play, adapted for high school students in 1938, was called "Lightning in the Laboratory." John A. Miller's 1958 biography was entitled

Modern Jupiter, Floyd Miller's 1965 juvenile biography was called *The Man Who Tamed Lightning,* and the *Life* magazine piece was subtitled "The Thunderer's Legacy."[104]

Although these later efforts failed to make Steinmetz as well known after World War II as he had been in the 1920s and 1930s, he has not dropped out of sight. Electrical engineers, who use much of his technical work, still honor him, several juvenile biographies published in the 1950s and 1960s have kept the legend alive, and he has appeared in some recent books. Kurt Vonnegut, who worked as a publicity writer for GE, named the protagonist Paul Proteus of *Player Piano* (1952) after Steinmetz. E. L. Doctorow mentioned him in his historical novel *Ragtime* (1975). And in 1976 William Stevenson claimed that Steinmetz, whom he described as a Jewish refugee scientist from Germany, was a close friend and colleague of *A Man Called Intrepid*: Sir William Stephenson, the World War I flying ace, radio experimenter, and head of the secret British Security Coordination group in New York City during World War II.[105]

But Steinmetz has certainly not kept pace with Edison, Ford, Babe Ruth, and the other folk heroes of the "lowbrow" newspaper story. One reason is that his publicity rose to their level in only two years. The number of newspaper and magazine stories on Steinmetz increased rapidly in 1922, skyrocketed in 1923 when he died on Jupiter's Throne, then dropped precipitously. The pattern is similar to that for Edison, but the Wizard of Menlo Park maintained higher publicity levels over his fifty-year career (see the Appendix).

Another reason for Steinmetz's decline in popularity has to do with the shift in hero worship in the twentieth century from idols of production (in industry, politics, and the military) to idols of consumption (movie stars and sports figures, for example) noted by Wyn Wachhorst and others. All of the "lowbrows" fall into these categories. Edison, Ford, Steinmetz, and Harding were idols of production; Ruth and Dempsey were idols of consumption. One reason Edison and Ford were so popular is that they easily became idols of consumption as well. Edison's picture and name appeared on such everyday items as phonograph cylinder cases and light bulb advertisements. Ford's name was stamped on every Model T, in town and on the farm.[106] But what was there to consume in the case of Steinmetz? The square root of minus one? Lightning in the laboratory?

Mythmakers tackled the problem in 1983 when the government issued postage stamps of prominent electrical inventors: Steinmetz, Nikola Tesla, Edwin Armstrong, and Philo Farnsworth.[107] None is a household name today. Few people associate electric motors with Tesla, FM radio with Armstrong, or television with Farnsworth. Part of the reason is that they were not the sole inventors of these complex technologies, and since the Manhattan Project ended World War II the public has increasingly come to see invention as a large-scale cooperative effort. The stamp designer chose a sine wave of alternating current to represent Steinmetz's "inventions." The image was appropriate—and better than a lightning flash. But its abstruseness symbolizes the difficulty of popularizing his highly mathematical work, which previous imagemakers overcame by writing of the gnarled genius throwing thunderbolts about the GE laboratory.

Conclusion

$\mathbf{B}$ehind the Steinmetz legend, but much harder to popularize, were substantial contributions to the science of electrical engineering, the emergence of corporate capitalism, and the Progressives' attempt to engineer society—legacies more lasting than the fleeting fame of Modern Jove. Steinmetz was in a unique position to participate in these movements in the late nineteenth and early twentieth centuries.

When he decided to give up a career in mathematics for one in engineering in the late 1880s, electrical engineering was changing from a craft into a scientific discipline. Steinmetz joined other engineers trained in science to "translate" electrophysics into a form useful to engineers and to create new engineering knowledge. The favorable reception of his books and articles testifies to his success and to the advantage he had in being able to combine the skills of a mathematician with the practical experience of a corporate engineer who had taken out nearly two hundred patents. His complex-number method and theories of apparatus based on equivalent circuits are widely used today, because they are applicable to electronic circuits as well as to electric machines. His work on magnetic hysteresis and transient phenomena laid the foundation for future research. A personal characteristic that did not bode well for the longevity of his work—without the intervention of others—was a stubborn determination to stick to his own way of doing things. The trait led Arthur Kennelly to standardize a different vector diagram and Ernst Berg to return to Heaviside's mathematics after Steinmetz's death.

Steinmetz helped make electrical engineering more scientific by these technical contributions and also by introducing more rigorous mathematics and a scientific methodology into the profession. From this perspective his work illustrates the nature of engineering science described by Edwin Layton. Although Steinmetz often considered his theories to be applied science, my analysis shows a more complex Laytonian relationship between science and technology. His investigation of magnetic hysteresis is a good example of autonomous engineering research, his complex-number method represents the translation from scientific to technical knowledge, and his theories of apparatus were a synthesis of applied science, engineering research, and practical design experience in the best tradition of scientist-engineers like John Hopkinson. In these respects—and in the institutional environment of professional societies and industrial research laboratories that he helped create—his work was a twin of basic science that formed a connecting link between ideal physics and the artifacts of electrical engineering. But it was a fraternal rather than a mirror-image twin, because Steinmetz, like "pure" scientists, valued general theories more than specific knowledge. He placed his theory of induction machines, for example, above empirical design equations for individual devices. While both types of engineering knowledge were necessary for designing commercial equipment, academic and practicing engineers honored Steinmetz as a theorist. The international electrical engineering community held him and such colleagues as Kennelly and Gisbert Kapp in such high esteem precisely because of their many theoretical books and articles.

His technical contributions were closely linked to the incorporation of science and engineering at General Electric. Moving from the inventor-entrepreneur Rudolf Eickemeyer to GE symbolized a process occurring in American industry at the turn of the century. The industrial scientist and the design engineer helped institutionalize innovation in the corporation. Despite being absent from the plant a great deal, he was a major force in establishing the General Electric Research Laboratory and organized the Consulting Engineering Department, both of which set precedents followed by other corporations. GE provided a wealth of facilities, interesting work, trained colleagues, and an opportunity to make a technical reputation as high as (and usually greater than) that of electrical engineers in academia. The corporation became the home of Steinmetz and many top electrical engineers, including Benjamin Lamme and Ernst F. W. Alexanderson.

GE treated Steinmetz very well, of course, by funding a home laboratory and allowing him to teach at Union College and work summers at Camp Mohawk. He returned the favor in his efforts to engineer society. For Steinmetz, technical and social engineering were not separate endeavors. They had been inextricably linked in his thinking since his conversations with Heinrich Lux in the Breslau days, particularly the symbiosis between electricity and socialism. His reading of utopian literature and the writings of American Socialists who looked for a gradual evolution from proprietary capitalism to huge trusts on the road to the Cooperative Commonwealth supported this view, as did his experiences at GE, Union College, the American Institute of Electrical Engineers, and the Schenectady common council. The resulting theory in *American and the New Epoch* was clearly a product of its times. As a liberal form of corporatism, it embraced the welfare policies of a prevailing corporate liberalism without extending as far to the right as the Fascist corporate state. Herbert Hoover considered the theory—as embodied in the Swope Plan—to be a Fascist proposal, yet Hoover's own proposal of an "associative state" was a voluntary rather than a technocratic form of corporatism.

As we gain a better understanding of how corporatism and corporate liberalism have shaped the American political economy in the twentieth century, the debt of these ideologies to corporate engineers and managers like Steinmetz, Gerard Swope, and Owen Young should be considered more fully. Swope and Young understood the value of nurturing science and engineering in a corporate setting, of working toward a less radical version of *America and the New Epoch,* and of using Modern Jove to humanize the corporation in the face of antitrust charges. But they did not exploit an unwilling Steinmetz in accomplishing these goals, as was assumed by Eugene Debs and John Dos Passos. Steinmetz largely agreed with the philosophy of Swope, Young, and their predecessors. He instituted corporate reforms in engineering education, standards, and ethics, and worked for an improved welfare capitalism and universal electrification—in the United States and Russia—to reach the technocratic corporate utopia of the New Epoch.

Playing the part of the Electrical Wizard and writing popular articles allowed him to bring this message to a wide audience. When interviewed by newspapers Steinmetz rarely let an opportunity go by to illustrate how municipal socialism and electrification were leading the country toward a corporate state. Some of the earliest explanations of his corporatism (as the Corporation of the United States, for example), appeared in interviews on Schenectady politics. His technological predictions described an all-electric utopia that required a federally regulated electrical grid and government-owned power sources. Pastoral pictures of him working in a canoe at Camp Mohawk humanized the corporation, and lightning in the laboratory reinforced the image of the altruistic corporation. Wizards of Science were creating a better life for humankind by taming one of nature's most mysterious forces in a controlled corporate setting.

The connection between playing the Electrical Wizard and promoting corporate socialism illustrates one of the themes that dominated Steinmetz's later years and provided consistency in his thought and actions. Some other themes were popularization, technological determinism, efficiency, and the application of scientific methods to technical and social engineering. Most of these beliefs and proclivities—as well as a certain stubbornness and eccentric habits—have roots in his Breslau days. At the university he first considered socialism as a means to make a better world and also started to deal with his deformity. His handicap and political beliefs were probably not unrelated, as some magazine writers suggested, but archival sources do not adequately answer this question. What is clear is that he developed a public persona based on the incongruity between his outstanding intellect and his deformed body, a persona that worked to his advantage while protecting him much better from the harsh gaze of the public than the uncouth shell described by George Moser.

The contrast between his big brain and little body was at the heart of his public image. It was also indicative of the many contrasts in Steinmetz's life. He was the Socialist chief engineer of one of America's largest capitalist corporations, a rugged individual employed by a huge bureaucracy, an upholder of corporate values who worked at home or at a summer camp, an inveterate smoker who spent his summers swimming and canoeing. Steinmetz, of course, resolved many of these contradictions. He worked out a philosophy of autonomy through cooperation between his profession and the commercial demands of his employer. He created a theory of corporate socialism, gained freedom from bureaucratic constraints at home and at his camp, and favored welfare capitalism over labor unions. These contrasts and their resolution are what made him so enticing to mythmakers, who intuitively understood that a cultural hero embodying these paradoxical realities could resolve cultural contradictions. He did not resolve other contradictions, however, such as being honorary president of the Russian Chamber of Commerce during the war while he was supporting Germany.

Such a glaring human inconsistency counters Jonathan Leonard's statement that "such publicity-built characters are not human; they have printers' ink in their arteries instead of blood." Yet after reading through Steinmetz's large collection of newspaper clippings, I can sympathize with Leonard's characterization. Ironically, the Wizard of Schenectady, the Modern Jove created to humanize GE, does not appear to be human in these stories. He was a socially constructed figure, the Little Cripple with the Giant Mind sitting on high in Jupiter's Throne, hurling thunder-bolts for cameramen and newspaper reporters serving a soulless corpora-tion. His interviews with John Hammond and the many photographs he took at Liberty Hall, Wendell Avenue, and Camp Mohawk show a differ-ent side. He was close to Lux, Oskar Asmussen, and Edward Mueller, called the Bergs his family, took care of his stepsister Clara all her life, and worked hard for an improved school system in Schenectady. When he adopted the Haydens as his new family at the turn of the century, be became "Daddy" to the young engineer and "Grandfather" to his chil-dren. The photographs clearly show these close relationships (although not between Steinmetz and Mrs. Hayden). Hayden and his children did not think grandfather Steinmetz had printers' ink for blood.

Yet Steinmetz enjoyed being a public figure—as shown by his well-kept collection of newspaper clippings. His public image and his articles and books on technical and social engineering were vital to him. They were the primary means by which he worked as an engineer and a Social-ist to realize his vision of a technocratic corporate utopia.

Appendix

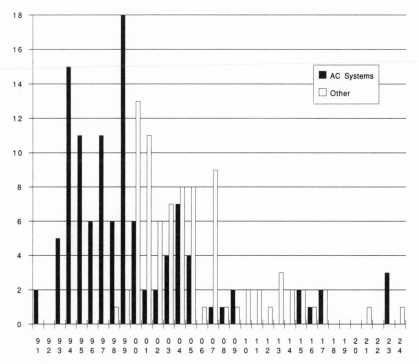

Figure A.1 Steinmetz's patents, number filed each year. The category "AC systems" covers electrical machinery, transformers, and transmission and distribution networks. The category "other" consists of protective devices, electric lighting, turbines, electrochemical subjects, and distribution systems for arc-lighting and mercury-arc rectifiers.

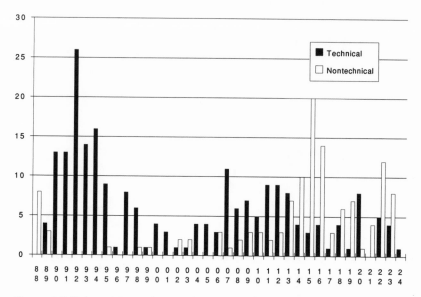

Figure A.2 Steinmetz's articles, number published each year. Articles on popular science, the engineering profession, engineering education, and engineering economics have been placed in the "nontechnical" category. Newspaper editorials have been included.

Table A.1 Steinmetz's books, number published each year

Year	Technical	Technical Revised	Nontechnical
1888	0	0	1
1897	1	0	0
1898	0	1	0
1900	0	1	0
1901	1	0	0
1902	0	1	0
1908	1	1	0
1909	2	1	0
1910	0	1	0
1911	2	1	0
1914	0	1	0
1915	0	2	0
1916	0	2	1

Table A.1　*(continued)*

Year	Technical	Technical Revised	Nontechnical
1917	2	1	0
1918	0	3	0
1920	0	1	0
1923	0	0	1
Total	9	17	3

Note: The 1923 book on the theory of relativity has been placed in the nontechnical category because it was a popularization. Translations are not included.

Table A.2　Number of Entries in Reader's Guide to Periodical Literature by or about Steinmetz and Edison

Year	Steinmetz	Edison
1920	3	4
1921	2	16
1922	13	10
1923	10	3
1924	8	5
1925	1	11
1926	0	6
1927	0	11
1928	1	12
1929	3	25
Total	41	103

Note: The figures include items about the Edison employment questionnaire (10 in 1921, 3 in 1922) that were listed under a separate heading. The table does not contain entries about Edison's companies or members of his family.

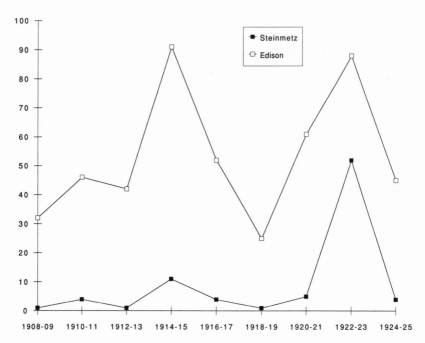

Figure A.3   Number of *New York Times* articles on Steinmetz and Edison, 1908–1925. *Source: New York Times Index*, for Steinmetz; and Wyn Wachhorst, *Thomas Alva Edison*, pp. 231–232, for Edison.

Notes

Abbreviations

AC   August Claessens Papers, Tamiment Library, New York University
AIEE   American Institute of Electrical Engineers
BL   Burndy Library, Norwalk, Connecticut
CPSC   C. P. Steinmetz Papers, History Center, City Hall, Schenectady, New York
CPSF   C. P. Steinmetz Papers, Henry Ford Museum, Dearborn, Michigan
CPSG   C. P. Steinmetz Papers, General Electric R&D Center, Schenectady, New York
CPSS   C. P. Steinmetz Papers, Schenectady County Historical Society, Schenectady, New York
CPSU   C. P. Steinmetz Papers, Schaffer Library, Union College, Schenectady, New York
CUA   Cornell University Archives, Ithaca, New York
CULA   Archives, Cooper Union Library, New York, New York
DAB   *Dictionary of American Biography*
DCJ   Dugald C. Jackson Papers, University Archives, Massachusetts Institute of Technology
EJB   Ernst J. Berg Papers, Schaffer Library, Union College, Schenectady, New York
ETA   Elihu Thomson Papers, American Philosophical Society, Philadelphia, Pennsylvania
ETG   Elihu Thomson Papers, General Electric Co., Lynn, Massachusetts
ETHA   Eidgenössisches Polytechnikum Rector Archives, Zurich, Switzerland
ETZ   *Elektrotechnische Zeitschrift*
EWR   Edwin W. Rice, Jr., Papers, Schaffer Library, Union College, Schenectady, New York
GEH   Hall of History Foundation, Schenectady, New York
GER   Historical Collection, General Electric R&D Center, Schenectady, New York
GS   Gerard Swope Papers, General Electric R&D Center, Schenectady, New York

HUA	Harvard University Archives, Cambridge, Massachusetts
HVV	Hawley B. Van Vechten Papers, History Center, City Hall, Schenectady, New York
HWB	H. W. Bibber Papers, Schaffer Library, Union College, Schenectady, New York
IEEE	Institute of Electrical and Electronics Engineers
IEEEA	Archives, Institue of Electrical and Electronics Engineers, Piscataway, New Jersey
ISS	Intercollegiate Socialist Society
ISSP	Intercollegiate Socialist Society Papers, Tamiment Library, New York University
JWH	John W. Hammond Collection, Hall of History Foundation, Schenectady, New York
MHA	McGraw-Hill Corporate Reference Center, New York, New York
NACS	National Association of Corporation Schools
NACB	*National Cyclopaedia of American Biography*
NYSL	New York State Library, Albany, New York
OCR	Otis Company Records, United Technology Corporate Archives, Hartford, Connecticut
ODY	Owen D. Young Papers, O. D. Young Library, St. Lawrence University, Canton, New York
RSR	Rand School Records, Tamiment Library, New York University (microfilm)
SC	Surrogates Court, Schenectady County Courthouse, Schenectady, New York
SCHS	Schenectady County Historical Society, Schenectady, New York
SL	Special Collections, Schaffer Library, Union College, Schenectady, New York
SPA	Socialist Party of America Papers, Duke University (microfilm)
SPN	S. P. Nixdorf Papers, Schaffer Library, Union College, Schenectady, New York
TAE	Thomas A. Edison Papers, Edison National Historic Site, West Orange, New Jersey
TAEF	Thomas A. Edison Papers, Henry Ford Museum, Dearborn, Michigan
TAEM	Thomas A. Edison Papers (microfilm)
TSF	Thomas S. Fiske Papers, Rare Book and Manuscript Library, Columbia University
Trans AIEE	American Institute of Electrical Engineers, *Transactions*
UTA	United Technology Archives, Hartford, Connecticut
VK	Vladimir Karapetoff Papers, Cornell University Archives,Ithaca, New York
WW	Woodrow Wilson Papers, Library of Congress (microfilm)

Chapter 1: German Origins

1. John W. Hammond, *Charles P. Steinmetz: A Biography* (New York: Century, 1924), pp. 7–17, 46–47; Hermann Markgraf, *Geschichte Breslaus in kurzer Ubersicht*, 2nd ed., ed. Otfried Schwarzer (Breslau, 1913); Baptism Certificate for Carl August Rudolph Steinmetz, 12 June 1888; Death Certificate for Carl August Steinmetz, 11 Jan. 1864; and Last Will of Carl Heinrich Steinmetz, 9 Oct. 1889, amended 10 Oct. 1889, item 4, CPSS.

2. *New York Times*, 1 Nov. 1923, p. 1.

3. Last Will of Carl Heinrich Steinmetz, items 2 and 3; and C. H. Steinmetz to C. P. Steinmetz, 7 Dec. 1888, 7 March, 21 July 1889, CPSS. Clara was emancipated as the ward of Carl Heinrich in 1878 at the age of twenty-one. See "Verfügung in der Vormundschaft Buchhalter Karl August Steinmetz," 23 Nov. 1878, CPSS.

4. G. J. Moser, "Charles Proteus Steinmetz: Personlicher Errinerlungen aus der Jugend des genialen Elektrikers," *Staats-Zeitung und Herold* (New York), 11 Nov. 1923. The translations are mine.

5. Königliche Prüfungskommission, "Zeugnis Reife" (school-leaving certificate), 29 Sept. 1883, CPSS.

6. Ibid. Hammond, or Steinmetz, confused first place in the mathematics and science examinations with first in the entire class. Compare the school-leaving certificate with Hammond, *Steinmetz*, pp. 40–41.

7. Georg Kaufmann, ed. *Festschrift zur Feier des hundertjährigen Bestehens der Universität Breslau*, 2 vols. (Breslau, 1911), 2: 633–634; and *Dictionary of Scientific Biography*, s.v. "Galle, Johann Gottfried."

8. Karl Steinmetz, "Universität Breslau, Anmeldungs-Buch," 22 Oct. 1883 to Summer Semester 1888, CPSS. All further references to Steinmetz's lecture courses, unless otherwise noted, are from this source.

9. Kaufmann, *Festschrift*, 2: 434–440. On Schröter, see also Florian Cajori, *A History of Mathematics*, 2nd ed. rev. (New York: Macmillan, 1931), pp. 314, 416.

10. Kaufmann, *Festschrift*, 2: 440–448. On Kirchhoff and Julius Meyer, see *Dictionary of Scientific Biography*. On Oskar Meyer, Weber, and Auerbach, see Poggendorff's *Biographisch-Literarisches Handwörterbuch der exakten Naturwissenschaften*.

11. Karl Steinmetz, "Mathematische Theorie der Elektricität und des Magnetismus von Dr. Auerbach, Breslau," Winter Semester 1884–1885; and Steinmetz, "Magnetismus—X," 2 Oct. 1884, CPSS.

12. Georg Dettmar, *Die Entwicklung der Starkstromtechnik in Deutschland*, 2 vols. (Berlin: ETZ, 1940), 1: 185–193, 280; Thomas P. Hughes, *Networks of Power: Electrification in Western Society, 1880–1930* (Baltimore: Johns Hopkins Univ. Press, 1983), pp. 66–72, 144–145; Kittler, *Handbuch der Elektrotechnik*, 2 vols. (Stuttgart, 1886–1890); and *Deutsches Biographisches Jahrbuch*, 1929, s.v. "Kittler, Erasmus."

13. Loose pages of notebook, undated, in CPSS; O. E. Meyer and F. Auerbach, "Ueber die Theorie der dynamoelektrischen Maschine," *ETZ*, 7 (1886): 240–244; and Auerbach, *Die Wirkungsgesetze der dynamo-elektrische Maschinen* (Vienna, 1887).

14. Otto Mahr, *Enstehung der Dynamomaschine* (Berlin: Springer, 1941), pp. 53–56, 100; Silvanus P. Thompson, *Dynamo-Electric Machinery*, 2 vols., 7th ed. rev. (London: E. and F. Spon, 1904–1905), 1: 31–34; J. Clerk Maxwell, "On the Theory of the Maintenance of Electric Currents by Mechanical Work without the Use of

Permanent Magnets," *Proceedings of the Royal Society,* 15 (1867): 397–402; and R. Clausius, "Zur Theorie der dynamoelektrische Maschine," *Annalen der Physik und Chemie,* new series, 20 (1883): 354–390. Frölich's theories are discussed in Kittler, *Handbuch,* 1: 368–397.

15. Hammond, *Steinmetz,* ch. 4, esp. pp. 57–58. Jonathan Leonard, *Loki: The Life of Charles Proteus Steinmetz* (Garden City, N.Y.: Doubleday, 1929), pp. 35–37, says that Steinmetz knew from the *Odyssey* that Proteus was hunchbacked. None of the versions I consulted described Proteus in this manner. The copy of the *Bier Zeitung* is in CPSS. The deformity did exempt him permanently from military service; see Königliche Ober-Ersatzkommission im Bezirt der 21th Infanteriebrigade, "Ausmusterungs-Schein," 3 July 1885, CPSS.

16. The certificates, notes, and papers are in CPSS. The *Stipendium* was awarded for continual residence for four years and for obeying the academic regulations. No monetary value is given on the certificate.

17. Steinmetz, "Ueber die durch ein lineares Flachensystem n^{ter} Ordnung definirten mehrdeutigen involutorischen Raumverwandtschaften," *Zeitschrift für Mathematik und Physik,* 35 (1890): 219–236, 272–292, 354–375; Steinmetz, "Multivalent and Univalent Involutory Correspondences in a Plane Determined by a Net of Curves of nth Order," *American Journal of Mathematics,* 14 (1892): 39–66; and Steinmetz, "On the Curves Which are Self-Reciprocal in a Linear Nulsystem, and Their Configurations in Space," ibid., pp. 161–186.

18. Steinmetz, "Anmeldungs-Buch," pp. 16–21.

19. Proteus [Steinmetz] to Hinz and Mrs. Hinz [Heinrich Lux], 25 Oct. 1893, in Justin G. Turner, "Steinmetz," *Manuscripts,* 15 (Fall 1963): 27–34, on p. 31.

20. Guenther Roth, *The Social Democrats in Imperial Germany: A Study in Working-Class Isolation and Rational Integration* (Totowa, N.J.: Bedminister Press, 1963), pp. 73, 267.

21. Sylvester A. Piotrowski, *Etienne Cabet and the Voyage en Icarie* (Washington, D.C.: Catholic Univ. of America, 1935), chs. 4–5; Christopher H. Johnson, *Utopian Communism in France: Cabet and the Icarians, 1839–1851* (Ithaca: Cornell Univ. Press, 1974), ch. 1; and Howard P. Segal, *Technological Utopianism in American Culture* (Chicago: Univ. of Chicago Press, 1985), pp. 72–73.

22. Heinz Lux, "Der Breslauer Sozialistenprozess," in *Mit Gerhart Hauptmann,* ed. Walter Heynen (Berlin, 1922), pp. 69–82; Theodor Müller, *Die Geschichte der Breslauer Sozialdemokratie* (Breslau: Sozialdemokratischen Vereins Breslau, 1925), Pt. 2, pp. 207–212, 214; C. F. W. Behl and Felix A. Voigt, *Chronik von Gerhart Hauptmanns Leben und Schaffen* (Munich: Korn, 1957), pp. 17–20; Eberhard Hilscher, *Gerhart Hauptmann* (Berlin: Nation, 1969), pp. 37–40; Frederick W. J. Heuser, "Gerhart Hauptmann's Trip to America in 1894," *Germanic Review,* 13 (1938): 3–31; and Charles Nordhoff, *The Communistic Societies of the United States* (New York, 1875), pp. 333–339.

23. Hammond, *Steinmetz,* pp. 66–89, quotation on p. 66. My dating of these events comes from Lux, "Breslauer Sozialistenprozess," who does not mention Steinmetz, rather than from Hammond.

24. Horst Bartel, *Das Sozialistengesetz, 1878–1890* (Berlin: Dietz, 1980).

25. Theodor Müller, ed., *Die sozialdemokratische Presse Schlesiens* (Breslau: Herrmann, 1925), pp. 21–27, 31–34, 38–41.

26. Ibid., pp. 27–31, 34–37; Leo Stern, ed., *Der Kampf der deutschen Sozialdemokratie in der Zeit des Sozialistengesetzes, 1878–1890: Die Tätigkeit der Reichs-Commission,* 2 vols. (Berlin: Rütten and Loening, 1956), 2: 842–861; and Vernon L. Lidtke, *The Outlawed Party: Social Democracy in Germany, 1878–1890* (Princeton: Princeton Univ. Press, 1966), p. 190.

27. *Der Sozialdemokrat,* 15 July and 25 Nov. 1887; Müller, *Breslauer Sozialdemokratie,* pp. 214–218; and Hammond, *Steinmetz,* pp. 84–87, 91–92. Hammond publishes Steinmetz's recollection that the photograph was displayed in the photographer's window in 1884, where it was seen by the police, which differs from Müller's history based on the trial records.

28. Müller, *Breslauer Sozialdemokratie,* pp. 217–218, 230–231, 233, 271; and Hammond, *Steinmetz,* pp. 92–97.

29. *Der Sozialdemokrat,* 1 July, 12 and 19 Aug., 2 and 30 Sept., 4 and 25 Nov. 1887; and Müller, *Breslauer Sozialdemokratie,* pp. 234, 245–247, 252–254, 263, 276, 281.

30. Müller, *Breslauer Sozialdemokratie,* pp. 260, 270–271; *Der Sozialdemokrat,* 12 Aug. 1887; Hammond, *Steinmetz,* pp. 97–111; and Lidtke, *Outlawed Party,* pp. 141–142, 164, 263–265, 283–285.

31. C. H. Steinmetz to Herr Lehmann, 3 June 1888; C. H. Steinmetz to C. P. Steinmetz, 28 May 1889, CPSS; and Müller, *Breslauer Sozialdemokratie,* p. 271. Trappe later emigrated to Cleveland, Ohio, where he was an active Socialist up to his death in 1890. See Müller, *Breslauer Sozialdemokratie,* and C. P. Steinmetz to C. H. Steinmetz, 7 June 1890, CPSS.

32. Steinmetz, "Anmeldungs-Buch," passim.

33. Steinmetz started at $12 (approximately 48 marks) per week in 1889, which impressed his family in Breslau. See C. H. Steinmetz to C. P. Steinmetz, 20 Oct. 1889, CPSS. The exchange rate comes from C. H. Steinmetz to C. P. Steinmetz, 3 June 1890, CPSS.

34. See C. H. Steinmetz to C. P. Steinmetz, 7 Dec. 1888, 7 March, 20 May, 25 Aug., and 20 Oct. 1889, CPSS.

35. Gerhart Hauptmann, *Das Abenteuer meiner Jugend,* 2 vols. (Berlin: S. Fischer, 1937), 2: 445–447; Lux, "Breslauer Sozialistenprozess," pp. 70–71; and Philip J. Pauly, *Controlling Life: Jacques Loeb and the Engineering Ideal in Biology* (New York: Oxford Univ. Press, 1987), pp. 45–46.

36. "Autobiography of Mr. Charles P. Steinmetz," *Electrical World,* 41 (1903): 524–525. The articles were published from 22 July 1888 to 31 March 1889. Copies are in CPSS. The book was *Astronomie, Astrophysik, und Meteorologie* (Volks-Bibliothek, n.d.), listed in Oskar Wekel, ed., *Vollständiges Bücher—Lexicon* (Leipzig, 1891), Pt. 2 of 10th Supplement, p. 538.

37. C. H. Steinmetz to C. P. Steinmetz, 7 Dec. 1888 and 7 Mar. 1889, CPSS; and Steinmetz to Clara Steinmetz, 27 Oct. 1890, CPSF.

38. "Autobiography of Steinmetz," p. 524.

39. Kgl. Regierungs-Präsident to C. H. Steinmetz, 2 Aug. 1888; and Die Gemeindrathskanzlei, "Schriften-Empfangschein," 22 Oct. 1888, CPSS.

40. Eidgenössisches Polytechnikum, "Matrikel-Auszug," 16 March 1889, CPSS; see also "Matrikel [for Karl Steinmetz]," ETHA.

41. Karl Steinmetz, Reading Notebook, June 1888 to Feb. 1889, pp. 1–3, 5, 19,

and last page; and C. H. Steinmetz to C. P. Steinmetz, 7 March 1889, CPSS. Class handouts on arc-lighting systems and his copy of Uppenborn's *Kalender* (1887) are also in CPSS. Although the *Kalender* has a datebook for 1887, Steinmetz probably used it in Zurich in 1888, because its back page contains names and addresses of Breslau friends, followed by those of people he met in Zurich.

42. C. H. Steinmetz to C. P. Steinmetz, 21 July and 1 Sept. 1889, CPSS.

43. Hughes, *Networks of Power*, pp. 86–105; Dettmar, *Starkstromtechnik*, 1: 83, 106–107; and Harold C. Passer, *The Electrical Manufacturers, 1875–1900* (Cambridge, Mass.: Harvard Univ. Press, 1953), pp. 121, 149.

44. G. B. Prescott, "Discussion," *Trans AIEE*, 5 (1888): 38; and J. Clerk Maxwell, "A Dynamical Theory of the Electromagnetic Field," *Philosophical Transactions of the Royal Society*, 155 (1865): 459–512.

45. This characterization of electrical engineering knowledge is an extended and modified version of that given by me in "Charles P. Steinmetz and the Development of Electrical Engineering Science" (Ph.D. diss., Univ. of Wisconsin—Madison, 1983), pp. 13–32. John Staudenmaier independently derived a similar but broader conception of technological knowledge. See his *Technology's Storytellers: Reweaving the Human Fabric* (Cambridge, Mass.: MIT Press, 1985), pp. 103–120.

46. Steinmetz, "Ueber den scheinbaren Widerstand stromdurchflossener Leitungsdräte," *Centralblatt für Elektrotechnik*, 11 (1889): 105–109; and "Grundzüge einer Theorie der Transformatoren," *Centralblatt für Elektrotechnik,*, 12 (1889): 171ff. Although published after Steinmetz arrived in America, both articles are signed "K. Steinmetz, Zurich." On his having written the article at the request of Uppenborn, see Steinmetz, "On the Efficiency of Thermo-magnetic Machines," *Electrical World*, 14 (1889): 48.

47. Maxwell, "Dynamical Theory," p. 473; John Ambrose Fleming, *The Alternate-Current Transformer in Theory and Practice*, 2 vols. (London, 1889–1892), 1: 268–269; John Hopkinson, "Note on Induction Coils or 'Transformers,'" *Proceedings of the Royal Society*, 42 (1887): 164–167; John David Miller, "Rowland's Magnetic Analogy to Ohm's Law," *Isis*, 66 (1975): 230–241; and D. W. Jordan, "The Magnetic-Circuit Model, 1850–1890: The Resisted Flow Image in Magnetostatics," *British Journal for the History of Science*, 23 (1990): 131–173.

48. James E. Brittain, "B. A. Behrend and the Beginnings of Electrical Engineering, 1870–1920" (Ph.D. diss., Case Western Reserve Univ., 1970), pp. 36–51; Hughes, *Networks of Power*, pp. 84, 146–147; Edison Notebook No. 8, 11 June 1879, pp. 212, 232, reel 29: 904, 914; and Edison Notebook No. 37, 25 May 1880, p. 197, reel 32: 344, TAEM.

49. Steinmetz, "Theorie der Transformatoren"; and Douglas S. Martin, "Steinmetz and His Discovery of the Hysteresis Law," *General Electric Review*, 15 (1912): 544–551.

50. "Autobiography of Steinmetz." On Lux's engineering career, see the obituary in *ETZ*, 66 (1945): 45. His Socialist writings include *Sozialpolitisches Handbuch* (Berlin, 1892) and *Etienne Cabet und der ikarische Kommunismus* (Stuttgart, 1894). See also Theodor Müller, *45 Führer aus den Anfängen und dem Geldenzeitalter der Breslauer Sozialdemokratie* (Breslau: Herrmann, 1926), pp. 112–114.

51. Wolfgang König, "Friedrich Engels und 'Die elektrotechnische Revolution': Technikutopie und Technikeuphorie im Sozialismus in den 1880er Jahren," *Tech-*

nikgeschichte, 56 (1989): 9–37; James W. Carey and John J. Quirk, "The Mythos of the Electronic Revolution," *American Quarterly*, 39 (1970): 219–241, 395–424; Segal, *Technological Utopianism*, pp. 23–26; Thomas P. Hughes, "The Industrial Revolution That Never Came," *American Heritage of Invention and Technology*, 3 (Winter 1989): 58–64; and Lux, *Die wirtschaftliche Bedeutung der Gas- und Elektrizitätwerke in Deutschland* (Leipzig, 1898).

52. C. H. Steinmetz to C. P. Steinmetz, 7 March, 28 May, and 17 June 1889; Eidgenössisches Polytechnikum, "Entlassungs-Zeugnis," 30 April 1889, CPSS; Steinmetz to Lux, 25 Oct. 1893, and "Autobiography of Steinmetz," p. 524. The date for his departure is in the daily-diary portion of his copy of F. Uppenborn, ed., *Kalender für Elektrotechniker* (Munich, 1889), CPSS.

Chapter 2: Eickemeyer's

1. Proteus [Steinmetz] to Hinz and Mrs. Hinz [Heinrich Lux], 25 Oct. 1893, in Justin G. Turner, "Steinmetz," *Manuscripts*, 15 (Fall 1963): 27–34, on p. 31. The arrival date is in Steinmetz to Clara Steinmetz, 20 May 1889, CPSF.

2. See, for example, John W. Hammond, *Charles P. Steinmetz: A Biography* (New York: Century, 1924), pp. 129–130.

3. Ernst J. Berg, "Charles Proteus Steinmetz," 23 April 1931, EJB. On the amount of the fare, see "Autobiography of Mr. Charles P. Steinmetz," *Electrical World*, 41 (1903): 524–525.

4. Hammond, *Steinmetz*, pp. 131, 136; and Steinmetz to Frank Cousins, 21 May 1914, CPSS. On Asmussen's profession, see his business card, printed "Mechanical Engineer," in CPSS.

5. Hans-Joachim Braun, "The National Association of German-American Technologists and Technology Transfer between Germany and the United States, 1884–1930," *History of Technology*, 8 (1983): 15–35; and *Der Techniker*, 11 (1889): 136; 12 (1889): 4. Braun mentions Steinmetz's membership in his "A Technological Community in the United States: The National Association of German-American Technologists, 1884–1941," *Amerikastudien*, 30 (1985): 447–463, but he cites an 1892 date for his membership. I would like to thank Gregory Dreicer for bringing this last article to my attention.

6. "Autobiography of Steinmetz," p. 525.

7. Fred H. Whipple, *The Electric Railway* (Detroit, 1889), pp. 95–113; and Harold C. Passer, *The Electrical Manufacturers, 1875–1900* (Cambridge, Mass.: Harvard Univ. Press, 1953), pp. 117, 121, 149.

8. Robert Rosenberg, "Test Men, Experts, Brother Engineers, and Members of the Fraternity: Whence the Early Electrical Work Force?" *IEEE Transactions on Education*, E-27 (1984): 203–210; Joseph W. Slade, "The Man behind the Killing Machine [Hiram Maxim]," *American Heritage of Invention and Technology*, 2 (Fall 1986): 18–26; William Le Roy Emmet, *Autobiography of an Engineer* (New York: ASME, 1940); James E. Brittain, ed., *Turning Points in American Electrical History* (New York: IEEE Press, 1977), pp. 102, 107, 112, 120, 135; and Passer, *Electrical Manufacturers*, pp. 14, 31, 129–130, 230.

9. Thomas P. Hughes, *Networks of Power: Electrification in Western Society* (Baltimore: Johns Hopkins Univ. Press, 1983), pp. 142–157; Robert Rosenberg, "The Origins of EE Education: A Matter of Degree," *IEEE Spectrum*, 21 (July 1984): 60–

68; Monte A. Calvert, *The Mechanical Engineer in America, 1830–1910: Professional Cultures in Conflict* (Baltimore: Johns Hopkins Press, 1967), pp. 54, 215–217, quotation on p. 215; A. Michal McMahon, *The Making of a Profession: A Century of Electrical Engineering in America* (New York: IEEE Press, 1984), chs. 1–2; and Benjamin Garver Lamme, *An Autobiography* (New York: G. P. Putnam's Sons, 1926), p. 21.

10. *Dictionary of Scientific Biography* (Pupin); *DAB* (Bell, Crocker, and Ryan); "Carl Hering," *Electrical Engineer*, 10 (1890): 329; Robert Friedel et al., *Edison's Electric Light: Biography of an Invention* (New Brunswick: Rutgers Univ. Press, 1986), pp. 36–37; Rosenberg, "Origins of EE Education," p. 65; and Brittain, *Turning Points*, pp. 79, 131.

11. John W. Servos, "Mathematics and the Physical Sciences in America, 1880–1930," *Isis*, 77 (1986): 611–629.

12. Steinmetz to Lux, 25 Oct. 1893, p. 31; and Steinmetz, "[Eickemeyer] Time Book," 10 June 1889, p. 1, BL. On his application to the Edison Machine Works, see the daily-diary portion of his copy of F. Uppenborn, ed., *Kalender für Elektrotechniker* (Munich, 1889), entry for 30 June (used as address space), CPSS; and Hammond, *Steinmetz*, p. 131. His new address is on a postcard from C. H. Steinmetz to C. P. Steinmetz, 21 July 1889, CPSS.

13. Steinmetz to Lux, 25 Oct. 1893, p. 31; and Hammond, *Steinmetz*, p. 148.

14. Steinmetz, "Rudolf Eickemeyer," *Electrical World*, 25 (1895): 331–332; "Rudolf Eickemeyer," *Trans. AIEE*, 12 (1895): 668–669; *DAB*, vol. 6 (1931), s.v. "Eickemeyer, Rudolf"; Charles E. Allison, *The History of Yonkers* (New York, 1896), pp. 185–186, 350; David B. Sicilia, "Technological Risk in Manufacturing, 1850–1880, Yonkers, New York: A Case Study," Brandeis Univ., Dept. of the History of American Civilization, n.d., p. 10; and a list of Eickemeyer's U.S. patents kindly provided by Mr. Sicilia.

15. "Rudolf Eickemeyer," *Trans AIEE*, 12 (1895): 668–669; *Electrical Engineer*, 7 (1888): 84–85; and *Electrical World*, 11 (1888): 213. On the armature winding, see *Electrical Engineer*, 10 (1890): 358–359; and Malcom Maclaren, *The Rise of the Electrical Industry during the Nineteenth Century* (Princeton: Princeton Univ. Press, 1943), p. 95.

16. U.S. Patent Nos. 358,340 (filed 2 March 1885, issued 22 Feb. 1887) and 377,996 (filed 11 May 1887, issued 14 Feb. 1888); *Electrical Review*, 13 (1888): 5; 14 (1889): 8; 15 (1890): 1; *Electrical World*, 15 (1890): 313; *Electrical Engineer*, 10 (1890): 669–670; and J. H. Bunnell and Co., *Eickemeyer Dynamos and Electric Motors* (New York, c. 1889), OCR.

17. Passer, *Electrical Manufacturers*, pp. 212, 216–218, 227, 258–259; and B. G. Lamme, *Electrical Engineering Papers* (Pittsburgh, 1919), pp. 721–753.

18. *Electrical Engineer*, 7 (1888): 84–85; 10 (1890): 691–692; 12 (1891): 619–622; Passer, *Electrical Manufacturers*, pp. 220–221; Agreement between Stephen D. Field and Thomas A. Edison, 26 April 1883, reel 86: 555–561; and Geo. A. Bliss to Edison, reel 70: 100, TAEM. On Field, see *DAB*, vol. 3 (1930).

19. *Electrical World*, 16 (1890): 3; 18 (1892): 79; Joseph Sachs, "Electric Elevators," *Cassier's Magazine*, 13 (1895): 387–408; and Thomas E. Brown, "Passenger Elevators," *Transactions of the American Society of Civil Engineers*, 52 (1904): 133–186, on pp. 134–136, 162. I would like to thank Anne Millbrooke for the last two references.

20. New York State Labor Dept., Inspection Bureau, *Annual Report* (New York, 1892), p. 325; *Turner's First Annual Directory of the City of Yonkers* (Yonkers, 1891/92); Steinmetz, "Eickemeyer's Differential Magnetometer," *Electrical Engineer*, 11 (1891): 353–355; Rudolf Eickemeyer, "Some Remarks on the Magnetic Circuit in Its Relation to a New Type of Dynamo Machine," 3 March 1887, in Bunnell, *Eickemeyer Dynamos*, pp. 29–48; and Andre Millard, *Edison and the Business of Innovation* (Baltimore: Johns Hopkins Univ. Press, 1990), chs. 1–2.

21. Steinmetz, "Time Book," passim.

22. *Electrical Engineer*, 12 (1891): 619–622; Passer, *Electrical Manufacturers*, pp. 243, 248, 254; and Steinmetz, "Time Book," 18 Oct. 1890, 15 April, 7 Dec. 1891. The Lynchburg cars climbed a grade of 11.25 percent, while the steepest grade of the Richmond system was 10 percent.

23. *Electrical World*, 19 (1892): 105; *Street Railway Journal*, 8 (1892): 136–138, 208, 391–393, 399–400; and *Electrical Review*, 20 (1892): 124; 21 (1892): 155; 22 (1893): 16.

24. Steinmetz, "Time Book," July to Sept. 1891; Steinmetz, "Discussion," *Trans AIEE*, 9 (1892): 156, 157, 162–163, 166; Steinmetz, Eickemeyer Abstracts Notebook, n.d., p. 28, CPSS; and Friedrich Tischendörfer, "Der Eickemeyer-Field-Strassenbahnmotor," *ETZ*, 13 (1892): 497–500.

25. Steinmetz, "Time Book," 15 and 26 May, 11 Sept. 1891; and Steinmetz, Eickemeyer Patent Notebook, 9 and 12 Jan., 5 July 1892, pp. 49–51, 58, CPSS.

26. Ronald Kline, "Science and Engineering Theory in the Invention and Development of the Induction Motor, 1880–1900," *Technology and Culture*, 28 (1987): 283–313.

27. U.S. Patent No. 448,326 (filed 20 July 1888, issued 17 March 1891).

28. C. P. Steinmetz to C. H. Steinmetz, 7 June 1890, CPSS.

29. On the books by Kapp and Uppenborn, see Accession Nos. 607468, 607063, and 607064, NYSL. Steinmetz ordered vol. 2 of Kittler's handbook through his father; see C. H. Steinmetz to C. P. Steinmetz, 3 June 1890, CPSS.

30. Steinmetz, "Time Book," 17–19 Dec. 1890, 9–14 March 1891, and passim, 1890–1891; and Steinmetz, Abstracts Notebook, 17 July 1890, 6 Feb., 24 April, May/June 1891, pp. 1–2, 12, 29–31.

31. See Eickemeyer, U.S. Patent Nos. 567,119 (filed 31 Dec. 1890, issued 1 Sept. 1896), 700,310 (filed 10 Aug. 1892, issued 20 May 1902), and 664,733 (filed 21 Nov. 1892, issued 25 Dec. 1900); Steinmetz, Patent Notebook, 8 and 22 Feb. 1891, pp. 2–22; Steinmetz, Abstracts Notebook, n.d. and 8 Oct. 1892, pp. 34–35, 16; and Steinmetz, "The Alternating-Current Railway Motor," *Trans AIEE*, 23 (1904): 9–25, quotations on pp. 10–12.

32. Compare Steinmetz, Patent Notebook, 23 Oct. 1891, pp. 44–45, with U.S. Patent No. 700,310. On Steinmetz's pointing out the problem of self-induction, see a loose page of a patent statement in Steinmetz's handwriting, n.d., in the 1889 Folder, CPSS. This document gives 19 July 1889 as the date that Steinmetz conceived of the idea of exciting the compensating coil by induction.

33. Compare the description of the first motor in Steinmetz to A. G. Davis, 12 Jan. 1899, L5470–73, JWH, with U.S. Patent No. 541,604.

34. Filed on 31 Aug. 1891, the application was divided into two parts on 13 May 1897, representing phase transformation and the induction motor. See U.S. Patent

Nos. 630,418 and 630,419 (both issued 8 Aug. 1899); and *Electrical World*, 34 (1899): 276. A rough draft of the patent application, dated 22 Nov. 1890, is in CPSS.

35. Steinmetz to A. G. Davis, 12 Jan. 1899, CPSS; Steinmetz, Abstracts Notebook, n.d. and 30 Dec. 1892, pp. 24–27, 43–44; Arthur E. Kennelly and Edwin Houston, *Recent Types of Dynamo-Electric Machinery* (New York, 1895), pp. 517–519; and Kline, "Science and Engineering Theory," pp. 292–297.

36. "Thomas Holmes Blakesley," *Proceedings of the Royal Society*, 41 (1928–1929): 594–596; and D. G. Tucker, *Gisbert Kapp (1852–1922)* (Birmingham, England: Univ. of Birmingham Press, 1973).

37. J. Clerk Maxwell, "A Dynamical Theory of the Electromagnetic Field," *Philosophical Transactions of the Royal Society*, 155 (1865): 459–512, on pp. 469, 473.

38. Thomas Blakesley, *Papers on Alternating Currents of Electricity* (London, 1889), which was based on a series of articles in the *Electrician* in 1885; and Gisbert Kapp, "Induction Coils Graphically Treated," *Electrician*, 18 (1887): 502–504, 524–525, 568–571. As a vector revolves counterclockwise around the origin of a clock diagram with rectangular coordinates (x–y axes), its projection on the vertical axis corresponds to a sine curve, representing the alternating voltage or current. The length of the vector represents the magnitude of the sine wave. The vector's inclination to the horizontal axis represents the phase angle of the sine wave (where it starts relative to a reference voltage or current), and the speed of the vector's rotation represents the frequency of the sine wave. Since all voltages or currents in an AC power circuit have the same frequency, their rotating vectors are drawn as stationary vectors from a common origin. Their phases are shown in reference to the zero-phase line (the right half of the horizontal axis). To add voltages or currents, Blakesley and Kapp simply combined vectors graphically, using the well-known rules common to engineering graphical analysis. The rotating vector was called a *phasor* after World War II.

39. Thomas Blakesley, "Alternating Currents," *Electrician*, 14 (1885): 199–200; and Kapp, "Induction Coils," pp. 502–503, 524, 569. On the history of engineering graphics, see Hans Straub, *A History of Civil Engineering*, trans. E. Rockwell (London: L. Hill, 1952), pp. 197–201.

40. Steinmetz, "Elementary Geometrical Theory of the Alternate-Current Transformer," *Electrical Engineer*, 11 (1891): 627ff. on p. 627; 12 (1891): 12ff.

41. Francis Caldwell, "A Comparative Study of the Electrical Engineering Courses Given at Different Institutions," Society for the Promotion of Engineering Education, *Proceedings*, 7 (1899): 127–129.

42. Steinmetz, "Das Transformatorenproblem in elementar-geometrischer Behandlungsweise," *ETZ*, 11 (1890): 185–186, 205–206, 225–227, 233–234, 345–348; Steinmetz, "Elementary Geometrical Theory," pp. 650, 671; and Steinmetz, "Anwendung des Polardiagrams der Wechselstrom für inductive Widerstande," *ETZ*, 12 (1891): 394–396, 405–407.

43. Eidenössisches Polytechnikum, "Matrikel-Auszug [for Karl Steinmetz]," 16 March 1889, CPSS. The course was "Schiebersteuerungen," by Prof. Fliegner. Zeuner, a professor of applied mechanics at the Zurich polytechnic, had introduced the diagram as early as 1856. See Gustav Zeuner, "Ueber Coulissensteuerungen," *Civilingenieur Zeitschrift*, 2 (1856): 202–222.

44. Kapp, "Induction Coils"; and Kapp, "On Alternate-Current Trans-

formers . . . ," *Journal of the Society of Telegraph Engineers*, 17 (1888): 96–119, on p. 106.

45. Steinmetz, "Das Transformatorenproblem."

46. Steinmetz, "Elementary Geometrical Theory," pp. 698, 27–28, 277. Compare Steinmetz's theory with those by Kapp, George Forbes, and Fleming in John Ambrose Fleming, *The Alternate Current Transformer in Theory and Practice*, 2 vols. (London, 1889–1892), 1: 268–302.

47. *Electrical Engineer*, 11 (1891): 630; and *Electrical Review*, 20 (1892): 135.

48. Steinmetz, "Notes on Transformers," *Electrical World*, 14 (1889): 168; and Steinmetz, "Influence of Frequency on the Working of Alternate Current Transformers," *Electrical Engineer*, 15 (1893): 184–185, 207–208, 260–261, on p. 261. Steinmetz performed the experiments for the latter paper in May 1891; see Steinmetz, Patent Notebook, pp. 28–35.

49. See, for example, the front covers of his student notebooks entitled "Magnetismus" and "Turbinen Wasserräder," CPSS; Steinmetz's copy of Felix Klein, *Ueber Riemann's Theorie der algebraischen Funktionen* . . . (Leipzig, 1882), NYSL, Accession No. 608061; C. H. Steinmetz to C. P. Steinmetz, 21 July and 1 Sept. 1889, CPSS; Steinmetz, "Time Book," front cover; and Steinmetz, "Influence of Frequency," p. 261.

50. See the Appendix. He did write one paper and a letter to the editor on electrical interference with signals on telegraph and telephone lines caused by nearby power lines. See "Das Verhältniss von Starkstrom zu Schwachstrom in der Vereinigten Staaten," *ETZ*, 13 (1892): 271–272; and "Starkstrom und Schwachstrom," ibid., p. 414. The six papers were on DC electric railways, batteries, the effects of electric light on plants, and spark lengths.

51. Lain's reminiscences are in *Schenectady Gazette*, 9 April 1936, p. 7. For the published version of Steinmetz's comments on Reid's paper, see *Trans AIEE*, 7 (1890): 395–396, 401.

52. Minutes of the New York Electrical Society, 30 April 1891, CULA.

53. C. H. Steinmetz to C. P. Steinmetz, 20 Oct. 1889, and 21 Feb. 1890, CPSS; Steinmetz, "Ueber die durch ein lineares Flachensystem n^{ter} Ordnung definirten mehrdeutigen involutorischen Raumverwandtschaften," *Zeitschrift für Mathematik und Physik*, 35 (1890): 219–236, 272–292, 354–375; K. Steinmetz, "Universität Breslau, Anmeldungs-Buch," 22 July 1890, p. 23 (quotation), CPSS.

54. Steinmetz to Thomas Fiske, 26 Feb., 24 March (quotation), 22 April, 6, 10, and 17 Aug. 1891, TSF; Steinmetz "Multivalent and Univalent Involutory Correspondences in a Plane Determined by a Net of Curves of nth Order," *American Journal of Mathematics*, 14 (1892): 39–66; Steinmetz, "On the Curves Which are Self-Reciprocal in a Linear Nulsystem, and Their Configurations in Space," ibid., pp. 161–186; and David E. Smith and Jekuthiel Ginsburg, *A History of Mathematics in America before 1900* (Chicago: Mathematical Association of America with the cooperation of Open Court, 1934), pp. 185, 188–191.

55. C. P. Steinmetz to Clara Steinmetz, 27 Oct. 1890, CPSF; Steinmetz to Thomas Fiske, 6 Aug. 1891, TSF; and Steinmetz to Lux, 25 Oct. 1893, p. 32. On comparable salary figures, see George Wise, "A New Role for Professional Scientists in Industry: Industrial Research at General Electric, 1900–1916," *Technology and Culture*, 21 (1980): 408–429.

56. C. P. Steinmetz to C. H. Steinmetz, 7 June 1890, CPSS; and Hammond, *Steinmetz*, p. 139. The last address I have been able to find for Asmussen was in a Brooklyn city directory for 1893.

57. Hammond, *Steinmetz*, pp. 140, 307–308; Steinmetz, "Rudolf Eickemeyer"; and Douglas S. Martin, "Steinmetz and His Discovery of the Hysteresis Law," *General Electric Review*, 15 (1912): 544–551. On using the library, see the stamp of "Sektion Yonkers" on the inside front cover of Steinmetz's copy of Karl Kautsky, *Das Erfurter Programm in seinem grundsätzlichen Theil*, 3rd ed. (Stuttgart, 1892), Accession No. 607850, NYSL.

58. C. P. Steinmetz to C. H. Steinmetz, 7 June 1890, CPSS.

59. C. H. Steinmetz to C. P. Steinmetz, 8 Dec. 1889, CPSS. On his father's admonitions to stay away from socialism, see C. H. Steinmetz to C. P. Steinmetz, 20 Oct. 1889, CPSS.

60. Steinmetz to Lux, 25 Oct. 1893, p. 34. On the SLP in this period, see Ira Kipnis, *The American Socialist Movement, 1897–1912* (New York: Columbia Univ. Press, 1952), p. 20; and Howard H. Quint, *The Forging of American Socialism*, 2nd ed. (New York: Bobbs-Merrill, 1963), pp. 52–59 and ch. 5. The Marxists preferred to work through trade unions, while the Lassalleans favored political action.

61. Steinmetz, "The World Belongs to the Dissatisfied," *American Magazine*, 85 (May 1918): 38–40, 76, 79–80, quotation on p. 40.

62. C. P. Steinmetz to C. H. Steinmetz, 7 June 1890, trans. by Franz Neugebauer, n.d., in CPSG. The original is in CPSS.

63. See the letters from C. H. Steinmetz to C. P. Steinmetz, 7 March 1889 to 3 June 1890; and C. H. Steinmetz to Königliche Eisenbau-Direktion, 28 April 1889, CPSS. Mention of Steinmetz's coming home in two years is in the letter dated 25 Aug. 1889. Berthe Mache's telegram about Carl Heinrich's death is described in [Franz Neugebauer], "Excerpts from Steinmetz Family Letters," No. 29, p. 7, CPSG. The original telegram is not in CPSS.

64. Steinmetz to Lux, 25 Oct. 1893, pp. 30, 34; and U.S. Citizenship Papers, 31 July 1896, CPSS. Steinmetz said in 1903 that the "proceedings against me were continued on the court records for several years, but nothing was ever done, and the action was finally dropped for lack of evidence." See "Autobiography of Steinmetz," p. 524.

Chapter 3: Engineering Research

1. Gisbert Kapp, "Alternate-Current Machinery," *Proceedings of the Institution of Civil Engineers*, 97 (19 Feb. 1889): 1–42; John Ambrose Fleming, *The Alternate Current Transformer in Theory and Practice*, 2 vols., 3rd ed. rev. (London, 1900), 2: 109, 114–115; and Geoffrey Tweedale, "Metallurgy and Technological Change: A Case Study of Sheffield Specialty Steel and America, 1830–1930," *Technology and Culture*, 27 (1986): 189–222, on pp. 216–217.

2. J. A. Ewing, "On the Production of Transient Electric Currents in Iron and Steel Conductors . . . ," *Proceedings of the Royal Society*, 33 (17 Nov. 1881): 21–23, on p. 22.

3. Emil Warburg, "Magnetische Untersuchungen," *Annalen der Physik und Chemie*, new series, 13 (1881): 141–164; and J. A. Ewing, "On Effects of Retentive-

ness in the Magnetization of Iron and Steel," *Proceedings of the Royal Society*, 34 (25 May 1882): 39–45, on p. 40.

4. J. A. Ewing, "Experimental Researches in Magnetism," *Philosophical Transactions of the Royal Society*, 176 (1885): 523–640.

5. J. A. Ewing, "Contributions to the Molecular Theory of Induced Magnetism," *Proceedings of the Royal Society*, 48 (1890): 342–358; "A Method for Measuring the Heat Developed on Account of Magnetic Hysteresis in the Core of a Transformer," *Electrician*, 27 (1891): 631–632; "A Magnetic Tester for Measuring Hysteresis in Sheet Iron," *Journal of the Institution of Electrical Engineers*, 24 (1895): 398–440; *Magnetic Induction in Iron and Other Metals* (London, 1892); and P. Strange, "Two Electrical Periodicals: *The Electrician* and *The Electrical Review* 1880–1890," *IEE Proceedings*, 132, Pt. A (1985): 574–581.

6. Steinmetz, "Magnetismus—X," 1884–1885, pp. 28–29, CPSS; and Steinmetz, "Grundzüge einer Theorie der Transformatoren," *Centralblatt für Elektrotechnik*, 12 (1889): 171ff., on pp. 221–222.

7. S. P. Thompson, *Dynamo-Electric Machinery*, 4th ed. (London, 1892), pp. 202–203.

8. Gisbert Kapp, *Alternate-Current Machinery* (London, 1889), p. 65.

9. The equation was $Wh = 300 + .00032 B^{1.83}$, where Wh = hysteresis loss in watts per ton and B = magnetic flux density in lines per centimeter squared. See Steinmetz's copy of Kapp's *Alternate-Current Machinery*, p. 66, NYSL, Accession No. 607468. I calculated an error of -1.9 percent.

10. Erasmus Kittler, *Handbuch der Elektrotechnik*, 2 vols. (Stuttgart, 1886–1890), 2: 252–254; and C. H. Steinmetz to C. P. Steinmetz, 3 June 1890, CPSS.

11. The equation was $Wh = .002 B^{1.6}$, where hysteresis loss (Wh) was now in ergs per centimeter cubed per cycle. See Steinmetz, "Note on the Law of Hysteresis," *Electrical Engineer*, 10 (1890): 676–677. Steinmetz most likely derived the 1.6 law in November; see a plot of Ewing's data, 16 Nov. 1890, in CPSS. On the derivation, see Steinmetz, "Einige Bemerken über Hysteresis," *ETZ*, 12 (1891): 62–63; Steinmetz, "On the Law of Hysteresis (Part II) and Other Phenomena of the Magnetic Circuit," *Trans AIEE*, 9 (1892): 621–724, on p. 674; Douglas S. Martin, "Steinmetz and His Discovery of the Hysteresis Law," *General Electric Review*, 15 (1912): 544–551; and James E. Brittain, "B. A. Behrend and the Beginnings of Electrical Engineering, 1870–1920" (Ph.D. diss., Case Western Reserve Univ., 1970), pp. 79–85, 101–109, 111–118.

12. Steinmetz, "Note on the Law of Hysteresis," p. 677.

13. E. T. Bell, *The Development of Mathematics* (New York: McGraw-Hill, 1940), p. 539; Francis E. Nipher, *Theory of Magnetic Measurements with an Appendix on the Method of Least Squares* (New York, 1886), p. 85; and Steinmetz, "Universität Breslau, Anmeldungs-Buch," pp. 6–7, CPSS.

14. See, for example, Ewing, *Magnetic Induction in Iron*, pp. 59–72; John Hopkinson, "Magnetisation of Iron," in *Original Papers of the Late John Hopkinson*, 2 vols., ed. B. Hopkinson (Cambridge, England, 1901), 2: 154–177; and E. B. Vignoles, "Some Researches in Electro-magnetic Induction," *Electrician*, 27 (1891): 49–51, 77–80, 111–113.

15. Steinmetz, "Some Practical Methods for the Determination of Inductive

Resistances and Self-Induction of Alternate-Current Instruments," *Electrical Engineer*, 10 (1890): 470–472, 508–509, quotation on p. 470; and Andre Millard, *Edison and the Business of Innovation* (Baltimore: Johns Hopkins Univ. Press, 1990), p. 12.

16. Steinmetz, "On the Law of Hysteresis," *Trans AIEE*, 9 (1892): 3–51, on pp. 5–11; and Steinmetz, Eickemeyer Patent Notebook, Jan. to Feb. 1891, pp. 2–8, 12–18, CPSS.

17. Steinmetz, "Some Practical Methods"; and C. P. Steinmetz to C. H. Steinmetz, 7 June 1890, CPSS.

18. Steinmetz, "Eickemeyer's Differential Magnetometer," *Electrical Engineer*, 11 (1891): 353–355; and Steinmetz, "On the Law of Hysteresis," pp. 28–34.

19. Steinmetz, "On the Law of Hysteresis—II," pp. 725–726.

20. Steinmetz, "On the Law of Hysteresis," pp. 5–6, 28, 30.

21. Proteus [Steinmetz] to Hinz and Mrs. Hinz [Heinrich Lux], 25 Oct. 1893, in Justin G. Turner, "Steinmetz," *Manuscripts*, 15 (Fall 1963): 27–34, on p. 31. He did most of the tests in June, and some at irregular intervals from Aug. to Dec. 1891. See Steinmetz, "[Eickemeyer] Time Book," BL.

22. The equation was $W = n B^{1.6} + e N B^2$, where W = energy loss in ergs per centimeter cubed per cycle, n = the coefficient of hysteresis, e = the coefficient of eddy currents, N = frequency of alternations of magnetism, and B = magnetic flux density in lines per centimeter squared. See Steinmetz, "On the Law of Hysteresis," pp. 10–28, 35–42.

23. *Electrical Engineer*, 13 (1892): 86–87, on p. 86. The journal had also praised Steinmetz's first note on hysteresis; see *Electrical Engineer*, 10 (1890): 673.

24. *Electrician*, 28 (1892): 371.

25. "Discussion," *Trans AIEE*, 9 (1892): 61, 52.

26. Ibid., p. 56. On the debates between Steinmetz and Pupin, see Brittain, "B. A. Behrend," pp. 98–99. On his career, see Michael Pupin, *From Immigrant to Inventor* (New York: Scribner, 1922).

27. Steinmetz, "On the Law of Hysteresis," p. 28.

28. "Discussion," *Trans AIEE*, 9 (1892): 54–55; Pupin agreed with Steinmetz that hysteresis was only a "special case of the general law, which was first formulated by Clausius," but he questioned the validity of the 1.6 law. The quotation on experimental physics is in Steinmetz, "Experimentelle Bestimmungen der Energieverlustes durch Hysteresis und seine Abhängigkeit von der Intensität der Magnetisirung, ausgeführt in der Eickemeyer-Dynamomaschinenfabrik, Yonkers, N.Y.," *ETZ*, 13 (1892): 43–48, 55–59, on p. 43.

29. "Discussion," *Trans AIEE*, 9 (1892): 58.

30. Steinmetz, "Appendix," *Trans AIEE*, 9 (1892): 63–64, on p. 64. See also Steinmetz, "Das Gesetz der Hysteresis," *ETZ*, 13 (1892): 136–137.

31. Steinmetz, "Time Book," Feb. to Aug. 1892; and Steinmetz, "On the Law of Hysteresis—II."

32. Steinmetz, "Discussion," *Trans AIEE*, 9 (1892): 739.

33. Steinmetz, "Appendix," *Trans AIEE*, 9 (1892): 725–729.

34. Steinmetz, "On the Law of Hysteresis—II," pp. 622–625, 634, 644, 693–694.

35. Steinmetz, "On the Law of Hysteresis," p. 46.

36. Steinmetz, "On the Law of Hysteresis—II," pp. 626–632, 637–679.

37. Ibid., pp. 622, 626–633.

38. "Discussion," *Trans AIEE*, 9 (1892): 730.

39. Ewing, *Magnetic Induction in Iron*, pp. 7–13, in which intensity of magnetization (*I*) was related to flux density (*B*) and field intensity (*H*) by the vector equation $B = 4\pi I + H$.

40. Steinmetz, "On the Law of Hysteresis—II," pp. 634–636, 678–679, 691–693; and A. E. Kennelly, "Magnetic Reluctance," *Trans AIEE*, 8 (1891): 485–517. Steinmetz defined metallic induction (*L*) in terms of flux density (*B*) and field intensity (*H*) by the vector equation, $B = L + H$.

41. Steinmetz, "Appendix—II," *Trans AIEE*, 9 (1892): 745–747, quotation on p. 747. The equation was $Bo = (n/po)^{2.5}$, where n = the hysteresis coefficient and *po* = the limiting value of permeability.

42. Ewing, *Magnetic Induction in Iron*, pp. 291–294.

43. Steinmetz, "On the Law of Hysteresis—II," pp. 718–719. The italics are in the original.

44. Ibid., pp. 719–724.

45. "Discussion," *Trans AIEE*, 9 (1892): 711–712. The italics are in the original.

46. Sydney Evershed, "A Key to All Magnetism," *Electrician*, 29 (1892): 668–671, on p. 670. Lacking formal training in science and mathematics, Evershed was manager of the firm Goolden and Trotter, which had been established in 1885 to made the Cardew voltmeter. A. P. Trotter was editor of the *Electrician* and published Evershed's researches. In 1895 Evershed set up his own instrument-making firm, Evershed and Vignoles. The remark about his humor comes from an obituary notice in *Journal of the Institution of Electrical Engineers*, 85 (1939): 775–776.

47. Evershed, "A Key to All Magnetism." The italics are in the original.

48. J. A. Ewing and Helen G. Klassen, "Magnetic Qualities of Iron," *Philosophical Transactions of the Royal Society*, 184A (1893): 985–1040, on p. 1017.

49. Ibid., p. 1018.

50. Reginald A. Fessenden, "A Formula for the Area of the Hysteresis Curve," *Electrical World*, 23 (1894): 768–770; Frank Holden, "Some Work on Magnetic Hysteresis," *Electrical World*, 25 (1895): 687–689; J. W. L. Gill, "On a New Method of Measuring Hysteresis in Iron," *Electrician*, 39 (1897): 718–722, a reprint of a paper read before the British Association for the Advancement of Science; and H. Maurach, "Ueber die Abhängigkeit des durch Hysteresis bedingten Effectverlustes im Eisen von der Stärke der Magnetisirung," *Annalen der Physik und Chemie*, 4th series, 6 (1901): 580–589. *Electrical World* defended Steinmetz against Fessenden by stating that Steinmetz had not claimed the law of hysteresis to be a physical law. See *Electrical World*, 23 (1894): 762.

51. Francis G. Bailey, "On Hysteresis in Iron and Steel in a Rotating Magnetic Field," BAAS, *Report*, 64 (1894): 576–577; and Bailey, "The Hysteresis of Iron in an Alternating Magnetic Field," *Electrician*, 36 (1895): 116–119, on p. 118. Two students of J. A. Fleming's confirmed Bailey's results; see R. Beattie and R. C. Clinker, "Magnetic Hysteresis in a Rotating Magnetic Field," *Electrician*, 37 (1896): 723–728.

52. At least three experimenters did note Steinmetz's use of the intensity of magnetization. See Evershed, "Key to All Magnetism," p. 669; J. A. Fleming et al., "On the Magnetic Hysteresis of Cobalt," *Philosophical Magazine*, 5th series, 48

(1899): 271–279; and Maurach, "Ueber die Abhängigkeit des Effectverlustes im Eisen," p. 580.

53. S. P. Thompson, "Discussion on Magnetic Testing," *Proceedings of the Institution of Civil Engineers*, 126 (1896): 256.

54. *Electrician*, 36 (1895): 107–108.

55. Steinmetz, "On the Law of Hysteresis (Part III) and the Theory of Ferric Inductances," *Trans AIEE*, 11 (1894): 570–608, on p. 573.

56. Steinmetz, "Discussion," *Trans AIEE*, 17 (1900): 330–332, on p. 332.

57. Bailey, "Hysteresis of Iron in an Alternating Magnetic Field," p. 116.

58. Applying the method of least squares to the data on this sample (taken from 342 to 2,600 lines per centimeter squared), I found the hysteresis exponent to be 1.79, with a small curve-fitting error (± 3.4 percent). If Steinmetz had assumed the exponent to be 1.6, as in his other tests, he would have obtained a large error (± 9.5 percent). Instead, he concluded that the horseshoe magnet had considerable eddy currents—a very questionable assumption about a laminated sample built up of three hundred sheets of well-insulated iron! He then calculated the total energy loss using a formula similar to his original law of hysteresis, which gave an error comparable to that for the 1.79 exponent (± 2.8 percent vs. ± 3.4 percent). See Steinmetz, "On the Law of Hysteresis," pp. 26–27. Ewing found a similar error (± 14 percent, according to my calculations) at 150 to 3,000 lines/cm^2. See Ewing and Klassen, "Magnetic Qualities of Iron," pp. 1017–1020.

59. Steinmetz, "On the Law of Hysteresis—II," pp. 704–705. Steinmetz also used data from other researchers, notably from John Hopkinson, which unfortunately did not extend to low magnetizations. See Steinmetz, "On the Law of Hysteresis," p. 47; and Hopkinson, "Magnetisation in Iron," in *Original Papers*, 2: 171.

60. Alexander Siemens, "Discussion," *Journal of the Institution of Electrical Engineers*, 23 (1894): 227–229; Horace F. Parshall, "Magnetic Data of Iron and Steel," *Proceedings of the Institution of Civil Engineers*, 126 (1896): 220–244, on p. 232; J. A. Fleming, "A Method of Determining Magnetic Hysteresis Loss in Straight Iron Strips," *Philosophical Magazine*, 5th series, 44 (1897): 262–282; and Clarence P. Feldmann, "Testing Iron of Transformers by a Workshop Method," *Electrician*, 32 (1894): 581–582. Ewing pointed out further practical applications for Steinmetz's law in 1894 and 1895. See Ewing, "Discussion," *Journal of the Institution of Electrical Engineers*, 23 (1894): 259; and Ewing, "A Magnetic Tester for Measuring Hysteresis in Sheet Iron," *Journal of the Institution of Electrical Engineers*, 24 (1895): 398–440, on pp. 402–403.

61. A. E. Kennelly, "Measurements of the Hysteresis Loss of Magnetic Energy in Nickel," *Electrical Engineer*, 13 (1892): 349–350; and Fleming, "On the Magnetic Hysteresis of Cobalt." The publication of this paper by Kennelly and the one cited in note 40 indicates that Edison's publication ban was not as strict as that described in Millard, *Edison*, p. 21.

62. See, for example, Kittler, *Handbuch der Elektrotechnik*, 3rd ed. rev. (1892), 1: 43–45; Gisbert Kapp, *Alternating Currents of Electricity* (New York, 1893); Dugald C. Jackson, *A Textbook on Electro-magnetism and the Construction of Dynamos*, 2 vols. (New York, 1893–1896), 1: 74–75; F. Loppe and R. Boquet, *Traité théorique et pratique des courants alternatifs industriels*, 2 vols. (Paris, 1894), 1: 270; C. P. Feldmann, *Wirkungsweise, Prüfung und Berechnung der Wechselstrom-Transformatoren für die Praxis*

bearbeitet, 2 vols. (Leipzig, 1894), 1: 142–143; Frederick Bedell, *The Principles of the Transformer* (New York, 1896), p. 4; and Eric Gérard, *Leçons sur l'electricité*, 2 vols., 6th ed. rev. (Paris, 1899), 1: 69.

63. Ewing, *Magnetic Induction in Iron*, 3rd ed. rev. (1900), p. 111. See, for example, Gisbert Kapp, *Dynamos, Motors, Alternators, and Rotary Converters*, trans. from 3rd German ed. rev. by Harold H. Simmons (London, 1902), p. 98; Englebert Arnold, ed., *Die Wechselstromtechnik*, 5 vols. (Berlin, 1902–1909), 1: 350, 2: 68–69, 4: 103–104, 5: 201; S. P. Thompson, *Dynamo-Electric Machinery*, 2 vols., 7th ed. rev. (London: E. and F. Spon, 1904–1905), 1: 99–100; Gisbert Kapp, *Transformers for Single and Multiphase Currents*, 2nd ed. rev. (New York, 1908), p. 17; C. Heinke, *Handbuch der Elektrotechnik*, 5 vols., ed. rev. (Leipzig, 1908), 5: 421–422; W. S. Franklin and W. Esty, *The Elements of Electrical Engineering*, 2 vols., 3rd ed. rev. (New York, 1912), 1: 380; and Horatio A. Foster, ed., *Electrical Engineers Pocketbook*, 7th ed. rev. (New York, 1913), pp. 98–99.

64. Kittler, *Handbuch der Elektrotechnik* (1886–1890), 2: 252–254.

Chapter 4: General Electric

1. Jonathan Leonard, *Loki: The Life of Charles Proteus Steinmetz* (Garden City, N.Y.: Doubleday, 1929), p. 124.

2. Alfred D. Chandler, Jr., *The Visible Hand: The Managerial Revolution in American Business* (Cambridge, Mass.: Harvard Univ. Press, 1977), chs. 9–11.

3. Harold C. Passer, *The Electrical Manufacturers, 1875–1900* (Cambridge, Mass.: Harvard Univ. Press, 1953), pp. 150, 321–329.

4. Monte A. Calvert, *The Mechanical Engineer in America, 1830–1910: Professional Cultures in Conflict* (Baltimore: Johns Hopkins Press, 1967), p. 74; David F. Noble, *America by Design: Science, Technology, and the Rise of Corporate Capitalism* (New York: Knopf, 1977), pp. 37–44, 171–176; and George Wise, "'On Test': Post Graduate Training of Engineers at General Electric," *IEEE Transactions on Education*, E-22 (1979): 171–177.

5. David O. Woodbury, *Beloved Scientist: Elihu Thomson* (New York: McGraw-Hill, 1944), p. 203; and John W. Hammond, *Charles Proteus Steinmetz: A Biography* (New York: Century, 1924), p. 204. Thomson did say in 1923 that after he and a patent attorney visited Eickemeyer's in 1892, "we both agreed that by all odds the most valuable asset that Eickemeyer had was Steinmetz." See Thomson to E. W. Rice, Jr., 29 Oct. 1923, in *Selections from the Scientific Correspondence of Elihu Thomson*, ed. Harold J. Abrahams and Marion B. Savin (Cambridge, Mass.: MIT Press, 1971), pp. 431–432, on p. 431. A copy of the original letter is in L1012–13, JWH.

6. E. W. Rice, Jr., "Charles Proteus Steinmetz," *Science*, 58 (1923): 388–390, on p. 388.

7. *Trans AIEE*, 8 (1891): 594, 616. This citation corrects Thomson's faulty 1923 recollection, in which he said that he first met Steinmetz in 1892 at Eickemeyer's. See Elihu Thomson to E. W. Rice, Jr., 29 Oct. 1923.

8. *Trans AIEE*, 9 (1892): 151–153, 156–157, 162–163. Also, Parshall became a full member of the AIEE at the same meeting (18 March 1890) at which Steinmetz became an associate member.

9. *Street Railway Journal*, 8 (1892): 399–400.

10. Charles W. Cheape, *Moving the Masses: Urban Public Transit in New York*,

Boston, and Philadelphia, 1880–1912 (Cambridge, Mass.: Harvard Univ. Press, 1980), pp. 112–120, quotation on p. 116; Passer, *Electrical Manufacturers*, pp. 228, 247–248, 252–253; W. Bernard Carlson, "Invention, Science, and Business: The Professional Career of Elihu Thomson, 1870–1900" (Ph.D. diss., Univ. of Pennsylvania, 1984), p. 409; and Richard H. Schallenberg, "The Anomalous Storage Battery: An American Lag in Electrical Engineering," *Technology and Culture*, 22 (1981): 725–752, on pp. 744–745.

11. Passer, *Electrical Manufacturers*, pp. 52–53; and *Annual Statement of the Thomson-Houston Electric Company*, 1891, pp. 4–5, GEH. I would like to thank W. Bernard Carlson for providing me with a copy of this document.

12. Agreements between Otis Brothers and Co., General Electric Co., Thomson-Houston Co., and Edison General Electric Co., 23 Aug. and [20 Dec.] 1892, in General Electric Company File, Patent Dept., Otis Elevator Co., OCR; and Charles E. Allison, *The History of Yonkers* (New York, 1896), p. 350.

13. *Electrical Engineer*, 11 (1891): 289; Notice of Stockholders Meeting for 5 May 1892 of Yonkers Machine Co., 26 April 1892, in 1890 Otis File, OCR; *Electrical World*, 20 (1892): 305–306; and *Otis Bulletin*, Nov. 1948, p. 16, OCR.

14. Steinmetz to Herr Uppenborn, 19 Feb. 1893, CPSS; and Edward Adler to Gerard Swope, 16 April 1936, GS.

15. New York State Labor Department, Inspection Bureau, *Annual Reports*, 1893–1900; Rudolph Eickemeyer, *Letters from the Southwest* (New York, 1894), pp. 5–7; Proteus [C. P. Steinmetz] to Hinz and Mrs. Hinz [Heinrich Lux], 25 Oct. 1893, in Justin G. Turner, "Steinmetz," *Manuscripts*, 15 (Fall 1963), 27–34, on p. 32; *NCAB*, vol. 11 (1909), s.v. "Eickemeyer, Carl"; and Carl Eickemeyer Alumni Folder, CUA. An obituary of Carl (Class of '92) said that he "organized the Eickemeyer-Field Manufacturing Company, which was later merged with the General Electric Company." See Cornell Univ., *Alumni News*, 15 Dec. 1927, p. 145.

16. *Otis Bulletin*, Nov. 1948, p. 14; John Diedrich Ihlder Alumni Folder, CUA; and *NCAB*, vol. E (1937–1938), s.v. "Ihlder, John [William]."

17. Steinmetz, "[Eickemeyer] Time Book," BL; Steinmetz, "Note on the Disruptive Strength of Dielectrics," *Trans AIEE*, 10 (1893): 85–105; Steinmetz, Eickemeyer Abstracts Notebook, 30 Dec. 1892, p. 43, CPSS; Steinmetz, "Influence of Frequency on the Working of Alternate Current Transformers," *Electrical Engineer*, 15 (1893): 184ff.; and Steinmetz to Lux, 25 Oct. 1893, quotation on p. 32.

18. Steinmetz to Lux, 25 Oct. 1893, p. 32; Steinmetz, *America and the New Epoch* (New York: Harper and Brothers, 1916), n.p. (introduction); and Steinmetz to F. C. Pratt, 5 March 1909, CPSS.

19. Passer, *Electrical Manufacturers*, pp. 228, 250; and Thomas P. Hughes, *Elmer Sperry: Inventor and Engineer* (Baltimore: Johns Hopkins Press, 1971), pp. 61, 71.

20. E. W. Rice, Jr., to S. D. Greene, 18 March 1893, EWR; *DAB* (Bell); *American Men of Science* (1921) (Reist, Foster, Emmet); *Who Was Who in America* (Moody); *NCAB*, vol. 35 (1949) (Berg); "E. Danielson," *ETZ*, 28 (1907): 1064; and Wise, "On Test," p. 172. Berg came to Thomson-Houston in the fall of 1892; see Ernst Berg to J. L. Hall, 6 July 1914, EJB.

21. Horace Field Parshall, *The Parshall Family, A.D. 870–1913* (London: Francis Edwards, 1915), pp. 163–170; and Steinmetz to A. G. Davis, 12 Jan. 1899, L5470–74, JWH. Parshall was an assistant to Knight in the Railway Department in

Sept. 1892, before the formation of the Calculating Department, sometime prior to Jan. 1893. See "Shop-Practice As Found at the Thomson-Houston Electric Co.'s Works . . . ," *Street Railway Journal*, 8 (1892): 522–530, on p. 530; and E. W. Rice, Jr., to S. D. Greene, 4 Jan. 1893, EWR.

22. *Electrical Review* 21 (1892): 155; E. W. Rice, Jr., to N. P. Otis, 11 Jan. 1893; Rice to Henry Osterheld, 4 Feb. 1893; Rice to H. F. Parshall, 6 March 1893; Rice to Henry Reist, 8 April 1893; Rice to Mr. Fiske, 9 May 1893, EWR; and Elihu Thomson to E. M. Bentley, 19 July 1893, ETA.

23. *Electrical Review*, 22 (1893): 197; Killingsworth Hedges, *American Electric Street Railways . . .* (London, 1894), pp. 75–76; Passer, *Electrical Manufacturers*, pp. 260–269; and Oscar T. Crosby and Louis Bell, *The Electric Railway in Theory and Practice* (New York, 1892), pp. 94–98.

24. *Electrical Engineer*, 10 (1890): 358–359; and Eugene Griffin to Department and Local Office Managers, 4 Aug. 1896, L1183–89, on L1189, JWH. I would like to thank W. Bernard Carlson for providing me with a copy of this letter.

25. Steinmetz to Lux, 25 Oct. 1893, p. 32; Hughes, *Elmer Sperry*, p. 71; and George Wise, "A New Role for Professional Scientists in Industry: Industrial Research at General Electric, 1900–1916," *Technology and Culture*, 21 (1980): 408–429, on p. 418.

26. E. W. Rice, Jr., to F. P. Fish, 1 March 1893, EWR; Steinmetz, Lynn Notebook, 18 May 1893, p. 16, CPSS; and Kennelly's AIEE Personal Classification Sheet, 13 Nov. 1917, IEEEA.

27. E. W. Rice, Jr., to Axel Ekström, 6 Feb. and 6 March 1893, EWR. Hammond, *Steinmetz*, p. 214, puts the number of Calculating Department staff at that time at nine.

28. Ronald Kline, "Science and Engineering Theory in the Invention and Development of the Induction Motor, 1880–1900," *Technology and Culture*, 28 (1987): 283–313.

29. W. J. Foster to J. T. Stockdale, 2 July 1925, L1014–16, JWH; John A. McManus, "Fifty Years of Induction Motor Manufacture at the Lynn Works of the General Electric Company," 25 Nov. 1941, GER, which cites entries from a work journal kept in the 1890s by Henry Hobart; and Edward L. Owen, "Induction Motor Developments at General Electric 1891 to Present," unpublished paper presented at the International Conference on Evolution and Modern Aspects of Induction Machines, Turin, Italy, July 1986.

30. *Electrical World*, 20 (1893): 325–326; Louis Bell, "Practical Properties of Polyphase Apparatus," *Trans AIEE*, 11 (1894): 3–31; and Kline "Science and Engineering Theory," p. 292.

31. Steinmetz, Lynn Notebook, passim. On the inductor, fan motor, and polyphase motors, see pp. 3–9, 49, 76–79.

32. Ibid., 17–18 May 1893, pp. 14–17; Glenn Weaver, *The Hartford Electric Light Company* (Hartford, Conn.: Hartford Electric Light Co., 1969), pp. 65–67; and Owen, "Induction Motor Developments," p. 3. The line also had a problem of burning out transformers, which was solved by removing a ground wire. See E. W. Rice, Jr., to H. M. Byllesby, 2 and 18 March 1893, EWR.

33. Steinmetz, U.S. Patent Nos. 513,370, 543,907, and 596,186 (with E. W. Rice, Jr.), all filed 9 Sept. 1893; Steinmetz, "Discussion," *Trans AIEE*, 10 (1893):

227–231, on p. 229; Steinmetz, Lynn Notebook, 17–18 June and 17 July 1893, pp. 42–43, 56; and Silvanus P. Thompson, "Some Advantages of Alternate Currents," *Electrician*, 33 (1894): 481–484, quotation on p. 484.

34. Carlson, "Thomson," pp. 407–409, 552–554; and Robert Friedel et al., *Edison's Electric Light: Biography of an Invention* (New Brunswick: Rutgers Univ. Press, 1986), ch. 2.

35. Weaver, *Hartford Electric Light Company,* p. 67, quoting A. C. Dunham, president of the utility; and Clyde D. Wagoner, "Steinmetz Revisited: The Myth and the Man," *General Electric Review,* 60 (1957): 16–21, on p. 16.

36. GE Power and Mining Dept., "Semimonthly Bulletin Letters Nos. 18 and 19," 1 and 15 Jan. 1896, p. 3, Commercial File, GEH, Box 19. The same department's Bulletin No. 4,123, 25 Oct. 1897, lists the Redlands plant and not the one at Hartford.

37. Steinmetz, Lynn Notebook, 18 June and 10 July 1893, pp. 43, 54; and Steinmetz, "On the Law of Hysteresis (Part III) and the Theory of Ferric Inductances," *Trans AIEE*, 11 (1894): 570–608, on p. 572. For a description of the Redlands plant, see "The First Three-Phase Plant in the United States," *Electrical World*, 22 (1893): 433–434, a title indicative of how little known the Hartford plant was in the technical press.

38. *Cornell Daily Sun*, 29 Oct. 1923, copy in VK. For earlier studies of the development of the complex-quantity method for electrical engineering, see G. Windred, "Complex Quantities: Theory and Application—Past and Present Developments," *Electrician*, 115 (1935): 339–340; Windred, "Early Developments in A.C. Circuit Theory . . . ," *Philosophical Magazine*, 7th series, 10 (1930): 905–916; and James E. Brittain, "B. A. Behrend and the Beginnings of Electrical Engineering, 1870–1920" (Ph.D. diss., Case Western Reserve Univ., 1970), pp. 121–132.

39. J. Clerk Maxwell, Balfour Stewart, and Fleeming Jenkin, "Description of an Experimental Measurement of Electrical Resistance . . . ," *Report of the British Association for the Advancement of Science*, 33 (1863): 163–176; Maxwell, "A Dynamical Theory of the Electromagnetic Field," *Philosophical Transactions of the Royal Society,* 155 (1865): 459–512, on pp. 475–477; and Victor Wietlisbach, "Ueber die Anwendung die Telephons zu elektrischen und galvanischen Messungen," *Monatsberichte der Königlich Preussischen Akademie der Wissenschaften zu Berlin*, 1879, pp. 278–283, on pp. 280–281. Wietlisbach's equations, in modern symbols, were $E = IZ$ and $Z = R + jwL + 1/jwC$, where $E, I, Z, R, L,$ and C stood, respectively, for the voltage, current, impedance, resistance, inductance, and capacitance of any leg of the bridge (with the resistance, inductance, and capacitance connected in series); $w = 2\pi$ times the frequency of the A.C. source; and $j = \sqrt{-1}$.

40. J. W. Strutt [Lord Rayleigh], *Theory of Sound*, 2 vols. (London, 1877), 1: 106–107; Rayleigh, *Scientific Papers*, 3 vols. (Cambridge, England, 1899), 1: 97ff., 126ff., 145ff., 365ff.; and Hermann von Helmholtz, "Telephon und Klangfarbe," *Annalen der Physik und Chemie*, 241 (1878): 448–460. Euler's identity, $e^{jwt} = \cos(wt) + j \sin(wt)$, was expressed by $A\,e^{jwt}$, where e was the base of the natural logarithms and A was the amplitude of the function. Wietlisbach's final equations did not contain the term e^{jwt}; it canceled out, because both the derivative and integral of e^{jwt} are functions of e^{jwt}.

41. A. Oberbeck, "Ueber elektrische Schwingungen mit besonderer

Berücksichtigung ihrer Phasen," *Annalen der Physik und Chemie*, new series, 17 (1882): 816–841, on pp. 819–822.

42. Lord Rayleigh, "The Reaction upon the Driving-Point of a System Executing Forced Harmonic Oscillations of Various Periods, with Applications to Electricity," *Philosophical Magazine*, 5th series, 21 (1886): 369–381, on pp. 372–373, 380; "On the Self-Induction and Resistance of Straight Conductors," ibid., pp. 381–394; and "On the Sensitiveness of the Bridge Method in Its Application to Periodic Electric Currents," *Proceedings of the Royal Society*, 49 (1891): 203–217. On the history of the skin effect and the impedance bridge, see D. W. Jordan, "D. E. Hughes, Self-Induction, and the Skin-Effect," *Centaurus*, 26 (1982): 123–153; and S. J. Zammataro, "Development of the Impedance Bridge," *Bell Laboratories Record*, 7 (Dec. 1928): 150–154.

43. Paul J. Nahin, *Oliver Heaviside: Sage in Solitude* . . . (New York: IEEE Press, 1987).

44. Oliver Heaviside, *Electrical Papers*, 2 vols. (New York, 1892), 2: 355–357. The resistance operator (Z) was a "function of the electrical constants of the [circuit] combination and of d/dt, the operator of time-differentiation," which was represented by the letter p. For a coil of wire, $Z = R + pL$, where R is its resistance and L its inductance.

45. For sinusoidal functions, $p^2 = -w^2$, because the second derivative of sin (wt) $= -w^2 \sin (wt)$. Therefore, $p = jw$, where $j = \sqrt{-1}$.

46. Heaviside, *Electrical Papers*, 2: 247; and Heaviside, "Electromagnetic Theory—LIX," *Electrician*, 32 (1893): 57–59. For instances in which he set $p^2 = -w^2$ before the paper on resistance operators, see *Electrical Papers*, 2: 62, 63, 71, 106, 114, 118, 192–193, 217, 228, 251, 270. For another instance of writing $p = jw$, see ibid., p. 199. But in 1898 he called j a "spurious imaginary" when it was used in equations such as $E = I (R + jwL)$; see Heaviside, *Electromagnetic Theory*, 3 vols. (1893–1912; rpt. New York: Chelsea, 1971), 2: 459.

47. Heaviside, *Electrical Papers*, 2: 355.

48. Max Wien, "Das Telephon als optischer Apparat zur Strommessung," *Annalen der Physik und Chemie*, new series, 42 (1891): 593–621, on p. 604; Wien, "Messung der Inductionconstanten mit dem 'optischer Telephon,'" *Annalen der Physik und Chemie*, 44 (1891): 689–712, on p. 693; J. V. Wietlisbach, "Fernsprechen," *Jahrbuch für Elektrotechnik*, 1887, pp. 205–270, on pp. 216–218, 231–234; Wietlisbach, "Theorie der Telephonkabel," *Elektrotechnische Rundschau*, 4 (Feb. 1887): 13–18; Wietlisbach, "Die Verwendung des Uebertragers (Translators) zu Fernsprecheinrichtungen," *ETZ*, 8 (1887): 238–241; J. H. W., "Dr. J. Wietlisbach," *ETZ*, 18 (1897): 753; and Adolf Franke, "Die elektrischen Vorgänge in Fernsprechleitungen und Apparaten," *ETZ*, 12 (1891): 447–452, 458–463, on pp. 451, 458. On Franke's career, see Georg Siemens, *Geschichte des Hauses Siemens*, 3 vols. 2nd ed. rev. (Munich: K. Alber, 1949–1953), 1: 209–210.

49. Lord Rayleigh, "The Self-Induction and Resistance of Compound Conductors," *Electrician*, 18 (1886–1887): 170–172, 222–223, 240–243; Rayleigh, "On the Sensitiveness of the Bridge Method in its Application to Periodic Electric Currents," *Electrician*, 26 (1891): 791–793; *ETZ*, 13 (1892): 61; C. Raveau, "Nouvelles applications scientifiques du téléphone—IV," *La lumière électrique*, 43 (1892): 563–570; J. V. Wietlisbach, "Theory of Telephone Circuits—Part I," *Western Electrician*, 1

(1887): 237; and, e.g., Heaviside, *Electrical Papers*, 2: 61–66. The *Electrician* also abstracted Wien's papers, but it did not include any mathematics; see *Electrician*, 27 (1891): 71.

50. *Electrician*, 26 (1891): 778.

51. Nahin, *Heaviside*, pp. 154, 182 (n. 57); and Raveau, "Nouvelles applications scientifiques du téléphone," pp. 564–566.

52. Steinmetz, "Complex Quantities and Their Use in Electrical Engineering," *Proceedings of the International Electrical Congress . . . 1893* (New York, 1894), pp. 37–74. Volume hereafter cited as *Proc. IEC, 1893*.

53. Steinmetz, Lynn Notebook, 18 May 1893, pp. 16–17. On the AC systems at West Orange, see Andre Millard, *Edison and the Business of Innovation* (Baltimore: Johns Hopkins Univ. Press, 1990), pp. 109, 120, 154.

54. Arthur E. Kennelly, "Impedance," *Trans AIEE*, 10 (1893): 175–216, on p. 186.

55. John Ambrose Fleming, *The Alternate Current Transformer in Theory and Practice*, 2 vols. (London: Electrician, 1889–1892), 1: 116.

56. Kennelly, "Impedance," p. 186.

57. Ibid., pp. 200–201, 210.

58. A. E. Kennelly to Ernst Berg, 27 Nov. 1923 and 4 April 1928, EJB. However, after 1893 Kennelly regularly substituted *jw* for *p* and referred to Heaviside's 1887 paper containing complex-number impedance. See Kennelly, "On the Fall of Pressure in Long Distance Alternating Current Conductors," *Electrical World*, 23 (1894): 17–19, on p. 19; Kennelly and Edwin J. Houston, "Resonance in Alternating Current Lines," *Trans AIEE*, 12 (1895): 133–157, on p. 157; Kennelly, *The Application of Hyperbolic Functions to Electrical Engineering Problems* (New York: McGraw-Hill, 1912), p. 333; and Kennelly to E. W. Rice, Jr., 28 Dec. 1923, in "Charles Proteus Steinmetz and Complex Quantities," *General Electric Review*, 27 (1924): 132–133.

59. Frank Sprague, "Discussion," *Trans AIEE*, 10 (1893): 227.

60. Ibid., pp. 217–227; *Electrical World*, 21 (1893): 340–341; and *Electrical Engineer*, 15 (1893): 438. One French journal did reprint the paragraph on complex impedance. See *La lumière électrique*, 48 (1893): 430–433, on pp. 432–433.

61. Steinmetz, "Discussion," *Trans AIEE*, 10 (1893): 227–231, on pp. 227, 228.

62. Ibid., p. 228.

63. Steinmetz, "Starkstrom und Schwachstrom," *ETZ*, 13 (1892): 414; C. Grawinkel, "Bemerkungen zur Abhandlung des Herrn Steinmetz . . . ," ibid., pp. 297–298; Adolph Franke, "Ueber die elektrischen Vorgänge in Fernsprechleitungen und Apparaten," ibid., pp. 295–296; and Steinmetz, "Das Verhältniss von Starkstrom zu Schwachstrom in den Vereinigten Staaten," ibid., pp. 271–272. In "Electromagnetic and Electrostatic Hysteresis," *Electrical World*, 22 (1893): 144–145, Steinmetz cited a paper by R. Arno in *La lumière électrique* for July 1892. Raveau's paper was published in March 1892.

64. Steinmetz, "The Determination of Inductance," *Electrician*, 26 (1891): 279; and Steinmetz, "Some Practical Methods for the Determination of Inductive Resistance and Self-Induction of Alternate Current Instruments—I," *Electrical Engineer*, 10 (1890): 470–472.

65. Frederick Bedell and Albert C. Crehore, "Derivation and Discussion of the General Solution for the Current Flowing in a Circuit Containing Resistance, Self-

Induction, and Capacity, with Any Impressed Electromotive Force," *Trans AIEE*, 9 (1892): 303–374, on pp. 310–314.

66. For a criticism of Steinmetz's lax citation practice, see Vladimir Karapetoff, "Discussion," *Trans AIEE*, 30 (1911): 396–397.

67. Steinmetz, "Complex Quantities," p. 37.

68. Ibid., pp. 38, 39.

69. Ibid., pp. 39–41. Steinmetz derived a minus sign because he represented vector A lagging behind vector B—both of which rotate in the counterclockwise direction—by placing A "ahead" of B in the clock diagram, while others would place A "behind" B in a counterclockwise system.

70. Steinmetz, "Discussion," *Proc. IEC, 1893*, p. 75.

71. *Electrical World*, 22 (1893): 170–175, on p. 171.

72. Compare Steinmetz, "Complex Quantities," pp. 41–74, with Steinmetz, *Theory and Calculation of Alternating Current Phenomena* (New York, 1897), chs. 8–30.

73. E. W. Rice, Jr., to Louis Bell, 19 Jan. 1893; Rice to H. M. Byllesby, 2 March 1893; Rice to Axel Ekström, 6 March 1893; and Rice to S. D. Greene, 6 March 1893, EWR; *Electrical World*, 21 (1893): 352, 371; George Forbes, "Discussion," *Proc. IEC, 1893*, pp. 433–436; Robert Belfield, "The Niagara System: The Evolution of the Electric Power Complex at Niagara Falls," *Proceedings of the Institute of Electrical and Electronics Engineers*, 64 (1976): 1344–1350; and Edward D. Adams, *Niagara Power: History of the Niagara Falls Power Company, 1886–1918*, 2 vols. (Niagara Falls, 1927), 2: 248, 276.

74. Steinmetz, "Complex Quantities," p. 74.

75. Louis Bell, "Discussion," *Proc. IEC, 1893*, pp. 423–428; and Adams, *Niagara Power*, 2: 270–278.

76. Steinmetz, "On the Law of Hysteresis (Part III)," pp. 578–584; and Heaviside, *Electrical Papers*, 2: 357. On the adoption of the term *impedance*, see M. J. Joubert, ed., *Congrés international des électriciens Paris 1889, comptes rendus des travaux* (Paris, 1890), p. 297.

77. Steinmetz, "Discussion," *Trans AIEE*, 17 (1900): 64. In the same paper in which he used Y for admittance, Steinmetz switched from I to U for impedance, while Heaviside continued to use I for impedance, J for admittance, and Z for the resistance operator. Electrical engineers later chose Z to represent impedance. Heaviside coined the following terms, which were later adopted by electrical engineers: *admittance* (Dec. 1887), *conductance* (Sept. 1885), *impedance* (July 1886), *inductance* (Feb. 1886), *permeability* (Sept. 1885), and *reluctance* (May 1888). See Heaviside, *Electrical Papers*, 2: 357, 24, 64, 28, 24, 168, respectively. Heaviside's equivalent term for *susceptance* was *permittance*, which he coined in June 1887. See *Electrical Papers*, 2: 120.

78. John Ambrose Fleming, "On Some Effects of Alternating-Current Flow in Circuits Having Capacity and Self-Induction," *Journal of the Institution of Electrical Engineers*, 20 (1891): 362–408, on pp. 400–401; "Discussion," *Trans AIEE*, 10 (1893): 226, 230; and Steinmetz, "Capacity and Self-Induction in High Potential Circuits," *Electrical World*, 21 (1893): 318–319. An earlier paper on the effect used exponential complex quantities to represent voltages along the line; see R. T. Glazebrook, "On the Rise of Electromotive Force Observed in the Deptford Mains,"

Electrician, 26 (1890): 232–233. On the Ferranti effect, see Percy Dunsheath, *A History of Electrical Power Engineering* (Cambridge, Mass.: MIT Press, 1962), pp. 167–168; and J. F. Wilson, *Ferranti and the British Electrical Industry* (Manchester: The University Press, 1988), p. 44.

79. Ernst J. Berg, "Reminiscences of Heaviside and Steinmetz," 22 April 1927, p. 6, EJB.

80. Heaviside, "Electromagnetic Theory—LIX"; and Steinmetz's copy of Heaviside, *Electrical Papers,* 1: 113, Accession No. 587141-2, NYSL. A further indication that Steinmetz was familiar with the November article is that it was immediately followed (on p. 59) by an editorial comment on a recent letter to the editor by Steinmetz on the candlepower of searchlights.

81. Ernst Berg, "Charles Proteus Steinmetz," 25 Jan. 1924, p. 5, EJB.

82. *Electrical Engineer,* 16 (1893): 206; *Electrical World,* 22 (1893): 171; and *Electrician,* 31 (1893): 522.

83. Steinmetz, "Anwendung complexer Grössen in der Elektrotechnik," *ETZ,* 14 (1893): 597–599, 631–635, 641–643, 653–654; and C. F. Guilbert, "La méthode de M. Steinmetz pour le calcul des courants alternatifs," *La lumière électrique,* 50 (1893): 451–458, 554–563, on p. 554. The IEC proceedings were published in Dec. 1894; see AIEE BoD minutes, 16 Jan. 1895, IEEEA.

84. Alexander Macfarlane, "On the Analytical Treatment of Alternating Currents," *Proc. IEC, 1893,* pp. 24–31; and "Discussion," ibid., pp. 74–75. On Macfarlane's career, see *Electrical World,* 24 (1889): 357; and Michael J. Crowe, *A History of Vector Analysis* (Notre Dame: Univ. of Notre Dame Press, 1967), pp. 190–219.

85. Edwin T. Layton, Jr., "Mirror-Image Twins: The Communities of Science and Technology in Nineteenth-Century America," *Technology and Culture,* 12 (1971): 562–580.

86. Vannevar Bush, "Arthur Edwin Kennelly, 1861–1939," National Academy of Sciences, *Biographical Memoirs,* 22 (1943): 83–119.

Chapter 5: Theory and Practice

1. George Wise, "The Shoemakers," chapter in an unpublished ms. on the history of General Electric in the author's possession, pp. 17–21; Harold C. Passer, *The Electrical Manufacturers, 1875–1900* (Cambridge, Mass.: Harvard Univ. Press, 1953), pp. 327–329; and John T. Broderick, *Forty Years with General Electric* (Albany, N.Y.: Fort Orange Press, 1929), pp. 30–31.

2. Broderick, *Forty Years,* pp. 36–37; and Proteus [Steinmetz] to Hinz and Mrs. Hinz [Heinrich Lux], 25 Oct. 1893, in Justin G. Turner, "Steinmetz," *Manuscripts,* 15 (Fall 1963): 27–34, on p. 32.

3. Wise, "Shoemakers," pp. 20–22; Harold C. Passer, "Development of Large-Scale Organization: Electrical Manufacturers around 1900," *Journal of Economic History,* 12 (1952): 378–395; Alfred D. Chandler, Jr., *The Visible Hand: The Managerial Revolution in American Business* (Cambridge, Mass.: Harvard Univ. Press, 1977), pp. 426–432; and Passer, *Electrical Manufacturers,* pp. 291, 323.

4. John W. Hammond, *Charles Proteus Steinmetz: A Biography* (New York: Century, 1924), pp. 216, 218–219; Horace Field Parshall, *The Parshall Family, A.D. 870–1913* (London: Francis Edwards, 1915), pp. 166–167; and Steinmetz, "Relation of

the Consulting Engineering Department to the Organization of the General Electric Company," June 1914, p. 2, CPSS.

5. Larry Hart, *Schenectady's Golden Era, 1880–1930*, 3rd ed. (Schenectady: Old Dorp Books, 1974), pp. ix-x, 1–7, 97, 291.

6. Hammond, *Steinmetz*, p. 218; and F. Tischendörfer to Steinmetz, 26 June 1894, CPSS. The quotation is in H. W. Bibber, "Personal," Steinmetz Centennial File, HWB.

7. Ernst Berg to Horatio Glenn, 2 April 1898, EJB; Steinmetz to Eskil Berg, 9 Jan. 1895; Steinmetz, Eickemeyer Patent Notebook, 20 Jan. 1900, p. 61, CPSS; Larry Hart, *Steinmetz in Schenectady: A Picture Story of Three Memorable Decades* (Schenectady: Old Dorp Books, 1978), sections 1–3, "family" photograph on p. 20; Hammond, *Steinmetz*, ch. 14; Floyd Miller, *The Electrical Genius of Liberty Hall: Charles Proteus Steinmetz, 1865–1923* (New York: McGraw-Hill, 1962); and George Wise, *Willis R. Whitney, General Electric, and the Origins of U.S. Industrial Research* (New York: Columbia Univ. Press, 1985), pp. 85–88.

8. Ernst Berg to J. Osky, 1 June 1898; Berg to Eskil Berg, 25 Nov. 1898; Berg to H. F. Barney and Co., 11 May 1899; Berg to William Close, 10 Aug. 1899, EJB; Steinmetz to J. H. Arnold, 29 March 1899; Steinmetz to W. Stoffregn, 17 April 1899; and Steinmetz to C. Hurd, 11 Sept. 1899, GEH; and William Le Roy Emmet, *Autobiography of an Engineer* (New York: ASME, 1940), p. 99.

9. "Proceedings of the Society for the Adjustment of Differences in Salaries," 1900 Folder, Box 5, CPSS.

10. Ernst Berg to "Sister," 3 Nov. 1899, EJB.

11. Ernst Berg to H. C. Senior, 22 Oct. 1912, EJB. Berg had written Steinmetz earlier, "I would ask as a special favor that Mr. Hayden will punch you every day or two until you have answered this letter." See Berg to Steinmetz, 17 March 1910, EJB.

12. Steinmetz, "Relation of the Consulting Engineering Department," p. 2.

13. Steinmetz to W. S. Moody, 26 March and 13 Sept. 1894, L5451–55, JWH; *Electrical World*, 26 (1895): 414–415; 27 (1896): 532–533; 33 (1899): 76–82; and "Walter Sherman Moody," *PTM Magazine*, Autumn 1931, copy in L5004–05, JWH.

14. On the Westinghouse system, see *Electrical Engineer*, 10 (1890): 153–154; and *Electrical World*, 19 (1892): 419–421; 20 (1892): 76–77.

15. See GE Power and Mining Department bulletins, 1896–1897, in GEH; and Edward L. Owen, "Induction Motor Developments at General Electric 1891 to Present," unpublished paper presented at the International Conference on Evolution and Modern Aspects of Induction Machines, Turin, Italy, July 1986. On the universal system, see Thomas P. Hughes, *Networks of Power: Electrification in Western Society, 1880–1930* (Baltimore: Johns Hopkins Univ. Press, 1983), pp. 121–139.

16. Steinmetz, "Discussion," *Trans AIEE*, 11 (1894): 387–392; Steinmetz, "Dielectric Strength of Air," *Trans AIEE*, 15 (1898): 281–326, on p. 293; Steinmetz, "Multiple Groove High Potential Insulator," 31 Jan. 1896, GE Report No. 4,515 (microfilm), GER; Lewis B. Stillwell, "The Electric Transmission of Power from Niagara Falls," *Trans AIEE*, 18 (1901): 445–531, on p. 502; and Steinmetz, "Review of Engineering Progress during November and December 1898," 5 Jan. 1899, L5458–67, JWH.

17. Steinmetz, Lynn Notebook, 19 July 1893, p. 56, CPSS; Elihu Thomson to

Steinmetz, 7 May 1894, ETA; Steinmetz, "Discussion," *Trans AIEE*, 11 (1894): 40–42; and Steinmetz to E. W. Rice, Jr., 14 Dec. 1896, L5772–78, JWH; and Steinmetz, "Discussion," *Trans AIEE*, 18 (1901): 425–428, quotation on p. 425.

18. Elihu Thomson to E. M. Bentley, 9 June and 12 Oct. 1893, ETA.

19. Steinmetz, Lynn Notebook, 6 Sept. 1893; "T.I. Motors [data sheet]," 1 Sept. 1893, L6671, JWH; *Electrical Engineer*, 17 (1894): 419; *Electrical World*, 23 (1894): 656–657; Louis Bell, "Electricity in Textile Manufacturing," *Cassier's Magazine*, 7 (1895): 275–284; Bell, U.S. Patent No. 505,505 (filed 14 June 1893); Steinmetz to A. G. Davis, 12 Jan. 1899, L5470–74, JWH; and Passer, *Electrical Manufacturers*, pp. 302–305.

20. Steinmetz to A. G. Davis, 12 Jan. 1899, L5470–74; John W. Hammond, "Notes from Old Papers of Charles P. Steinmetz," n.d., L5915–17, on L5916, JWH; and Steinmetz, "Discussion," *Trans AIEE*, 18 (1901): 425–428, quotation on p. 425.

21. U.S. Patent No. 587,340 (filed 20 Nov. 1893, issued 3 Aug. 1897); *Engineering News*, 38 (1897): 209; Steinmetz, "On the Use of Polyphase Motors on Electric Railroads," *Electrical World*, 31 (1898): 20–21; 44 (1904): 331–334; and A. H. Armstrong, "The Development of the Alternating-Current Railway Motor," *Street Railway Journal*, 24 (1904): 1111–1112.

22. *Electrical Engineer*, 18 (1894): 256–257; *Electrical World*, 24 (1894): 371–372; Steinmetz, "The Polycyclic System," 26 Nov. 1894, L5639–46; Steinmetz, "The Monocyclic System," 13 July 1895, L5435–37, JWH; Steinmetz, "The Monocyclic As Distinguished from the Polyphase System" (letter to the editor), *Electrical Engineer*, 18 (1894): 317; and, e.g., U.S. Patent No. 533, 244 (filed 2 April 1894, issued 29 Jan. 1895).

23. Louis Bell, "The Monocyclic System," *Electrical World*, 25 (1895): 302–306. On the debate, see *Electrical World*, 25 (1895): 296, 358, 371, 398–399.

24. *Electrical World*, 25 (1895): 611, 665; 26 (1895): 48; and D. C. Jackson and J. P. Jackson, *Alternating Currents and Alternating Current Machinery* (New York, 1896), pp. 673–678, quotation on p. 676. But the Jackson brothers also said that the monocyclic system was "essentially a single-phase system" (p. 677).

25. Steinmetz, "Some Features of Alternating Current Systems," *Trans AIEE*, 12 (1895): 326–349, on pp. 330–331; and Steinmetz, *Theory and Calculation of Alternating Current Phenomena*, 3rd ed. rev. (New York: W. J. Johnston, 1900), p. 247.

26. Steinmetz, *Theoretical Elements of Electrical Engineering* (New York: Electrical World and Engineer, 1901), p. 293.

27. Steinmetz, "Discussion," *Proceedings of the International Electrical Congress . . . , 1893* (New York, 1894), pp. 464–466; and Passer, *Electrical Manufacturers*, pp. 297, 314. For a comparison of the two systems, see Steinmetz to E. W. Rice, Jr., 9 Aug. 1895, L5438–40, JWH.

28. In 1898 Steinmetz could trace the invention of the monocyclic system only back to March 1894, but he thought it probably dated to his work at Lynn. See Steinmetz to A. G. Davis, 12 March 1898, L5750, JWH.

29. *Electrical World*, 24 (1894): 652; 26 (1895): 355; D. C. Jackson and S. B. Fortenbaugh, "Some Observations on a Direct-Connected 300 K.W. Monocyclic Alternator," *Trans AIEE*, 12 (1895): 350–353; Supplement to GE Power and Mining Dept., "Semimonthly Bulletin Letters Nos. 18 and 19," 1 and 15 Jan. 1896, p. 14, Commercial File, Box 19–3, GEH; *Electrical Engineer*, 21 (1896): 83, 482–486;

Steinmetz to Rice, 26 Feb. 1898, L5751–61, on L5754, JWH; Edwin J. Houston and Arthur E. Kennelly, *Recent Types of Dynamo-Electric Machinery* (New York: Collier, 1902), pp. 305–317; and W. J. Foster, "Early Days in Alternator Design," *General Electric Review,* 23 (1920), 81–90, on p. 85.

30. Frank H. Ray, Miami Cycle and Mfg. Co., to GE, 21 Feb. 1896, rpt. in Supplement to "Semimonthly Bulletin Letters Nos. 18 and 19," p. 12a. Jackson, *Alternating Currents,* p. 677, said, however, that most monocyclic equipment was sold for lighting purposes only.

31. Benjamin Lamme, "The Story of the Induction Motor," *Journal of the AIEE,* 40 (1921): 203–223, on p. 208.

32. Passer, *Electrical Manufacturers,* pp. 297, 314, 331–334.

33. *Electrical World,* 27 (1896): 428–430; GE, *Instructions for Installing and Operating Three-Phase and Monocyclic Apparatus* (Schenectady, 1897), copy in Commercial File, Box 22–10, GEH; *Electrical Engineer,* 25 (1898): 344–345; 26 (1898): 49–51; *Electrical Review,* 32 (1898): 220–221; and GE Commercial Dept., "Circular Letter No. 249," 10 Oct. 1898, p. 4, Commercial File, Box 19–3, GEH.

34. Steinmetz, "Systems of Electrical Distribution—IV," *Electrical Review,* 38 (1901): 528–529; and Foster, "Early Days," p. 85. Foster said the line was discontinued because lights were beginning to be placed on polyphase systems and that a "rating that could be given to a machine was much higher if it were polyphase than monocyclic" (pp. 85–86). Lamme noted the growing use of polyphase loads plus the "elimination of the patent situation" as the causes for the system's fading away. See Lamme, *Electrical Papers* (Pittsburgh, 1919), p. 604.

35. E. J. Berg, "Long-Distance Transmission of Power," *Electrical World,* 32 (1898): 206; E. W. Rice, Jr., to Steinmetz, 19 Jan. 1899, L5485; and Steinmetz to Rice, 14 Feb. 1899, L5486–95, JWH. On the stopper lamp, see Passer, *Electrical Manufacturers,* pp. 142–143, 160–161.

36. U.S. Patent Nos. 620,988 and 620,989 (filed 10 Sept. 1895 and 26 Jan. 1897, issued 14 March 1899); Steinmetz to A. G. Davis, 12 Jan. 1899, L5470–74, on L5473, JWH; Steinmetz, *AC Phenomena,* 5th ed. rev. (1916), p. 246; and Steinmetz, "Single-Phase Induction Motor," *Trans AIEE,* 15 (1898): 35–106, quotation on p. 84.

37. U.S. Patent Nos. 602,920 and 602,921 (filed 12 Feb. 1897, issued 26 April 1898); Steinmetz, "Review of Engineering Progress," L5461; and Philip Alger and Robert E. Arnold, "The History of Induction Motors in America," *Proceedings of the IEEE,* 64 (1976): 1380–1383.

38. James E. Brittain, "B. A. Behrend and the Beginnings of Electrical Engineering, 1870–1920" (Ph.D. diss., Case Western Reserve Univ., 1970); *Männer der Technik,* (Arnold); and *Dictionary of Scientific Biography* (Blondel and Fleming). On Hopkinson and Kittler, see Chapter 1 above. On Kennelly, Pupin, Thompson, and Ryan, see Chapter 2 above. Ryan consulted for the Los Angeles power bureau, Hopkinson for Edison, and Fleming for Edison and Marconi.

39. B. G. Lamme, "Induction Motor Characteristics and Curves," *Electric Journal,* 23 (1926): 451–454, 614–620; Lamme, *An Autobiography* (New York: G. P. Putnam's Sons, 1926), pp. 143–144; and "Inventor Tesla Replies to Dr. Louis Duncan, Explaining His Alternating Current Motor," *Electrical Review,* 12 (1888): 5.

40. Bernard Behrend, *The Induction Motor* (New York: Electrical World and Engineer, 1901), p. 2n.

41. Rice Circular Letter Nos. 85, 14 Dec. 1897, and 90, 8 April 1898, L5891–92, L5900, JWH; David E. Nye, *Image Worlds: Corporate Identities at General Electric, 1890–1930* (Cambridge, Mass.: MIT Press, 1985), pp. 14–15, 35–36; and Steinmetz, "[Review of D. C. Jackson's] *Alternating Currents . . . ,*" *Physical Review,* 4 (1897): 423–426, quotation on p. 424.

42. Gisbert Kapp, "Modern Continuous-Current Dynamo-Electric Machines and their Engines," *Proceedings of the Institution of Civil Engineers,* 83 (24 Nov. 1885): 123–154, quotation on p. 136.

43. Lamme, "Story of the Induction Motor," p. 216. Unless otherwise noted, this section is based on Ronald Kline, "Science and Engineering Theory in the Invention and Development of the Induction Motor, 1880–1900," *Technology and Culture,* 28 (1987): 283–313.

44. André Blondel, "Einige Bemerkungen über den Streuungs-Köefficienten bei Mehrphasenstrommotoren," *ETZ,* 16 (1895): 625–627; Gisbert Kapp, *Electric Transmission of Energy,* 4th ed. rev. (London, 1894), pp. 328–353; and Steinmetz, "Theory of the Induction Motor," *Trans AIEE,* 11 (1894): 754–760.

45. Kapp and Steinmetz defined leakage inductance (representing the decrease in induced voltage resulting from magnetic leakage) as the product of the number of coil turns and leakage flux, divided by current. This was similar to Maxwell's definition of the coefficient of self-induction, except that Maxwell's term referred to all flux produced by current in a coil: that surrounding only the coil (magnetic leakage) and that extending to another coil (mutual flux). Steinmetz separated these two fluxes into two parameters: leakage inductance (representing leakage flux) and primary admittance (representing mutual flux). Because the latter term was defined as the ratio of primary current to primary voltage at no load, it also took into account energy losses due to hysteresis and eddy currents. Although Duncan, Hutin, and Leblanc had obtained the correct general shape of the speed-torque curve by including the term self-inductance in their equations, Kapp and Steinmetz achieved more accurate curves by replacing this term with leakage inductance. On these relationships, see A. E. Fitzgerald and Charles Kingsley, Jr., *Electric Machinery* (New York: McGraw-Hill, 1961), pp. 13–26.

46. A. E. Kennelly, "Discussion," *Trans AIEE,* 11 (1894): 760; Rankin Kennedy, "The Early History of the Polyphase Motor," *Electrical Review* (London), 36 (1895): 409–410, on p. 409; and Michael Pupin, "Discussion," *Trans AIEE,* 37 (1918): 685–691, on p. 685.

47. *Electrical World,* 24 (1894): 25.

48. See, for example, Frederick Terman, *Radio Engineering* (New York: McGraw-Hill, 1932), pp. 123–135, which cites a 1919 paper on the equivalent circuit of the vacuum tube; *Reference Data for Radio Engineers,* 4th ed. rev. (New York: IT&T, 1956), pp. 500–502; and Joan Lisa Bromberg, "Engineering Knowledge in the Laser Field," *Technology and Culture,* 27 (1986): 798–818, on pp. 809–810.

49. Steinmetz, "Theory of the Induction Motor," p. 758.

50. Steinmetz, "The Alternating Current Induction Motor," *Trans AIEE,* 14 (1897): 185–217, on p. 193; and Steinmetz, "Discussion," ibid., pp. 220–221. The quotation is on p. 191.

51. "Discussion," *Trans AIEE*, 24 (1905): 685–686.

52. Steinmetz, *AC Phenomena*, 4th ed. (1908), pp. 9–11, 258–261; and Steinmetz, *Theory and Calculation of Electric Circuits* (New York: McGraw-Hill, 1917), pp. 216–231.

53. Steinmetz, "Alternating Current Induction Motor." This version appeared in Steinmetz, *AC Phenomena*, 3rd ed. (1900), pp. 262–265 and then in other textbooks and handbooks. See, for example, Harold Pender et al., eds., *Electrical Engineers Handbook: Electric Power*, 3rd ed. rev. (New York: Wiley, 1936), pp. 9-49–9-51.

54. Steinmetz, "Single Phase Induction Motor"; Galileo Ferraris, "A Method for the Treatment of Rotating or Alternating Vectors, with an Application to Alternating-Current Motors," *Electrician*, 33 (1894): 110ff.; and Steinmetz, "Notes on Single Phase Induction Motors and the Self-Starting Condenser Motor," *Trans AIEE*, 17 (1900): 25–61.

55. "Memorandum of Agreement [between W. J. Johnston and Steinmetz]," 4 April 1894, MHA; A. E. Kennelly to W. D. Weaver, 28 June 1895, cited in Hammond, "Notes," L5916; and Steinmetz to Weaver, 10 July 1895, GEH.

56. Steinmetz, *AC Phenomena*, 3rd ed. (1900), pp. 234–247, 274–281, 354–370. On the power curves, compare *AC Phenomena* (1897), pp. 275–290, with W. M. Mordey, "On Testing and Working Alternators," *Journal of the Institution of Electrical Engineers*, 22 (1893): 116–134. On the previous work on parallel alternators (of Hopkinson, Kapp, and Mordey), see John A. Fleming, *The Alternate Current Transformer in Theory and Practice*, 2 vols., 3rd ed. rev. (London, 1900), 2: 351–364.

57. Steinmetz "Alternating Current Motors," *Transactions of the International Electrical Congress . . . 1904*, 3 vols. (New York: IEC, 1905), 2: 76–128; and Philip Alger, *The Life and Times of Gabriel Kron* (Schenectady: Mohawk Development, 1969).

58. Steinmetz, *AC Phenomena*, 3rd ed. (1900), pp. 383–409.

59. Silvanus P. Thompson, *Polyphase Electric Currents and Alternate Current Motors*, 3rd ed. rev. (London: E. and F. Spon, 1901), p. 262; and Thompson, *Dynamo-Electric Machinery*, 2 vols., 7th ed. rev. (London: E. and F. Spon, 1904–1905), 1: 538–544, 2: 354–356. Steinmetz probably told Thompson about the rule in private correspondence or during a visit to America, because he had supplied Thompson with information on GE's AC machines in the 1890s. See Thompson to Steinmetz, 12 May 1898, EJB. Another rule was that the impedance of a long-distance transmission line should not be larger than twice its resistance in order not to have too large a voltage drop along the line. See Steinmetz to S. B. Paine, 17 Oct. 1893, Box 11, TAEF. Edward Owen gave me a tour of the electrical machinery engineering section of GE in Schenectady in June 1986.

60. Steinmetz, *Engineering Mathematics* (New York: McGraw-Hill, 1911), pp. 200–205. More practically oriented engineers like Bernard Behrend were more interested in developing empirical equations than rational ones. See Brittain, "Behrend," pp. 149–150.

61. Steinmetz, "Discussion," *Trans AIEE*, 12 (1895): 470–473, quotation on p. 471. Steinmetz may have adopted this terminology from Michael Pupin, who spoke before him during the discussion on this paper, saying, "In electrical engineering we have no necessity for empirical formula[e], at any rate not in this case, and it is not a step forward, but a step backward to substitute an empirical formula in place of an

exact formula." He also cited the practice of using empirical formulae as a "favorite method of mechanical engineers" (see p. 470).

62. Steinmetz, "Discussion," *Trans AIEE*, 11 (1894): 782.

63. Steinmetz, "The Dying Out of Alternate Current Waves," *Electrical Engineer*, 16 (1893): 543–544, quotation on p. 543.

64. Steinmetz, "Polyphase Motors," *Electrical Engineer*, 17 (1894): 430–431, quotation on p. 430.

65. Steinmetz, "Discussion," *Trans AIEE*, 27 (1908): 1555–1558, quotation on p. 1556.

66. Steinmetz, "Discussion," *Trans AIEE*, 30 (1911): 411–413, quotation on p. 413.

67. Steinmetz, "Discussion," *Trans AIEE*, 35 (1916): 21–24, quotation on p. 22.

68. Steinmetz, "Discussion," *Trans AIEE*, 12 (1895): 257–258.

69. Eskil Berg, Graph of I-8-30-900-111 induction motor, 30 Oct. 1895, and eleven pages of complex-number calculations for the motor, loose in Steinmetz, Lynn Notebook. For an example of a specification sheet of an induction motor, with one column for calculated values and another for tested values, see one by E. J. Berg for a motor for a Russian dredger, 27 Oct. 1897, EJB.

70. Steinmetz, "Discussion," *Trans AIEE*, 11 (1894): 34–39, quotation on p. 37. Similarly, GE engineer P. T. Hanscom told Steinmetz in about 1893, "You were very close in your estimate" on the electrical characteristics of a rewound induction motor. See Hanscom to Steinmetz, n.d., CPSS.

71. Steinmetz, "Discussion," *Trans AIEE*, 14 (1897): 220.

72. Cited in John A. McManus, "Fifty Years of Induction Motor Manufacture at the Lynn Works of the General Electric Company," 25 Nov. 1941, p. 6, GER, which reprints entries from a work journal kept in the 1890s by Hobart.

73. H. F. Parshall to Editor, 19 Dec. 1923, in *General Electric Review*, 27 (1924): 134.

74. Steinmetz, "Discussion," *Trans AIEE*, 14 (1897): 223; and W. I. Slichter to Ernst Berg, 3 Feb. 1899, EJB.

75. Ernst Berg to Steinmetz, 10 Jan. 1899, L5519–21, quotation on L5520. An undated document lists the members of the department as the Bergs, Slichter, Averrett, A. H. Kruesi, and S. Ferguson. See "C. P. Steinmetz's Office," n.d., L5690, JWH.

76. Steinmetz, "Relation of the Consulting Engineering Department," pp. 2, 3.

77. Ibid., p. 3. Shortly after moving to Schenectady Steinmetz told Moody at Lynn that the "Calculating Dept. does not exist any more, and we have to shift for ourselves." See Steinmetz to W. S. Moody, 26 March 1894, L5454–55, on L5455, JWH. Since Hammond, Emmet, and Steinmetz say that the Calculating Department existed up to the late 1890s, I have interpreted this letter to mean that the department was temporarily without a home in early 1894.

78. Steinmetz to E. W. Rice, Jr., 12 May (quotation) and 19 May 1899, L5505, L5503–04, JWH.

79. Steinmetz to J. P. Ord, 25 Sept. 1899, L5522; W. J. Foster to Steinmetz, 14 Aug. 1899, L5507, JWH; and Steinmetz to E. W. Rice, Jr., 2 Nov. 1899, GEH. On Steinmetz's increased consulting work, see several letters from Rice to Elihu Thomson, 11 Nov. 1899 to 5 March 1900, ETA.

80. Rice Circular Letter, Nos. 130, 6 April 1899, and 133, 13 May 1899, L5496, L5506; H. F. T. Erben to Steinmetz, 15 May 1899, L5510; E. M. Hewlett to Steinmetz, 16 May 1899, L5511; and W. J. Foster to Steinmetz, 14 Aug. 1899, L5507, JWH.

81. Rice Circular Letter No. 97, 4 May 1898, L5894–96; and W. B. Potter to Steinmetz, 23 Jan. 1899, L5516–18, JWH.

82. See A. E. Averrett, "Power Factor in Induction Motors," *General Electric Review,* 5 (1905): 44–45.

83. Emmet, *Autobiography,* pp. 111–114, quotation on p. 114; Edward D. Adams, *Niagara Power: History of the Niagara Falls Power Company, 1886–1918,* 2 vols. (Niagara Falls, 1927), 2: 248–249; and George Wise, "Inventors and Corporations in the Maturing Electrical Industry, 1890–1940," n.d., unpublished ms. in the author's possession. In contrast to Steinmetz, Emmet was a political conservative and an outspoken critic of Steinmetz's socialism but a "radical" designer. See Emmet, *Autobiography,* pp. 127, 154–155, 190–193. Neither Adams nor Emmet reveals what Emmet's "radical procedure" was.

84. Steinmetz to E. W. Rice, Jr., 3 June 1895, L5425–26, JWH.

85. John W. Hammond, *Men and Volts: The Story of General Electric* (New York: Lippincott, 1941), ch. 31; Passer, *Electrical Manufacturers,* pp. 310–313; and Hughes, *Networks of Power,* pp. 209–212. On Steinmetz's turbine work, see Charles Curtis to E. W. Rice, Jr., 27 Feb. 1905, CPSU; Steinmetz, "The Steam Turbine," *General Electric Review,* 7 (1906): 18–19; and Steinmetz to Rice, 15 and 27 Sept. 1910, CPSS.

86. Steinmetz to E. W. Rice, Jr., 31 Oct. 1895, L5448–49, JWH.

87. Arthur A. Bright, Jr., *The Electric Lamp Industry: Technological Change and Economic Development from 1800 to 1947* (New York: Macmillan, 1949), pp. 221–224.

88. Steinmetz to E. W. Rice, Jr., 31 Oct. 1899, L5484 (quotation); J. R. McKee to Charles Coffin, 16 Oct. 1899, L5565–66, JWH; Steinmetz, "What Are the Limits of High Potential Transmission," *Electrical Engineer,* 11 (1891): 442; and Steinmetz, "Discussion," *Trans AIEE,* 13 (1896): 312–315.

89. Passer, *Electrical Manufacturers,* pp. 309–310; and Hughes, *Networks of Power,* ch. 10.

90. Steinmetz, "Preliminary Report on the State of Alternating Current Engineering in Europe," 13 April 1897, L5779–87, quotations on L5781 and L5787, JWH; and Steinmetz to Clara Steinmetz, 6, 24, 27, and 28 Feb., 6 March 1897, CPSF. On the agreement between AEG and GE, see Hughes, *Networks of Power,* p. 179.

91. Steinmetz, "Review of Engineering Progress."

92. Rice Circular Letter No. 255, 6 Sept. 1901, L5703. Steinmetz read the committee reports and made suggestions. See Steinmetz to E. W. Rice, Jr., 18 April 1902, L5593–94, JWH.

93. Alger and Arnold, "History of Induction Motors," p. 1381; and Emmet, *Autobiography,* p. 106. Despite their disagreement over the Niagara alternators, Steinmetz praised Emmet in 1910, when Berg had left GE, as the "most prominent of our engineers." See Steinmetz to G. Alexander, 16 Feb. 1910, CPSU.

94. *Electrical Review,* 32 (1898): 243–244; 38 (1901): 333–340, 499; and Daniel Nelson, *Managers and Workers: Origins of the New Factory System in the United States* (Madison: Univ. of Wisconsin Press, 1975), p. 7.

95. U.S. Bureau of the Census, *The Statistical History of the United States from Colonial Times to the Present*, 2 vols., rev. ed. (Stamford, Conn.: Fairfield, 1965), p. 510; Richard DuBoff, "The Introduction of Electric Power in American Manufacturing," *Economic History Review*, 2nd series, 20 (1967): 509–518; Richard H. Schallenberg, "The Anomalous Storage Battery: An American Lag in Early Electrical Engineering," *Technology and Culture*, 22 (1981): 725–752; Nadja Maril, *American Lighting, 1840–1940* (West Chester, Pa: Schiffer, 1989), ch. 10; Passer, *Electrical Manufacturers*, pp. 106–108, 120, 197; Forrest McDonald, *Insull* (Chicago: Univ. of Chicago Press, 1962), pp. 104–105; David E. Nye, *Electrifying America: Social Meanings of a New Technology, 1880–1940* (Cambridge, Mass.: MIT Press, 1990), chs. 5–6; and Bright, *Electric Lamp Industry*, pp. 126–127.

Chapter 6: New Settings for Research

1. Leonard S. Reich, *The Making of American Industrial Research: Science and Business at GE and Bell, 1876–1926* (New York: Cambridge Univ. Press, 1985); and George Wise, *Willis R. Whitney, General Electric, and the Origins of U.S. Industrial Research* (New York: Columbia Univ. Press, 1985), chs. 6–7.

2. Robert Friedel et al., *Edison's Electric Light: Biography of an Invention* (New Brunswick: Rutgers Univ. Press, 1986); W. Bernard Carlson, "Thomas Edison as a Manager of R&D: The Case of the Alkaline Storage Battery, 1898–1915," *IEEE Technology and Society Magazine*, 7 (Dec. 1988): 4–12; and Andre Millard, *Edison and the Business of Innovation* (Baltimore: Johns Hopkins Univ. Press, 1990), esp. pp. 157–159.

3. John A. Miller, *Workshop of Engineers: The Story of the General Engineering Laboratory of the General Electric Company* (Schenectady: GE, 1953); Steinmetz to E. W. Rice, Jr., 9 July 1897, L5791–93, JWH; and W. Bernard Carlson, "Invention, Science, and Business: The Professional Career of Elihu Thomson, 1870–1900" (Ph.D. diss., Univ. of Pennsylvania, 1984), pp. 258, 366, 370, 407–409.

4. Steinmetz, "Request for experimental mfg. order," 14 Feb. 1896, EJB; and Friedel, *Edison's Electric Light*, pp. 34–35.

5. Ronald Kline, "The Origins of Industrial Research at the Westinghouse Electric Company, 1886–1922," a chapter in a forthcoming book on the history of R&D in the United States, ed. David Hounshell, Princeton University Press.

6. Steinmetz to E. W. Rice, Jr., 21 Sept. 1900, ETG. I would like to thank W. Bernard Carlson for sending me a copy of this and related letters from the Thomson Papers.

7. Arthur A. Bright, Jr., *The Electric Lamp Industry: Technological Change and Economic Development from 1800 to 1947* (New York: Macmillan, 1949), pp. 122, 126–129, 168–176, 218–229.

8. Steinmetz to E. W. Rice, Jr., 18 Dec. 1894, L5450; Steinmetz to Rice, 18 Jan. 1896, excerpt from Hammond, "Notes from Old Papers of Charles P. Steinmetz," L5917; Steinmetz to A. G. Davis, 13 July 1898, L5475–80; and Steinmetz, "Report on Preliminary Investigation of Luminescent Lighting," n.d. [c. 1899], L5525–33. In 1898 GE abandoned the experiments on metallic oxides because the results did not look promising. See Steinmetz, "Review of Engineering Progress during November and December 1898," 5 Jan. 1899, L5458–67, on L5463, JWH.

9. Steinmetz to J. W. Howell, 23 July 1897, L5788–89, JWH.

10. E. W. Rice, Jr., to Elihu Thomson, 16, 17, and 27 March 1899, ETA; and Bright, *Electric Lamp Industry*, pp. 224–229.

11. Steinmetz, "Investigation of Luminescent Lighting," quotation on L5531; and Steinmetz, Eickemeyer Patent Notebook, 20 Jan. 1900, p. 61, CPSS. He had written two short articles on electrochemistry at Eickemeyer's; see "Hydrochloric Acid in Bichromate Cells,"*Electrical Engineer*, 11 (1891): 468, 544.

12. Bright, *Electric Lamp Industry*, p. 227.

13. Steinmetz to E. W. Rice, Jr., 9 July 1897, L5791–93; Steinmetz to J. W. Howell, 23 July 1897, L5788–89; Steinmetz, "Review of Engineering Progress," L5467; Steinmetz to Rice, 20 Sept. 1897, L5794–95; and Steinmetz to A. G. Davis, 13 July 1898, L5475–80, JWH.

14. Steinmetz, "Review of Engineering Progress," L5467, JWH; Elihu Thomson to E. W. Rice, Jr., 11 April 1899; and Rice to Thomson, 17 April 1899, ETA.

15. Bright, *Electric Lamp Industry*, pp. 172–173; Wise, *Whitney*, pp. 76–77; and Millard, *Edison*, pp. 154–156.

16. Steinmetz to E. W. Rice, Jr., 21 Sept. 1900, ETG.

17. Elihu Thomson to Steinmetz, 24 Sept. 1900, ETG. On Thomson's reputation as a scientist, see W. Bernard Carlson, "Elihu Thomson: Man of Many Facets," *IEEE Spectrum*, 20 (Oct. 1983): 72–75; and Harold J. Abrahams and Marion B. Savin, eds., *Selections from the Scientific Correspondence of Elihu Thomson* (Cambridge, Mass.: MIT Press, 1971). Whitney tried several times to lure Thomson to the GE Research Laboratory. In 1913, for example, Whitney complained about the need to do so much applied work and thought his researchers "would appreciate the presence of a dean" like Thomson to discuss basic science. See W. R. Whitney to Thomson, 9 Feb. 1913, in Abrahams and Savin, *Thomson Correspondence*, pp. 538–539.

18. Charles Cross to Elihu Thomson, 1 Oct. 1900; and Steinmetz to Thomson, 8 Oct. 1900, ETG. On Cross, see Robert Rosenberg, "The Origins of E.E. Education: A Matter of Degree," *IEEE Spectrum*, 21 (July 1984): 60–68.

19. Wise, *Whitney*, chs. 4–5.

20. Steinmetz report, n.d. [c. Oct.–Dec. 1900], L5699–701, JWH.

21. Wise, *Whitney*, p. 85.

22. Steinmetz to Elihu Thomson, n.d. [shortly after 9 Feb. 1901), rpt. in John A. Miller, *Modern Jupiter: The Story of Charles Proteus Steinmetz* (New York: ASME, 1958), p. 100. On the date of the fire, see "Proceedings of the Society for the Adjustment of Differences in Salaries," 22 Jan. 1901, 1900 Folder, Box 5, CPSS; and Larry Hart, *Steinmetz in Schenectady: A Picture Story of Three Memorable Decades* (Schenectady: Old Dorp Books, 1978), pp. 111, 127–128.

23. Steinmetz to J. Riddell, 8 Feb. 1901, L5697; Steinmetz to E. W. Rice, Jr., 19 Feb. 1901, L5695; Steinmetz to A. L. Rohrer, 25 Feb. 1901, L5696; Steinmetz to G. E. Emmons, 27 March 1901, L5698, JWH; and Steinmetz to E. J. Berggren, 4 March 1901, GEH.

24. Steinmetz, Patent Notebook, 20 Jan. 1900, p. 61; Hart, *Steinmetz*, pp. 134–139, 143–144; and Bruce Maston, *An Enclave of Elegance: A Survey of the Architecture, Development, and Personalities of the General Electric Realty Plot Historical District* (Schenectady: GERPA, 1983), pp. 9, 11, 28–29, 102. Steinmetz's research at Liberty

Street was not entirely independent of GE, since an experimental manufacturing order covered part of his work there. See, for example, Steinmetz to A. L. Rohrer, 30 March and 4 April 1899, GEH.

25. Ernst Berg to G. Travers, 29 Nov. 1899; H. F. Huntington to Berg, 12 March 1900; Berg to Messrs. Manning and Hardy, 22 May 1900; Berg to H. F. Stimpson, 11 July 1900; Receipt from Vale Cemetery, 13 Oct. 1900; Berg to Mary de Ikalac, 12 Feb. 1901; Berg to Oscar Peterson, 24 May 1901; Berg to Schenectady Realty Co., 4 June 1901, EJB; Steinmetz to Stimpson, 4 June 1901, GEH; "Proceedings of the Society for the Adjustment of Differences in Salaries," 1 Aug. 1901; Steinmetz to A. P. Strong, 2 Aug. and 21 Sept. 1901, GEH; John W. Hammond, *Charles Proteus Steinmetz: A Biography* (New York: Century, 1924), pp. 244–245; Hart, *Steinmetz*, pp. 143–145, 227, 229; and Hayden's obituary in the *Schenectady Gazette*, 13 Aug. 1951. In 1903 Berg agreed to sell his lot next to Steinmetz, who wanted more land. See Schenectady Realty Co. to Berg, 7 Jan. 1903; and Berg to H. W. Darling, 26 March 1903, EJB.

26. Willis Whitney, Laboratory Notebook, 20 May 1901, 8 Aug. 1901, 11 Jan. 1902, pp. 40, 86, 196, reel 665, Whitney Library, GER; Wise, *Whitney*, pp. 96 (quotation), 98–106; Wise, "Ionists in Industry: Physical Chemistry at General Electric, 1900–1915," *Isis*, 74 (1983): 7–21, on p. 15; and Reich, *Making of American Industrial Research*, p. 71.

27. U.S. Patent Nos. 897,801 (filed 28 Dec. 1900, issued 1 Sept. 1908), 806,758 (filed 21 Jan. 1901, issued 5 Dec. 1905), and 701,959 (filed 6 Nov. 1901, issued 10 June 1902). On the luminous arc lamp, see Bright, *Electric Lamp Industry*, pp. 214–216.

28. U.S. Patent No. 914,891 (filed 26 Feb. 1902, issued 9 March 1909); Steinmetz, "Magnetite Arc Lamp," *Electrical World*, 43 (1904): 974–975; Steinmetz to N. R. Birge, 29 May 1909; Steinmetz to W. C. Fish, 25 Jan. 1911, CPSS; Elihu Thomson, "Development of Arc Lighting," *Electrical World*, 80 (1922): 542–543; and Bright, *Electric Lamp Industry*, p. 217.

29. Whitney, Laboratory Notebook, 15 Feb., 3 and 12 April, 3 May, 3 June 1902, pp. 209, 228, 230, 239, 247–248; A. G. Davis to W. R. Whitney, 2 May 1902, ETA; E. Weintraub, "The Mercury Arc Lamp and Rectifier," *Electrical World*, 45 (1905): 1031–1034; Steinmetz, "Constant-Current Mercury-Arc Rectifier," *Trans AIEE*, 24 (1905): 371–393; Hammond, *Steinmetz*, pp. 303–310; and Kendall Birr, *Pioneering in Industrial Research: The Story of the General Electric Research Laboratory* (Washington, D.C.: Public Affairs Press, 1957), p. 36.

30. Reich, *Making of Industrial Research*, p. 72; and Elihu Thomson to E. W. Rice, Jr., 8 Oct. 1903, in Abrahams and Savin, *Thomson Correspondence*, pp. 422–424. The photograph is published in Wise, *Whitney*, p. 109.

31. Steinmetz, Patent Notebook, 1 Jan. 1911, quotation on p. 81 (shorthand entry); Ernst Berg to Eskil Berg, 5 May 1913, EJB; Hammond, *Steinmetz*, pp. 278–280, 426–433; and Hart, *Steinmetz*, p. 145, 149. The Schenectady city directory lists Clara at Park Avenue and Union Avenue from 1903 to 1906.

32. Charles P. Matthews, "Discussion," *Trans AIEE*, 19 (1902): 1188; Charles W. Eliot to Steinmetz, 17 May 1902, HUA; *Boston Evening Transcript*, 25 June 1902, p. 1 (quotation); Steinmetz, "Presidential Address," *Trans AIEE*, 19 (1902): 1145–1150; and E. W. Rice, Jr., to C. A. Coffin, 10 Nov. 1902, copy in Union College

Trustees Minutes, 11 Dec. 1902, Book F, pp. 162–163, SL. On building Camp Mohawk, see F. W. Hugo to Steinmetz, 6 March 1899; and Steinmetz to Hugo, 24 April 1899, GEH.

33. "Dr. Steinmetz's Own Estimate of His Work," Dec. 1923, rpt. in Emil J. Remscheid, *Recollections of Steinmetz* (Schenectady: GE, 1977), pp. 62–67.

34. Steinmetz, "Photographic Investigation of a 150,000-Volt Power Discharge," *Electrical World*, 31 (1898): 294–295.

35. Steinmetz, "Dielectric Strength of Air," *Trans AIEE*, 15 (1898): 281–326; and Steinmetz, "Note on the Disruptive Strength of Dielectrics," *Trans AIEE*, 10 (1893): 85–105, on pp. 91–93, 101. He had also performed pioneering research on dielectric hysteresis at Eickemeyer's. See Steinmetz, "Dielectric Hysteresis—The Loss of Energy in a Dielectric Medium under Alternating Electrostatic Strain," *Electrical Engineer*, 13 (1892): 272–273.

36. Steinmetz, "Photographic Investigation," p. 295.

37. Steinmetz, "Discussion," *Trans AIEE*, 8 (1891): 167–175; Oliver Lodge, "On Lightning, Lightning Conductors, and Lightning Protectors," *Journal of the Institution of Electrical Engineers*, 18 (1889): 386–430; Steinmetz, Patent Notebook, 13 May 1892; Steinmetz, "Discussion," *Trans AIEE*, 11 (1894): 387–392; and Steinmetz, "The Natural Frequency of a Transmission Line and the Frequency of Lightning Discharges Therefrom," *Electrical World*, 32 (1898): 203–205.

38. Lewis B. Stillwell, "The Electric Transmission of Power from Niagara Falls," *Trans AIEE*, 18 (1901): 445–531, on p. 495; *Electrical World*, 32 (1898): 577–584; 35 (1900): 541–544; Max Vogelsang, *Die geschichtliche Entwicklung der Hochspannungs-Schalttechnik* (Berlin: Springer, 1929), chs. 5–9; and Roy Wilkins and E. A. Crellin, *High Voltage Oil Circuit Breakers* (New York: McGraw-Hill, 1930), pp. 3–34.

39. Steinmetz, "Theoretical Investigations of Some Oscillations of Extremely High Potential in Alternating High Potential Transmissions," *Trans AIEE*, 18 (1901): 383–405.

40. Ibid., pp. 387–405.

41. Edward James Hart, "Kalamazoo Valley, Michigan, Transmission Plant," *Electrical World*, 34 (1899): 483–485; and E. Hardy Luther, "Early Developments in High Voltage Transmission," *Michigan History*, 51 (1967): 93–115.

42. Steinmetz, "Discussion," *Trans AIEE*, 19 (1902): 269–271; and E. W. Rice, Jr., "The Control of High Potential Systems of Large Power," *Trans AIEE*, 18 (1901): 407–420.

43. *Electrical World*, 37 (1901): 10–14, quotation on p. 10.

44. Ibid., pp. 89–93, quotation on p. 91. See also L. B. Stillwell, "The Electric Power Plant of the Manhattan Railway Company," *Street Railway Journal*, 17 (1901): 21–47.

45. *New York Times*, 24 July 1903, p. 1.

46. *Scientific American*, 86 (1902), 21–22; Harold C. Passer, *The Electrical Manufacturers, 1875–1900* (Cambridge, Mass.: Harvard Univ. Press, 1953), p. 312; Steinmetz, "High Power Surges in Electrical Distribution Systems of Great Magnitude," *Trans AIEE*, 24 (1905): 297–315; and H. G. Stott, "Incidents in the Operation of a Large Power Plant and Distribution System," *Electric Club Journal*, 2 (1905): 278–283.

47. *Electrical World*, 34 (1899): 890. On Stott's GE affiliation, see *Trans AIEE*, 14 (1897–1898): 634.

48. Percy Thomas, "An Experimental Study of the Rise of Potential on Commercial Transmission Lines due to Static Disturbances . . . ," *Trans AIEE*, 24 (1905): 317–354; Thomas, "Discussion," ibid., pp. 362–365; and "Proceedings of the Asheville Meeting of the A.I.E.E.," *Electrical World*, 45 (1905): 1156 (quotation).

49. Steinmetz, "High Power Surges"; H. G. Stott, "Discussion," *Trans AIEE*, 24 (1905): 355–357.

50. E. W. Rice, Jr., to J. R. McKee, 29 May 1901, ETA; and U.S. Patent Nos. 897,800 (filed 28 Dec. 1900, issued 1 Sept. 1908) and 860,997 (filed 28 Dec. 1900, issued 23 July 1907).

51. Steinmetz to George Alexander, 12 Aug. 1908, CPSU; *American Men of Science*, 3rd ed. (1921), s.v., "Creighton, Prof. Elmer E. F"; and Creighton, "Discussion," *Trans AIEE*, 25 (1906): 447.

52. A. J. Wurts, "The Engineering Evolution of Electrical Apparatus—XXI & XXII: The History of the Lightning Arrester," *Electric Journal*, 13 (1916): 187–191, 209–212; and Steinmetz, "Discussion," *Trans AIEE*, 25 (1906): 427–435.

53. Steinmetz, "Discussion," *Trans AIEE*, 25 (1906): 431; Steinmetz, "Lightning Phenomena in the Clouds," *Electrical Review*, 51 (1907): 167–169; Steinmetz, *General Lectures on Electrical Engineering*, ed. J. L. R. Hayden, 2nd ed. (Schenectady: Robson and Adee, 1908), pp. 259–284; Bernhard Walter, "Ueber die Entstehungsweise des Blitzes," *Annalen der Physik*, 315 (1903): 393–407; and Alex Larson, "Photographing Lightning with a Moving Camera," Smithsonian Institution, *Annual Report*, 1905, 1: 119–127.

54. B. F. J. Schonland, *Flight of Thunderbolts* (Oxford: Clarendon Press, 1950); and Steinmetz, *General Lectures*, pp. 267–270.

55. Steinmetz, "Lightning Phenomena in Electric Circuits," *Trans AIEE*, 26 (1907): 401–423.

56. Steinmetz, *General Lectures*, pp. 137–146.

57. See, for example, Steinmetz, "Lightning Arresters and Lightning Protection," GE Technical Report, 1 March 1905, EJB.

58. U.S. Patent No. 923,024 (filed 9 July 1906, issued 29 May 1909); *Electrical World*, 51 (1908): 783, 990–991; R. P. Jackson, "The Engineering Evolution of Electrical Apparatus—XXIII: The History of the Lightning Arrester," *Electric Journal*, 13 (1916): 299–301; Creighton, "Methods of Testing Protective Apparatus," *Trans AIEE*, 25 (1906): 365–397; Creighton, "New Principles in the Design of Lightning Arresters," *Trans AIEE*, 26 (1907): 461–486; Creighton, "Protective Apparatus Engineering," ibid., pp. 1049–1095; and Creighton, "Measurements of Lightning, Aluminum Lightning Arresters . . . ," *Trans AIEE*, 27 (1908): 669–740.

59. Steinmetz to E. W. Rice, Jr., 7 Oct. 1908, CPSS.

60. Percy Thomas, "Static Strains in High Tension Circuits and the Protection of Apparatus," *Trans AIEE*, 19 (1902): 213–264; Steinmetz, "High Power Surges," pp. 305–309; and Steinmetz, "Lightning Phenomena in Electric Circuits," p. 411.

61. Steinmetz, "Complex Quantities and Their Use in Electrical Engineering," *Proceedings of the International Electrical Congress . . . , 1893* (New York, 1894), pp. 33–74, on pp. 57–68.

62. Steinmetz, "The General Equations of the Electric Circuit," *Trans AIEE*, 27 (1908): 1231–1305.

63. Oliver Heaviside, *Electrical Papers*, 2 vols. (London, 1892), 1: 139–140, 2: 91, 105, 123, 194–195, 247; Heaviside, *Electromagnetic Theory*, 3 vols. (London, 1893–1912), 1: 449–452; and Steinmetz, "General Equations," pp. 1282–1305.

64. D. C. Jackson, "Discussion," *Trans AIEE*, 27 (1908): 1307–1310, quotation on p. 1307. On Kelvin's and Heaviside's equations, see D. W. Jordan, "The Adoption of Self-Induction by Telephony, 1886–1889," *Annals of Science*, 39 (1982): 433–461.

65. James E. Brittain, "The Introduction of the Loading Coil: George A. Campbell and Michael I. Pupin," *Technology and Culture*, 11 (1970): 36–57; and Neil H. Wasserman, *From Invention to Innovation: Long-Distance Telephony at the Turn of the Century* (Baltimore: Johns Hopkins Univ. Press, 1985).

66. Steinmetz to O. A. Kenyon, 4 May, 17 and 23 Nov., 31 Dec. 1908, CPSS.

67. Ernst J. Berg, *Electrical Engineering Advanced Course* (New York: McGraw-Hill, 1916), p. v; *Electrical World*, 52 (1909): 131; *ETZ*, 30 (1909): 590; *Electrical Engineering* (London), 5 (1909): 719; and *Electrician*, 64 (1909): 399–400, on p. 399.

68. Steinmetz, "Surges and Oscillations," *General Electric Review*, 10 (1907): 91–94; Steinmetz, "Transient Electric Phenomena," ibid., pp. 38–44; Steinmetz, "Electric Transients," *Journal of the Franklin Institute*, 172 (July 1911), 39–54; and Steinmetz, *Elementary Lectures on Electric Discharges, Waves, and Impulses and Other Transients* (New York: McGraw-Hill, 1911), quotation on p. vi.

69. Steinmetz, Patent Notebook, 1 Jan. 1911, p. 81 (quotation) (shorthand entry); and Steinmetz to Paul Spencer, 19 Nov. 1908, CPSS.

70. Steinmetz, Patent Notebook, 1 Jan. 1911, p. 81 (shorthand entry).

71. Steinmetz to E. W. Rice, Jr., 7 Oct. 1908, CPSS. On the size of Whitney's staff, see Wise, *Whitney*, p. 125.

72. Steinmetz to E. W. Rice, Jr., 7 Oct. 1908, CPSS.

73. Wise, *Whitney*, pp. 128–138.

74. Steinmetz to E. W. Rice, Jr., 7 Oct. 1908, CPSS.

75. [Steinmetz], "Consulting Department," 27 Dec. 1909, CPSS.

76. Rice Circular Letter, 9 May 1910, extract from "General Electric's Departmental System," L6189, JWH.

77. James E. Brittain, "C. P. Steinmetz and E. F. W. Alexanderson: Creative Engineering in a Corporate Setting," *Proceedings of the IEEE*, 64 (1976): 1413–1417; and Hugh G. J. Aitken, *The Continuous Wave: Technology and American Radio, 1900–1932* (Princeton: Princeton Univ. Press, 1985), pp. 64–68.

78. Steinmetz to E. W. Rice, Jr., 29 July and 18 Sept. 1910; H. M. Hobart to Steinmetz, 30 Sept. 1910; John Taylor to W. S. Moody, 2 Dec. 1910; Steinmetz, Patent Notebook, 1 Jan. 1911, p. 81; Steinmetz to E. F. Collins, 25 Feb. and 13 March 1911, CPSS; and Miller, *Workshop of Engineers*, pp. 16–29. On Slichter's going to Columbia, see *Citizen*, 23 Sept. 1910.

79. See, for example, E. E. F. Creighton to C. W. Stone, 16 Nov. 1910, CPSS; and F. W. Peek, Jr., "Consulting Engineering Dept. Council Meeting," 13 Jan. 1912, SPN.

80. Steinmetz, "Organization of [the] Consulting Engineering Department and of Laboratories Connected Therewith . . . ," 10 July 1912, p. 7, SPN.

81. Steinmetz to W. R. Whitney, 26 April 1910, CPSS. Most of the administrative work was done by C. W. Stone. When Stone left to become head of the Lighting Department in 1912, he was replaced by Hayden and then G. R. Barksdale. See H. C. Senior to J. T. Broderick, 30 Sept. 1912, CPSS; and *General Electric Review,* 15 (1912): 345.

82. Steinmetz to E. W. Rice, Jr., 3 May 1912, CPSS.

83. Steinmetz to C. A. Richmond, 4 Jan. 1911, CPSU.

84. Steinmetz, "Review of Some Work in Progress in Dr. Steinmetz's Laboratory," 7 April 1909; "Report of Work of Dr. Steinmetz's Laboratory during 1910," 2 Feb. 1911; J. L. R. Hayden to W. B. Curtiss, 22 Sept. 1910, CPSS; and Steinmetz, "Power Characteristics of the Tungsten Filament," *Electrical World,* 54 (1909): 89.

85. Remscheid, *Recollections,* pp. 4–14; Steinmetz to J. T. Broderick, 21 Dec. 1908, 23 Jan. 1909, 8 April 1911; Steinmetz to W. B. Curtiss, 6 July 1908 and 15 Jan. 1909; Steinmetz to J. Riddell, 23 March 1908; Steinmetz to L. T. Robinson, 30 Oct. 1908; Steinmetz to C. C. Chesney, 9 Aug. 1909; Steinmetz to E. F. Peck, 11 June 1908; Steinmetz to E. J. Berggren, 22 June and 27 Nov. 1908; J. T. Broderick to Steinmetz, 26 Aug. 1910; and Steinmetz to Schenectady Illuminating Company, 11 Nov. 1916, CPSS.

86. Steinmetz, Patent Notebook, 28 Feb. 1911, p. 85. In comparison, Whitney, who ran a much larger facility, made $20,000 in 1925. See Wise, *Whitney,* pp. 240–241. But Owen Young started at $25,000 per year as vice president in 1913. See Josephine Y. Case and Everett N. Case, *Owen D. Young and American Enterprise: A Biography* (Boston: David R. Godine, 1982), pp. 103–106. Steinmetz said in 1911 that Hayden "has been in charge of my electrochemical laboratory since some years." See Steinmetz to W. E. Holland, 21 Jan. 1911, CPSS. Hayden's salary was quite low in comparison with Steinmetz's: $25 per week in 1905, $35 per week in 1907, and $45 per week in 1910; see his work contracts in CPSS.

87. Steinmetz to C. A. Coffin, 9 Jan. 1913, CPSS; and Miller, *Workshop of Engineers,* pp. 25–27.

88. "Review of Some Work in Steinmetz's Laboratory," pp. 6, 14; Steinmetz to E. W. Rice, Jr., 13 Sept. 1910, CPSS; Bryan M. Vanderbilt, *Thomas A. Edison, Chemist* (Washington, D.C.: American Chemical Society, 1971), pp. 238–240; Millard, *Edison,* pp. 213, 305; and Robert Friedel, *Pioneer Plastic: The Making and Selling of Celluloid* (Madison: Univ. of Wisconsin Press, 1983), pp. 103–107. Edison organized the Condensite Company of America in Sept. 1909 to manufacture the product, which found wide use in Edison's disc records.

89. Steinmetz to W. R. Whitney, 13 and 19 Sept. 1910; and Whitney to Steinmetz, 16 Sept. 1910, CPSS.

90. J. L. R. Hayden to A. McKay Gifford, 5 Oct. 1910; Steinmetz to E. W. Rice, Jr., 19 Nov. 1910; Steinmetz to C. C. Chesney, 16 Jan. 1911; Chesney to Steinmetz, 18 Jan. 1911; C. W. Stone to W. S. Moody, 19 Jan. 1911; Steinmetz to L. E. Barringer, 24 Jan. 1911; Steinmetz to H. L. Schermerhorn, 2 Feb. 1911; Barringer to Steinmetz, 10 Feb. 1911, CPSS; Steinmetz to Edison, 11 Oct. 1910 and 25 Jan. 1911; and Steinmetz to W. H. Meadowcroft, 21 Jan. 1911, TAE.

91. "Report of Steinmetz Laboratory, 1910," p. 2; C. E. Dept. Report, n.d. [c. 1912], p. 5, SPN; Steinmetz to C. C. Chesney, 21 May 1912, rpt. in Herman Liebhafsky et al., *Silicones under the Monogram* (New York: John Wiley, 1978), pp. 33–

34; U.S. Patent No. 1,215,072 (filed 2 Jan. 1915, issued 6 Feb. 1917); L. E. Barringer, "A Revolutionary Development in Mica Insulation," *General Electric Review,* 29 (1926): 757–762; and Birr, *Pioneering in Industrial Research,* pp. 114–116. Introduced in the mid-1920s as a bonding agent for high-voltage mica insulation and a general-purpose plastic for such novelty items as cigar holders, Glyptal later found much more use as an automotive paint.

92. Steinmetz to W. R. Whitney, 18 Feb. 1909; J. L. R. Hayden to Whitney, 1 March 1909, CPSS; "Review of Some Work in Steinmetz's Laboratory," p. 13; Steinmetz to Whitney, 10 Sept. and 2 Dec. 1910; and Whitney to Steinmetz, 16 Sept., 5 Dec. 1910, 14 April 1911, CPSS.

93. Thomas P. Hughes, "The Science-Technology Interaction: The Case of High-Voltage Power Transmission Systems," *Technology and Culture,* 17 (1976): 646–662; A. B. Headrick, "Discussion," *Trans AIEE,* 30 (1911): 115–119; Steinmetz to E. W. Rice, Jr., 16 Nov. 1908; and Steinmetz to Guiseppi Faccioli, 9 May 1910, CPSS.

94. John W. Hammond, "A Tribute to F. W. Peek, Jr.," *General Electric Review,* 36 (1933): 383–384; and Steinmetz to C. A. Richmond, 4 Jan. 1911, CPSS.

95. F. W. Peek, Jr., "The Law of Corona and the Dielectric Strength of Air," *Trans AIEE,* 30 (1911): 1889–1965, on pp. 1901–1906; Steinmetz to E. J. Berg, 22 Nov. 1910; Peek to Giuseppe Faccioli, 13 Dec. 1910; and Steinmetz to D. C. Jackson, 18 May 1911; CPSS.

96. F. W. Peek, Jr., "The Law of Corona and the Dielectric Strength of Air [Parts I, II, III]," *Trans AIEE,* 30 (1911), 1889–1965; 31 (1912): 1051–1092; 32 (1913): 1767–1785.

97. Hughes, "Science-Technology Interaction," 649–654; Hughes, *Networks of Power: Electrification in Western Society, 1880–1930* (Baltimore: Johns Hopkins Univ. Press, 1983), pp. 380–384; and John Whitehead "The Electric Strength of Air," *Trans AIEE,* 29 (1910): 1159–1187.

98. H. J. Ryan, "Discussion," *Trans AIEE,* 30 (1911): 1979, 1983; Peek, *Dielectric Phenomena in High-Voltage Engineering* (New York: McGraw-Hill, 1915); and Hughes, "Science-Technology Interaction," p. 655.

99. Peek, "Law of Corona—I," pp. 1904, 1910, 1941, 1955; and Steinmetz, "Scientific Research in Its Relation to the Industries," *Journal of the Franklin Institute,* 182 (1916): 711–718.

100. L. T. Robinson to F. C. Pratt, 9 Dec. 1919, CPSS; Miller, *Workshop of Engineers,* pp. 27, 43; and conversation with George Wise.

101. Wise, *Whitney,* pp. 149–158.

102. Reich, Making of Industrial Research, pp. 205–213; and Kline, "Origins of Industrial Research at Westinghouse."

103. Steinmetz to O. A. Kenyon, 23 March 1908, CPSS.

104. J. M. Cattell to Steinmetz, 27 July 1909, CPSS; Ernst Berg to Steinmetz, 1 Feb. 1913, EJB; and Steinmetz, "Scientific Research in Its Relation to the Industries".

Chapter 7: Reforming a Profession

1. Robert H. Wiebe, *The Search for Order, 1877–1920* (New York: Hill and Wang, 1967), ch. 5. For an excellent account of the relevant literature, see Daniel T.

Rodgers, "In Search of Progressivism," *Reviews in American History*, 10 (1982): 113–132.

2. Nathan O. Hatch, ed., *The Professions in American History* (Notre Dame: Univ. of Notre Dame Press, 1988); Edwin T. Layton, Jr., *The Revolt of the Engineers: Social Responsibility and the American Engineering Profession* (1971; rpt. Baltimore: Johns Hopkins Univ. Press, 1986); and A. Michal McMahon, *The Making of a Profession: A Century of Electrical Engineering in America* (New York: IEEE Press, 1984).

3. David F. Noble, *America by Design: Science, Technology, and the Rise of Corporate Capitalism* (New York: Knopf, 1977), pp. xxiii–xxv, 167–202, 321.

4. This section is based on Ronald Kline, "The General Electric Professorship at Union College, 1903–1941," *IEEE Transactions on Education*, 31 (1987): 141–147. References are given only for quotations and added material. On Nott, see George Wise, "Reckless Pioneer," *American Heritage of Invention and Technology*, 6 (Spring/Summer 1990): 26–33.

5. "Union's New Era," *Concordiensis*, 26 (17 Dec. 1902): 5–8, quotation on p. 6, SL.

6. Ernst Berg to A. L. Rohrer, 15 Nov. 1912, EJB.

7. Steinmetz to F. C. Pratt, 5 March 1909 (quotation); Steinmetz to E. W. Rice, Jr., 24 Oct. 1910; and Carl Magnusson to Steinmetz, n.d., CPSS. His stay at GE is mentioned in *NCAB*, vol. 14, s.v. "Magnusson, Carl."

8. Union College Trustees Minutes, 11 Dec. 1902, Book F, pp. 161–167, SL. In 1909 Berg said that GE gave its "leading engineers" such contracts because of "patent matters." See Ernst Berg to William Goss, 27 Sept. 1909, draft of letter, EJB.

9. Ernst Berg to F. Ginsburg, 8 Aug. 1914, EJB.

10. Steinmetz to George Alexander, 12 Aug. 1908 and 16 Feb. 1910, CPSU.

11. Steinmetz, "The New Electrical Course," *Concordiensis*, 26 (18 Feb. 1903): 5–6.

12. Steinmetz to C. A. Coffin, 9 Jan. 1913, CPSS.

13. Steinmetz to C. A. Richmond, 21 May 1910, 4 Jan. 1911, 12 Feb. 1912. C. M. Davis, in the Consulting Engineering Department, also entered the program. See Steinmetz to Richmond, 7 Dec. 1910, CPSS. On Steinmetz's teaching, see Morland King to Ernst Berg, 26 April 1939; Walter Upson to Berg, 30 April 1939; and O. J. Ferguson to Berg, 26 April 1939, Steinmetz Faculty File, SL.

14. M. W. Alexander to D. C. Jackson, 21 May 1908, cited in W. Bernard Carlson, "Academic Entrepreneurship and Engineering Education: Dugald C. Jackson and the MIT-GE Cooperative Engineering Course, 1907–1932," *Technology and Culture*, 29 (1988): 536–567, on p. 551.

15. M. W. Alexander to Steinmetz, 19 Sept. 1908, CPSS.

16. See, for example, Steinmetz, "Engineering Schools of Electrical Manufacturing Companies," NACS, *Bulletin*, 1 (March 1914): 23–28.

17. Carlson, "Academic Entrepreneurship."

18. J. P. Jackson, "College and Apprenticeship Training: The Relation of the Student Engineering Courses in the Industries to the College Technical Courses," *Transactions of the American Society of Mechanical Engineers*, 29 (1907): 473–490, quotation on p. 477. On his visit to Schenectady, see *General Electric Review*, 8 (1907): 156.

19. Edwin T. Layton, Jr., "Science, Business, and American Engineering," in

The Engineers and the Social System, ed. Robert E. Perrucci and Joel E. Gerstl (New York: Wiley, 1969), pp. 51–72, quotation on p. 54.

20. This section is based on Ronald Kline, "Professionalism and the Corporate Engineer: Charles P. Steinmetz and the American Institute of Electrical Engineers," *IEEE Transactions on Education,* 23 (1980): 144–150. References are given only for quotations and added material.

21. See *Trans AIEE,* passim.

22. Steinmetz to Elihu Thomson, 23 April 1900, ETA. On the nomination procedure, see Layton, *Revolt of the Engineers,* p. 81.

23. "Rules," *Trans AIEE,* 1 (1884): appendix, pp. 1–5, quotation on p. 4; "Constitution," *Trans AIEE,* 18 (1901): 923–936, quotation on p. 924.

24. Charles Scott, "Presidential Address," *Trans AIEE,* 22 (1903): 3–15, on p. 7.

25. Layton, *Revolt of the Engineers,* pp. 79–93, quotation on p. 79; and Peter Meiksins, "The 'Revolt of the Engineers' Reconsidered," *Technology and Culture,* 29 (1988): 219–246.

26. BoD Minutes, 23 March 1906, IEEEA; and *Electrical World,* 47 (1906): 310, 398, 650, 831, 879, 906, 976, 1020, 1083. H. C. Wirt at GE was doing some of the campaigning for Rice; see Wirt to D. C. Jackson, 22 Jan. 1906, DCJ. On Mailloux, Mershon, and Sheldon, see *NCAB,* 26: 428–429, 15: 225, and 14: 208.

27. R. W. Pope to the Intermediate Grade of Membership Committee, 17 Aug. 1910; and F. T. Hutchinson to the committee, 18 Aug. 1910, DCJ.

28. Steinmetz to C. W. Stone, 1 April 1911, CPSS.

29. Steinmetz to H. G. Stott, 29 March 1911, CPSS.

30. Steinmetz's letter to the board of directors, dated 13 Jan. 1913, is mentioned in Secretary of AIEE to Steinmetz, 18 Jan. 1913, Steinmetz Folder, IEEEA. But no copy of it has been located.

31. A. G. Langsdorf to Steinmetz, 6 Dec. 1910; and Steinmetz to Langsdorf, 9 Dec. 1910, CPSS.

32. Steinmetz, "Enhancing the Prestige of the Institute," *Electrical World,* 74 (1919): 243–246, quotation on p. 246.

33. See, for example, "Schenectady Branch of A.I.E.E.," *General Electric Review,* 8 (1907): 36–39; and Steinmetz to D. B. Rushmore, 25 April 1908, CPSS.

34. Steinmetz to E. W. Rice, Jr., 25 July 1910, CPSS.

35. Steinmetz to E. W. Rice, Jr., 27 Feb. 1911, CPSS.

36. Steinmetz to A. H. Armstrong, 8 Feb. 1911, CPSS.

37. Steinmetz to H. G. Stott, 29 March 1911, CPSS.

38. Steinmetz to Edison and Edison to Steinmetz, 17 Jan. 1911. Without knowing of Steinmetz's request, Rice also telegraphed Edison and asked him to support Dunn. See E. W. Rice, Jr., to Edison, 19 Jan. 1911, TAE. On Dunn, see *NCAB,* 39: 603–604.

39. A. H. Armstrong to Steinmetz, 6 Feb. 1911; and Steinmetz to Armstrong, 8 Feb. 1911, CPSS. Armstrong told him that F. L. Hutchinson, the assistant AIEE secretary, thought the Schenectady section was acting out of "revenge."

40. Steinmetz to Lewis Stillwell, 29 Jan. 1912, CPSS. Steinmetz also circularized for Stott as president in 1907 and Berg as manager in 1912. See W. C. Andrews to H. G. Stott, 1 Feb. 1907, DCJ; and Steinmetz to P. Junkersfeld, 18 Jan. 1912, CPSS. Berg was defeated and, according to a colleague at GE, was not fairly

treated by the AIEE. See Edward Berry to Steinmetz, 11 March 1916, CPSS.

41. H. C. Senior to Ernst Berg, 13 Jan. 1913, EJB.

42. Rice and Steinmetz often discussed highly technical topics. See, for example, one on the nature of high-voltage discharges in E. W. Rice, Jr., to Steinmetz, 24 Feb. 1898, L5906–07, JWH. GE engineers planned in 1911 to run Rice for president again in the future; see A. H. Armstrong to Steinmetz, 6 Feb. 1911, CPSS. On Buck, Carty, and Lincoln, see *American Men of Science*, 3rd ed., 1921.

43. Bruce Sinclair, *A Centennial History of the American Society of Mechanical Engineers, 1880–1980* (Toronto: Univ. of Toronto Press, 1980), pp. 46–60.

44. Steinmetz to E. W. Rice, Jr., 26 Feb. 1898, L5751–61, quotation on L5758, JWH.

45. See, for example, Larry May, "Professional Action and the Liabilities of Professional Associations: A.S.M.E. v. Hydrolevel Corp.," *Business and Professional Ethics Journal*, 2 (1982): 1–14.

46. C. L. de Muralt, "Discussion," *Trans AIEE*, 32 (1913): 141.

47. "The Standardization of Generators, Motors, and Transformers (A Topical Discussion)," *Trans AIEE*, 15 (1898): 3–32, quotation on p. 14. On Crocker and Lieb, see *DAB*.

48. "Standardization of Generators," p. 20.

49. Compare Rice's discussion in *Trans AIEE*, 15 (1898): 4–8, with "Report of the Committee on Standardization," *Trans AIEE*, 16 (1899): 255–268. Rice based his remarks on a paper by GE engineer S. Dana Greene. See Greene's "The Relations between the Customer, Consulting Engineer, and the Electrical Manufacturer," *Electricity*, 14 (19 Jan. 1898): 20–22.

50. Steinmetz, "Standardization Rules of the A.I.E.E.," *General Electric Review*, 13 (1910): 34–39, quotation on pp. 37–38.

51. F. B. Crocker to Samuel Sheldon, 7 June 1907, DCJ.

52. "History of the Standardization Rules," *Trans AIEE*, 35 (1916): 1551–1554, quotation on p. 1552. GE engineers W. S. Moody and E. B. Raymond also reviewed changes to the standards in 1909. See Steinmetz to Moody, 8 and 26 Feb. 1909; and A. E. Kennelly to Steinmetz, 8 March 1909, CPSS.

53. E. W. Rice, Jr., to Elihu Thomson, 2 May 1899, ETA; Steinmetz to Rice, 18 April 1902, L5593–94, JWH; Steinmetz to H. F. T. Erben, 14 Nov. 1908; Steinmetz to J. W. Ham, 17 Nov. 1908; and Steinmetz to C. W. Stone, 16 Dec. 1910, CPSS.

54. Steinmetz to E. W. Rice, Jr., 16 Dec. 1910, CPSS. The AIEE adopted *transformer, converter,* and *cycles per second,* used by GE, instead of Westinghouse's *converter, rotary transformer,* and *alternations per minute.*

55. A. H. Moore to M. P. Rice, 19 Dec. 1910, CPSS.

56. B. A. Behrend, "Discussion," *Trans AIEE*, 32 (1913): 143.

57. "History of the Standardization Rules," p. 1552.

58. See the lengthy correspondence between Steinmetz and B. G. Lamme, from 24 Feb. 1912 to 28 April 1914 in CPSS; and Steinmetz and Lamme, "Temperature and Electrical Insulation," *Trans AIEE*, 32 (1913): 79–81. Lamme later recalled that he suggested that they make the revisions on the basis of fundamental principles rather than codify best practice. See Benjamin Garver Lamme, *An Autobiography* (New York: G. P. Putnam's Sons, 1926), pp. 146–148.

59. B. G. Lamme to Steinmetz, 30 Oct. 1912; and Steinmetz to Lamme, 5 Nov. 1912, CPSS.

60. Steinmetz to B. G. Lamme, 8 July 1912, CPSS. For a long list of the books Lamme sent Steinmetz, see Lamme, *Autobiography*, pp. 229–232.

61. Ronald Kline, Joyce Bedi, and Thomas Lindblom, "Wheeler's Gift of Electrical Books at the Engineering Societies Library: A Legacy and a Responsibility," *Science and Technology Libraries,* 7 (Summer 1987): 63–80.

62. Schulyer Skaats Wheeler, "Engineering Honor," *Trans AIEE,* 25 (1906): 241–248.

63. "Discussion," *Trans AIEE,* 25 (1906): 266–268.

64. Editor, "Engineering and a Code of Ethics," *American Machinist,* 25 July 1907, quoted partially in Monte A. Calvert, *The Mechanical Engineer in America, 1830–1910: Professional Cultures in Conflict* (Baltimore: Johns Hopkins Press, 1967), p. 267.

65. BoD Minutes, 26 July, 30 Aug., 27 Sept. 1907, IEEEA; "Proposed Code of Ethics," *Trans AIEE,* 26 (1907): 1789–1793; S. S. Wheeler to D. C. Jackson, 3 March 1910; F. L. Hutchinson to Jackson, DCJ; and "History of the Code," *Trans AIEE,* 31 (1912): 2229–2230. For another account of the origins of the code, see McMahon, *Making of a Profession,* pp. 112–117.

66. S. S. Wheeler to Steinmetz, 17 Sept. 1908, CPSS.

67. Steinmetz to S. S. Wheeler, 15 Oct. 1908, CPSS. On Ferguson, see *NCAB,* vol. 39; and Forrest McDonald, *Insull* (Chicago: Univ. of Chicago Press, 1962), pp. 58, 68–70. The nature of the friction between Steinmetz and Ferguson, a major GE customer, is unknown.

68. S. S. Wheeler to D. C. Jackson, 3 March 1910, DCJ. On Stillwell, see *NCAB,* 14: 520–512. Ironically, Steinmetz supported Stillwell for president in 1909. See Steinmetz to H. G. Stott, 25 Jan. 1909, CPSS.

69. F. L. Hutchinson to D. C. Jackson, 29 Nov. 1910, DCJ. On Pope, see *NCAB,* 27: 469–470.

70. BoD Minutes, 23 Oct. 1895, IEEEA. The minutes do not reveal what the misconduct was. Also, the board voted not to admit as a member Harold Brown, infamous for his less than ethical role in promoting the DC system in the AC-DC controversy between Edison and Westinghouse. See BoD Minutes, 15 Nov. and 20 Dec. 1893. On Brown, see Terry Reynolds and Theodore Bernstein, "Edison and 'The Chair,'" *IEEE Technology and Society Magazine,* 8 (March 1989): 19–28.

71. D. C. Jackson to S. S. Wheeler, 24 Oct. 1907; and BoD Minutes, 11 March 1910, IEEEA.

72. See the correspondence from 3 March 1910 to 27 June 1911 between D. C. Jackson and S. S. Wheeler, Samuel Sheldon, William B. Jackson, H. G. Stott, Steinmetz, Lewis Stillwell, Frank Sprague, F. L. Hutchinson, and John W. Lieb, DCJ.

73. Steinmetz to D. C. Jackson, 6 Jan. 1911, CPSS; the original is in DCJ.

74. Ibid.

75. BoD Minutes, 27 June 1911, IEEEA. On the appointment of the advisory committee, see BoD Minutes, 10 Nov. 1911.

76. H. G. Stott to F. L. Hutchinson, 11 Dec. 1911, Stott Folder, IEEEA;

George Sever to Steinmetz, 9 Jan. 1912, CPSS; BoD Minutes, 9 Feb. 1912, IEEEA; and correspondence from 2 Jan. to 5 Feb. 1912 between Steinmetz and Sever, Frank Waterman, Hutchinson, S. S. Wheeler, Samuel Reber, H. W. Buck, C. F. Scott, Louis Bell, Elihu Thomson, and William Stanley, CPSS. Also see Acting Chairman [Steinmetz] to F. B. Crocker, 26 Jan. 1912, Crocker Folder, IEEEA; and D. C. Jackson to Steinmetz, 5 Feb. 1912, DCJ. On Sever, see *American Men of Science*, 8th ed., 1949.

77. George Sever, "Report of Committee on Code of Principles of Professional Conduct," *Proceedings of the AIEE*, 31 (April 1912): 148–151, quotation on p. 149; and "History of the Code," p. 2230.

78. See Layton, *Revolt of the Engineers*, pp. 84–85; and McMahon, *Making of a Profession*, pp. 112–117.

79. Compare *Trans AIEE*, 26 (1907): 1421–1425, and "Proposed Code of Ethics" [1907] with *Trans AIEE*, 31 (1912): 2227–2229.

80. "Proposed Code of Ethics" [1907], clause 5.

81. Steinmetz to Louis Bell, 1 Feb. 1912, CPSS.

82. Steinmetz, "Organization of [the] Consulting Engineering Department and of Laboratories Connected Therewith . . . ," 10 July 1912, p. 18, SPN.

83. He did, however, serve on the public policy committee in 1908. See Steinmetz to W. C. L. Elgin, 23 June 1908, CPSS.

84. Steinmetz, "Enhancing the Prestige of the Institute," p. 246.

85. Steinmetz, "Symbolic Representation of General Alternating Waves and of Double Frequency Vector Products," *Trans AIEE*, 16 (1899): 269–296. For criticisms of his method, see "Discussion," *Trans AIEE*, 17 (1900): 380–381, 384.

86. W. S. Franklin, "A Discussion on Some Points in Alternating Current Theory," *Trans AIEE*, 22 (1903): 589–601, quotation on pp. 591–592.

87. The books included substantially revised and enlarged editions of *Theory and Calculation of Alternating Current Phenomena* (1908) and *Theoretical Elements of Electrical Engineering* (1909), in addition to first editions of *Theory and Calculation of Transient Electric Phenomena and Oscillations* (1909), *General Lectures on Electrical Engineering* (1908), *Radiation, Light, and Illumination* (1909), *Elementary Lectures on Electric Discharges, Waves, and Impulses* (1911), and *Engineering Mathematics* (1911). The last four were based on lectures he had given at Union College.

88. A. E. Kennelly, "Vector Power in Alternating-Current Circuits," *Trans AIEE*, 29 (1910): 1233–1267, quotation on p. 1265; and "Discussion," ibid., pp. 1268–1272, 1274. Kennelly included only one of Steinmetz's three books in this survey and failed to list at least three other works that followed Steinmetz: C. P. Feldman, *Wirkungsweise, Prüfung, und Berechnung der Wechselstrom-Transformatoren* (Leipzig, 1894); J. Kramer, *Die Einfachen und Mehrphasen Elektrischen Wechselströme* (Geneva, 1896); and Silvanus P. Thompson, *Polyphase Electric Circuits*, 2nd ed. rev. (London, 1900).

89. F. L. Hutchinson to D. C. Jackson, 21 July 1910; and A. E. Kennelly to Jackson, 15 Aug. 1910, DCJ.

90. BoD Minutes, 11 Nov. and 9 Dec. 1910, IEEEA; R. W. Pope to U.S. National Committee, 17 Nov. 1910, DCJ; and Steinmetz to Ernst Berg, 22 Nov. 1910, CPSS. The BoD minutes do not support Steinmetz's understanding on sending out Kennelly's paper and his discussion of it.

91. *Trans AIEE*, 30 (1911): 575–596; and *Electrical World*, 57 (1911): 464. Although a "disciple of Steinmetz," as he told Kennelly, Berg did not present his paper in person because of a previous commitment (!) to give a lecture in Chicago. See Ernst Berg to A. E. Kennelly, 16 Nov. 1910; and Berg to Steinmetz, 31 Jan. 1911, EJB.

92. Ernst Berg to A. E. Kennelly, 16 Nov. 1910, EJB. Steinmetz did not attend the Turin meeting, but he did ask his GE colleague Guiseppi Faccioli, who was going, to keep an eye on the vector question for him. See Steinmetz to Faccioli, 20 March 1911, CPSS.

93. IEC, *Report of the Turin Meeting . . .* , pub. 12 (London, 1912), p. 78.

94. BoD Minutes, 13 Oct. 1911, IEEEA; A. E. Kennelly, "Turin Meeting of the [IEC]," *Trans AIEE*, 30 (1911): 2507–2518, on pp. 2514–2515; and "Addenda to the A.I.E.E. Standardization Rules," ibid., pp. 2570–2585, on p. 2571.

95. Ernst Berg to Steinmetz, 3 Oct. 1911, EJB.

96. Ernst Berg to Martin M. Foss, 26 Sept. 1912. Also see Berg to W. S. Franklin, 31 May 1913, EJB.

97. Steinmetz to Edward Caldwell, 5 March 1912; and Steinmetz to Martin M. Foss, 15 July 1913, CPSS.

98. *Electrician*, 79 (1917): 193. See also *Electrical World*, 67 (1916): 948; 71 (1918): 1087.

99. See, for example, D. G. Fink, ed., *Standard Handbook for Electrical Engineers*, 11th ed. (New York: McGraw-Hill, 1978), section 2, pp. 36–43.

100. On the use of the books in colleges, see, for example, O. A. Kenyon to Steinmetz, 15 June 1908, CPSS. On the translations, see Ernst Berg to C. L. von Muralt, 14 Oct. 1898, EJB; K. Tornberg to J. L. R. Hayden, 3 Feb. 1909; Steinmetz to L. Bunet, 21 March 1911; Steinmetz to G. Semenza, 20 May 1912; Steinmetz to Kimura Shunkkhi, 12 July 1912; Steinmetz to T. Nomura, 19 May 1913; and Steinmetz to Ricardo Capopte Valente, 12 Sept. 1916, CPSS.

101. The advertisement [c. 1920] is in EJB.

102. Carolyn Marvin, *When Old Technologies Were New: Thinking about Electric Communication in the Late Nineteenth Century* (New York: Oxford Univ. Press, 1988), pp. 56–62; and Marcel C. LaFollette, *Making Science Our Own: Public Images of Science, 1910–1955* (Chicago: Univ. of Chicago Press, 1990), pp. 98–100.

103. Wyn Wachhorst, *Thomas Alva Edison: An American Myth* (Cambridge, Mass.: MIT Press, 1981), pp. 50–51; and James O. Robertson, *American Myth, American Reality* (New York: Hill and Wang, 1980), esp. ch. 5.

104. On these concepts, see Trevor J. Pinch and Wiebe E. Bijker, "The Social Construction of Facts and Artifacts," in *The Social Construction of Technological Systems: New Directions in the Sociology and History of Technology*, ed. Bijker, Thomas P. Hughes, and Pinch (Cambridge, Mass.: MIT Press, 1987), pp. 17–50.

105. Carl Snyder, "American 'Captains of Industry' . . . ," *Review of Reviews*, 25 (1902): 417–432, on pp. 421–423. Steinmetz did not attend the luncheon with Prince Henry (*New York Times*, 27 Feb. 1902, p. 1), apparently because he still had an antipathy toward the Prussian government (see his letter to Clara Steinmetz, 2 July 1915, CPSF). On *Success*, see Frank Luther Mott, "The Magazine Revolution and Popular Ideas in the Nineties," *Proceedings of the American Antiquarian Society*, 64 (April 1954): 195–214.

106. Herbert Wallace, "A Man Who Knows—A Talk with Charles P. Steinmetz . . . ," *Success*, 6 (1903): 145–146. It was reprinted in *Concordiensis*, 36 (11 March 1903): 5–8; and in *Electrical World*, 41 (1903): 524–525.

107. "Edison's Most Important Discovery," *Harper's Weekly*, 45 (1901): 1302–1303; Wachhorst, *Edison*; "A New Edison on the Horizon," *Review of Reviews*, 9 (1894): 355; Margaret Cheney, *Tesla: Man Out of Time* (New York: Dell, 1981), chs. 6, 15–16; and Susan J. Douglas, *Inventing American Broadcasting, 1899–1922* (Baltimore: Johns Hopkins Univ. Press, 1987), chs. 1–2.

108. *New York Times*, 1 Nov. 1908, V, p. 9.

109. Ibid. For an earlier story along the same lines, see Arthur Goodrich, "Charles P. Steinmetz, Electrician," *World's Work*, 8 (1904): 4866–4869.

110. H. H. McClure, "Interesting People—Charles P. Steinmetz," *American Magazine*, 69 (Nov. 1909): 68 (quotation), 70; and Peter Lyon, *Success Story: The Life and Times of S. S. McClure* (New York: Scribner, 1963), pp. 147, 294–295. For an example of stories that follow this one, see *Baltimore Evening Sun*, 22 July 1915.

111. Arthur Gleason, "The Socialism of Steinmetz," *Metropolitan Magazine*, March 1914, partially rpt. in *Citizen*, 13 March 1914.

112. *New York Times*, 12 Nov. 1911, V, p. 4; and Steinmetz to E. W. Rice, Jr., 15 June 1908, CPSS. For an example of Tesla's predictions, see "Tesla Tells of Wonders, Says He Will Soon Be Able to Transmit Wireless Power across Seas," *New York Times*, 16 May 1911, p. 22. Steinmetz indirectly praised Tesla's induction motor in *Trans AIEE*, 8 (1891): 591.

113. David E. Nye, *Image Worlds: Corporate Identities at General Electric, 1890–1930* (Cambridge, Mass.: MIT Press, 1985).

114. J. L. R. Hayden to Albert Spies, 6 Nov. 1908; Hayden to E. D. Snow, 8 June 1909; John Phillips to Steinmetz, 29 Aug. 1912; Hayden to Phillips, 24 Sept. 1912; Phillips to Hayden, 25 Sept. 1912, CPSS; and Mary B. Mullett, "The Little Giant of Schenectady," *American Magazine*, 94 (22 Oct. 1922): 18–19, 123–130. The interviews were published in May 1918, April 1919, and July 1922.

115. M. P. Rice to J. L. R. Hayden, 8 May 1911, CPSS; Joseph R. Baker, "Charles Proteus Steinmetz: An Appreciation," *Scientific American*, 104 (1911): 443; and Douglas S. Martin, "Charles Proteus Steinmetz: One Man's Share in a Few Recent Years of Electrical Progress," *Scientific American Supplement*, 76 (1913): 28. Martin had earlier written "Steinmetz and His Discovery of the Hysteresis Law," *General Electric Review*, 15 (1912): 544–551.

116. See, for example, H. C. Senior to W. L. Chandler, 13 July 1914; Senior to Earle H. Eaton, 16 July 1914; and Senior to E. A. Roehty, 11 Jan. 1915, CPSS.

117. Arthur A. Bright, Jr., *The Electric Lamp Industry: Technological Change and Economic Development from 1800 to 1947* (New York: Macmillan, 1949), pp. 145–159; and Martin J. Sklar, *The Corporate Reconstruction of American Capitalism, 1890–1916: The Market, the Law, and Politics* (New York: Cambridge Univ. Press, 1988), chs. 3–4. For examples of the publicity surrounding the antitrust suit, see *New York Times*, 4 March, 29 July, 12 and 29 Aug., 13 Oct. 1911; *New York World*, 4 March and 13 Oct. 1911; and *Outlook*, 99 (1911): 440–441.

118. "GE's Third Generation: Wilson and Reed," *Fortune*, 21 (Jan. 1940): 101–104; David Loth, *Swope of GE* (New York: Simon and Schuster, 1958), pp. 130–131; and Josephine Y. Case and Everett N. Case, *Owen D. Young and American Enterprise: A*

Biography (Boston: David R. Godine, 1982), pp. 106, 112–113, 125.

119. J. L. R. Hayden to M. P. Rice, 10 May 1915; and H. C. Senior to M. P. Rice, 28 Oct. 1915, CPSS.

120. Baker, "Steinmetz"; Arthur E. Kennelly, "Charles Proteus Steinmetz (1865–1923)," *Proceedings of the American Academy of Arts and Sciences*, 59 (1924): 657–660; James E. Brittain, "B. A. Behrend and the Beginnings of Electrical Engineering, 1870–1920" (Ph.D. diss., Case Western Reserve Univ., 1970), p. 99; and Matthew Josephson, *Edison* (New York: McGraw-Hill, 1959), p. 434.

121. Steinmetz to Paul Spencer, 19 Nov. 1908; E. T. Lake to Steinmetz, 13 Sept. 1910; H. C. Feuers to Steinmetz, 1 Feb. 1909; Steinmetz to Feuers, 8 Feb. 1909; Chris Kraft to Steinmetz, 20 March 1914, quotation on p. 1, CPSS; and Tyler G. Price, "Steinmetz in Fullerton Hall," *General Electric Review*, 27 (1924): 648–649. The test course students at GE's plant in Erie, Pennsylvania, also renamed their club after Steinmetz. See E. C. Clarke to Steinmetz, 5 April 1915; and H. C. Senior to Clarke, 9 April 1915, CPSS.

122. E. W. Rice, Jr., to Steinmetz, 19 Dec. 1914; H. C. Senior to Mr. Scovil, 23 Dec. 1914; Robert B. Rifenberick to L. E. Gould, 22 June 1915; E. W. Allen to Steinmetz, 25 June 1915; C. A. Coffin to Steinmetz, 8 July 1915; and Senior to Rifenberick, 9 July 1915, CPSS.

Chapter 8: Corporate Socialism

1. Ellis W. Hawley, "The Discovery and Study of a 'Corporate Liberalism,'" *Business History Review*, 52 (1978): 309–320; R. Jeffrey Lustig, *Corporate Liberalism: The Origins of Modern American Political Theory, 1890–1920* (Berkeley: Univ. of California Press, 1982), ch. 1; Daniel T. Rodgers, "In Search of Progressivism," *Reviews in American History*, 10 (Dec. 1982): 113–132, esp. n. 27, p. 129; Louis Galambos, "Technology, Political Economy, and Professionalization: Central Themes of the Organizational Synthesis," *Business History Review*, 57 (1983): 471–493; and Martin J. Sklar, *The Corporate Reconstruction of American Capitalism, 1890–1916: The Market, the Law, and Politics* (New York: Cambridge Univ. Press, 1988), pp. 34–40, quotation on p. 35.

2. David A. Shannon, *The Socialist Party of America: A History* (New York: Macmillan, 1955), ch. 1; James Weinstein, *The Decline of Socialism in America, 1912–1925* (New York: Monthly Review Press, 1967), chs. 1–2; and W. E. Walling et al., eds., *The Socialism of Today* (New York, 1916), p. 191.

3. Arnold E. Kaltinick, "Socialist Administration in Four American Cities (Milwaukee, Schenectady, New Castle, Pennsylvania, and Conneaut, Ohio), 1910–1916" (Ph.D. diss., New York Univ., 1982), ch. 4; David Montgomery, *The Fall of the House of Labor: The Workplace, the State, and American Labor Activism, 1865–1925* (New York: Cambridge Univ. Press, 1987), pp. 313–314; and *Citizen*, 10 Feb. 1911.

4. Kaltinick, "Socialist Administration," ch. 6; and Kenneth E. Hendrickson, "Tribune of the People: George R. Lunn and the Rise and Fall of Christian Socialism in Schenectady," in *Socialism and the Cities*, ed. Bruce M. Stave, (Port Washington, N.Y.: Kennikat Press, 1975), pp. 72–98.

5. *Citizen*, 15 March and 20 Sept. 1912; Arthur Many, "A Visitor's Impression of Schenectady," *New York Call*, 13 Aug. 1912; and *New York Call*, 21 Sept. 1912.

6. Isador Ladoff, *The Passing of Capitalism and the Mission of Socialism* (Terre

Haute, Ind.: Debs, 1901), annotation in Steinmetz's copy of the book, Accession No. A607804A, NYSL. On Ladoff, see Howard H. Quint, *The Forging of American Socialism*, 2nd ed. (New York: Bobbs-Merrill, 1963), p. 376; Willis R. Whitney, Laboratory Notebook, 3 Oct. 1901 and 11 Jan. 1902, pp. 141, 196, GER; and George Wise, *Willis R. Whitney, General Electric, and the Origins of U.S. Industrial Research* (New York: Columbia Univ. Press, 1985), pp. 101, 146. Steinmetz was involved with state politics in 1902–1903, when he served on a three-person commission to study the feasibility of establishing a state electrical standards laboratory. See *Electrical World*, 41 (1903): 202, 387, 517.

7. Shannon, *Socialist Party*, pp. 55–56; "Socialists in Faculty, 1900–1920"; "Numbers of Chapters & Members," n.d., ISSP; Harry Laidler to Steinmetz, 14 Dec. 1911; Steinmetz to Laidler, 28 Dec. 1911, CPSS; Executive Committee Minutes, 30 Oct., 17 Nov., 11 Dec. 1911; "Third Annual Convention—Executive Sessions," 28–29 Dec. 1911; "Organizing Secretary's Report, 8 Jan. 1912, ISSP; and *Citizen*, 2 Feb., 22 March, 11 Oct. 1912.

8. Kaltinick, "Socialist Administration," pp. 62–64; *Citizen*, 29 Dec. 1911; and Steinmetz to George Lunn, 16 June 1910, CPSS.

9. Chad Gaffield, "Big Business, the Working-Class, and Socialism in Schenectady, 1911–1916," *Labor History*, 19 (1978): 350–372; and Weinstein, *Decline of Socialism*, quotation on p. 109.

10. *Citizen*, 19 July and 30 Aug. 1912 (quotations).

11. Samuel P. Hays, "The Politics of Reform in Municipal Government in the Progressive Era," *Pacific Northwest Quarterly*, 60 (Oct. 1964): 157–169; Edwin T. Layton, Jr., *The Revolt of the Engineers: Social Responsibility and the American Engineering Profession* (1971; rpt. Baltimore: Johns Hopkins Univ. Press, 1986), pp. 115–116, 160–165; Jean Christie, *Morris Llewellyn Cooke: Progressive Engineer* (New York: Garland, 1983); and Martin J. Schiesl, *The Politics of Efficiency: Municipal Administration and Reform in America, 1800–1920* (Berkeley: Univ. of California Press, 1977), pp. 163–165, 178–183.

12. Steinmetz, "Development of Schenectady's Educational System," 26 Jan. 1916, quotation on p. 4, CPSS.

13. Steinmetz to Thomas Wooley, 23 Nov. 1912, CPSS; *Citizen*, 10 and 31 Jan., 7 Feb., and 29 Aug. 1913; Steinmetz, "Schenectady's Educational System," p. 6; Steinmetz to Henry Berger, 27 Nov. 1916, CPSS; Kaltinick, "Socialist Administration," pp. 110, 144; and U.S. Commission on Industrial Relations, *Final Report and Testimony*, 11 vols. (Washington, D.C.: U.S. Govt. Printing Office, 1916), 2: 1828–1835.

14. Hawley B. Van Vechten, "Complete History of the Schenectady Board of Parks and City Planning Commission," *Citizen*, 30 Jan. 1914; and Kaltinick, "Socialist Administration," pp. 119–123.

15. *Citizen*, 21 and 28 Nov. 1913 (quotations).

16. *Citizen*, 5 Dec. 1913. The McClellan Street site became popular as the present "Central Park." See Larry Hart, *Schenectady's Golden Era, 1880–1930*, 3rd ed. (Schenectady: Old Dorp Books, 1974), pp. 45–48.

17. *Citizen*, 8 Aug. and 10 Oct. 1913.

18. Ronald Steel, *Walter Lippmann and the American Century* (Boston: Little,

Brown, 1980), ch. 4; *Schenectady Socialist Party Campaign Book* (Schenectady, 1913), HVV; and *New York Times*, 2 Nov. 1913, V, p. 3.

19. Kaltinick, "Socialist Administration," pp. 153–155; and *Citizen*, 14 Nov. 1913. The editorial, from the *Scranton Tribune-Republican*, is reprinted in ibid.

20. *Citizen*, 13 March 1914. The editorials are reprinted in *Citizen*, 20 March and 10 April 1914.

21. Van Vechten, "Complete History"; *Citizen*, 23 Jan. (quotation), 27 Feb., 13 and 31 July 1914.

22. *Citizen*, 13 July 1914 (quotation), 20 and 27 Nov. 1914, 14 and 28 May 1915 (quotation). Steinmetz also advised the landscape architect on how to deal with the "narrow mindedness and political partisanship" of the Schoolcraft administration. See Steinmetz to C. W. Leavitt, 30 July 1914, CPSS.

23. *Citizen*, 31 July 1914, 20 and 27 Aug., 3 Sept., 1 Oct. 1915 (headline). Compare the 1913 and 1915 platforms in ibid., 8 Aug. 1913 and 3 Sept. 1915. In an interview Steinmetz said he was a "constructive Socialist"; see *Knickerbocker Press*, 12 July 1914.

24. *Citizen*, 30 Oct. and 5 Nov. 1915; and *New York Call*, 3 Nov. 1915. He was probably misquoted as saying thirty-eight rather than twenty-eight years (i.e., 1887, the year of Lux's trial).

25. *Citizen*, 17 Dec. 1915, 7 Jan., 24 March 1916. Schoolcraft reappointed Dempster to the school board in Dec. 1915. See *Knickerbocker Press*, 13 Dec. 1915.

26. *Citizen*, 16 Jan., 6 March, 1 May 1914, 6 Aug. (quotation), 13 Aug. 1915.

27. *Citizen*, 11 and 18 Feb. 1916, 3, 17, and 31 March 1916; *Knickerbocker Press*, 29 March 1916; *New York Times*, 29 March 1916, p. 2; and *New York Call*, 31 March 1916.

28. Harry Laidler, Organizing Secretary's Report, Dec. 1914, ISSP, p. 3; *Citizen*, 1 Jan. and 12 Nov. 1915; and *Knickerbocker Press*, 1 April 1915.

29. *New York Call*, 6 and 13 April 1916; *Schenectady Gazette*, 12 April 1916; and *Citizen*, 14 and 21 April 1916.

30. *Citizen*, 17 Dec. 1915, 31 March, 23 and 30 June 1916; and *Schenectady Gazette*, 27 June and 4 Aug. 1916.

31. *Citizen*, 11 Aug., 15 and 29 Sept. 1916 (quotation); and *New York Call*, 3 Oct. 1916.

32. *Schenectady Union-Star*, 23 Oct. 1916; *Gloversville Herald*, 25 Oct. 1916; *Albany Times*, 26 Oct. 1916; *Citizen*, 27 Oct. 1916; *New York Call*, 29 Oct. 1916; Steinmetz to August Pellens, 31 Oct. 1916; and H. C. Senior to Mark Stern, 1 Nov. 1916, CPSS. Steinmetz also opposed Wilson's antitrust policy, which resulted in "industrial disorganization and interference, lowering our national efficiency at the time when the other nations are organizing for the higher efficiency of national cooperation." See an untitled signed statement, 21 Oct. 1916, in CPSU.

33. *Knickerbocker Press*, 13 and 14 Oct. 1916, 3 March, 26 Oct., 4 Nov. 1917 (quotation); *New York Call*, 16 and 17 Nov. 1916; *Appeal to Reason*, 6 Jan. 1917; *Citizen*, 12 Jan. and 9 March 1917; *Schenectady Union-Star*, 9, 10, 12, and 17 April 1917; *Schenectady Gazette*, 19 April 1917; Schenectady Common Council, *Proceedings* (Schenectady, 1917), minutes for 16 and 21 April 1917; and Steinmetz to Henry Berger, 27 Nov. 1916, CPSS.

34. Common Council, *Proceedings*, 1916–1917; *Knickerbocker Press*, 16 July, 23 Aug., 23 Nov. 1917; and Hendrickson, "Tribune of the People," p. 94. Lunn fired Hunt in April because of his feud with the city engineer; see *Schenectady Union-Star*, 9 April 1917. As a Democrat, Lunn beat the Socialist candidates for mayor by a margin of six to one in 1919 and 1921.

35. ISS press releases, May 1914 and 30 Dec. 1914; "Contributions Received in Response to Appeal, June 3–18, 1912," ISSP; Algernon Lee to Steinmetz, 8 July 1912 and 16 Jan. 1914; Harry Laidler to Steinmetz, 20 Sept. 1912 and 15 Sept. 1913; Steinmetz to Alice Kuebler, 11 Feb. 1914; and Steinmetz to Alice Boehme, 20 Feb. 1915, CPSS.

36. H. F. Simpson to Steinmetz, 19 Jan. 1914; W. E. Walling to Steinmetz, 5 May 1914; Steinmetz to Walling, 13 May 1914; H. C. Senior to Upton Sinclair, 8 July 1915; J. G. Phelps Stokes to Steinmetz, 21 June 1916; Algernon Lee to Steinmetz, 20 April 1916; Max Eastman to Steinmetz, 26 June 1916, CPSS; and *New Review*, May 1914, p. 257, and title pages for June and July 1914. On the journal, see Weinstein, *Decline of Socialism*, pp. 86–87.

37. Samuel Haber, *Efficiency and Uplift: Scientific Management in the Progressive Era, 1890–1920* (Chicago: Univ. of Chicago Press, 1964), p. 72; Steinmetz, *The Future of Electricity* (New York: New York Electrical School, 1911), quotation on p. 9; and Society for Electrical Development, *Camp Co-operation: Book of Proceedings* (Association Island, 1913), pp. 151–166.

38. Steinmetz, "The Future Development of the Electrical Business," in *Camp Co-operation*, pp. 53–72, quotation on p. 70. He expressed similar ideas in "Effect of Electrical Engineering on Modern Industry," *Journal of the Franklin Institute*, 177 (1914): 115–124; and "Electrifying America," *Collier's*, 56 (27 Nov. 1915): 12–13, 34–35. This view is closer to the position that artifacts have politics than to a Marxist technological determinism. See Langdon Winner, *The Whale and the Reactor: A Search for Limits in an Age of High Technology* (Chicago: Univ. of Chicago Press, 1986), ch. 2.

39. Steinmetz to Harry N. Slattery, 24 Dec. 1913, CPSS; and Steinmetz, "Commission Control," *Collier's*, 57 (8 April 1916): 17, 27 (quotation). *Collier's* asked him to write on this topic; see Steinmetz to Henry J. Forman, 3 Feb. 1914, CPSS.

40. Mary B. Mullet, "The Little Giant of Schenectady," *American Magazine*, 94 (22 Oct. 1922): 18–19, 123–124, 126, 128 (quotation), 130.

41. Steinmetz, "Discussion," *NACS Proc.*, 1 (1913): 406–408; "Address," ibid., pp. 424–425; "Opening Address," *NACS Proc.*, 2 (1914): 55–60; and "Discussion," ibid., pp. 676–678. Steinmetz even suggested that the AIEE and other engineering societies be admitted as corporate members of the NACS. See *NACS Proc.*, 1 (1913): 380–381.

42. Arthur H. Gleason, "The Socialism of Steinmetz," *Metropolitan Magazine*, March 1914, pp. 46, 48–50, quotation on p. 48. See also Steinmetz, "Commission Control."

43. *New York Times*, 2 Nov. 1913, V, p. 3. Interview by George McAdam.

44. Hays, "Politics of Reform in Municipal Government"; James Weinstein, *The Corporate Ideal in the Liberal State, 1900–1918* (Boston: Beacon Press, 1968), pp. 96–105; and Schiesl, *Politics of Efficiency*, passim.

45. Donald Wilhelm, "The 'Big Business' Man as a Social Worker . . . III—Dr. Steinmetz, of the General Electric Company," *Outlook*, 108 (1914): 496–500, quota-

tion on p. 500; Steinmetz, "Response and Annual Address," *NACS Proc.*, 3 (8–11 June 1915): 48–55, quotation on p. 54; and Steinmetz, "Address," ibid., pp. 839–842, quotation on p. 840. For other instances where he alluded to the Corporation of the United States, see *Cincinnati Times-Star*, 25 Aug. 1915; and Steinmetz, "Commission Control," p. 17. Steinmetz's views on the NACS are described well in James Gilbert, *Designing the Industrial State: The Intellectual Pursuit of Collectivism in America, 1880–1940* (Chicago: Quadrangle Books, 1972), pp. 195–198.

46. *Knickerbocker Press*, 12 July 1914; and Steinmetz, "Commission Government Offers No Cure for Municipal Ills," *Citizen*, 12 Feb. 1915. Parts of the interview were reprinted in *New York Times*, 12 July 1914, p. 3. On attitudes toward commission government, see Schiesl, *Politics of Efficiency*, ch. 7; and Stave, *Socialism and the Cities*, p. 7.

47. Gleason, "Socialism of Steinmetz," p. 50; W. S. English to Steinmetz, 4 Jan. 1915; Steinmetz to English, 6 Jan. 1915, CPSS; and *Citizen*, 10 Jan. 1913. On his view of the IWW before 1919, see, for example, Wilhelm, "Steinmetz," p. 498.

48. John M. Jordan, "Technic and Ideology: The Engineering Ideal and American Political Culture, 1892–1934" (Ph.D. diss., Univ. of Michigan, 1989), esp. pp. 161–170; Gilbert, *Designing the Industrial State*, passim; Kenneth M. Roemer, "Technology, Corporation, and Utopia: Gillette's Unity Regained," *Technology and Culture*, 26 (1985): 560–570; Ira Kipnis, *The American Socialist Movement, 1897–1912* (New York: Columbia Univ. Press, 1952), pp. 63–64, 110–113, 117–119, 222–223; Nick Salvatore, *Eugene V. Debs: Citizen and Socialist* (Urbana: Illinois Univ. Press, 1982), pp. 170–171, 193–194; and Walling, *Socialism of Today*, pp. 475–476.

49. Morris Hillquit, *Socialism in Theory and Practice* (New York, 1909), p. 113; Charles H. Vail, *Principles of Scientific Socialism* (New York, 1899), pp. 24–25; Donald Stabile, *Prophets of Order: The Rise of the New Class, Technocracy, and Socialism in America* (Boston: South End Press, 1984), chs. 4–5; Quint, *Forging of American Socialism*, pp. 75, 277; and Shannon, *Socialist Party*, p. 251.

50. *New York Call*, 5 Dec. 1911. The editorials appeared in the *Citizen* on 9, 16, and 23 June 1911. On Lunn's evolutionary socialism, see Gaffield, "Socialism in Schenectady," p. 357. Steinmetz owned an autographed copy of Gaylord Wilshire, *Socialism Inevitable* (New York, 1907), Accession No. A606987A, NYSL, but no presentation date is indicated in the book.

51. Gleason, "Socialism of Steinmetz"; Victor L. Berger, *Broadsides*, 3rd ed. (Milwaukee, 1913), Accession No. A607433A22, NYSL; *New York Call*, 30 Dec. 1911; Steinmetz to Rand School, 3 July 1912, CPSS; and John Spargo, *Applied Socialism* (New York, 1912), p. 118.

52. On the popularity of *Looking Backward*, see Howard P. Segal, *Technological Utopianism in American Culture* (Chicago: Univ. of Chicago Press, 1985), p. 47. John M. Jordan, "Society Improved the Way You Can Improve a Dynamo: Charles P. Steinmetz and the Politics of Efficiency," *Technology and Culture*, 30 (1989): 57–82, on pp. 74–77, describes well the similarity between *Looking Backward* and *America and the New Epoch*, including the quotation from Bellamy in the previous paragraph. Steinmetz owned numerous books by Wells and ordered *What Is Coming* just before he wrote *America and the New Epoch*. See Steinmetz to Robson and Adee, 24 April 1916, CPSS.

53. C. P. Steinmetz to Clara Steinmetz, 7 July 1914, CPSF; and *New York Times*,

13 Sept. 1914, IV, p. 2. Although outside the scope of this study, there was a strong connection between eugenics and theories of the corporative state in Germany in the nineteenth century. See Daniel Gasman, *The Scientific Origins of National Socialism: Social Darwinism in Ernst Haeckel and the German Monist League* (London: Mac-Donald, 1971), ch. 4.

54. Steinmetz, "Russia the Real Menace," *New Review,* 2 (1914): 702–703, quotation on p. 703; Steinmetz to Ivan Narodny, 16 Jan. 1915 (quotation); *New York Times,* 18 April and 7 May 1915; and Narodny to Steinmetz, 28 May 1915, CPSS. In 1916 the Russian government sanctioned the formation of another agency, the American-Russian Chamber of Commerce, which did foster much trade between the two countries. See *New York Times,* 28 Jan. 1916, p. 13.

55. Steinmetz to Clara Steinmetz, 30 Jan. 1915, CPSF; Steinmetz to C. H. Davis, 14 April 1915; Steinmetz to V. Ridder, 1 March 1915 (plus monthly contributions thereafter), CPSS; and Steinmetz to W. E. Walling, 8 Feb. 1915, published as a letter to the editor in *New Review,* 3 (1915): 149–150. He also told Clara that buying English war bonds was "another good chance for you to lose some money in foreign investments." See Steinmetz to Clara Steinmetz, 2 July 1915, CPSF.

56. Waldemar Kaempffert to Steinmetz, 30 July 1914; Atherton Brownell to Steinmetz, 13 Aug. 1914; H. C. Senior to Brownell, 21 Aug. 1914; and Brownell to Steinmetz, 23 Dec. 1914, CPSS. In an unpublished editorial for the *New Review,* dated 18 May 1915, he correctly called the *Lusitania* an "ammunitions carrier." On Brownell, see *Who Was Who in America* (Chicago, 1943), 1: 153.

57. "Neutrality Meeting," 2 May 1915, pamphlet, CPSC; and *Schenectady Gazette,* 3 May 1915. On O'Leary and the American Truth Society, see Mark Sullivan, *Our Times: The United States, 1900–1925,* 6 vols. (New York: Scribners, 1926–1935), 5: 140, 160, 236–237. The editorials are in CPSC.

58. *New York World,* 14 July 1915; *Knickerbocker Press,* 14 July 1915; and *New York World,* 8 Aug. 1915. For the favorable reaction of other newspapers, see *Baltimore Evening Sun,* 22 July 1915; *Lawrence* (Mass.) *Telegram,* 23 July 1915; *San Antonio Express,* 23 July 1915; and *New York Evening Mail,* 12 Aug. 1915. On the origins of the board, see Daniel J. Kevles, *The Physicists: The History of a Scientific Community in America* (New York: Knopf, 1977), pp. 105–109.

59. *New York Press,* 11 Aug. 1915, copy in Steinmetz Folder, IEEEA; and *New York Evening Mail,* 10 Aug. 1915. The editorial also appeared in the *St. Joseph* (Mo.) *News-Press,* 29 July 1915. On the New York papers, see Frank L. Mott, *American Journalism: A History, 1690–1960,* 3rd ed. (New York: Macmillan, 1962), pp. 616, 637–638.

60. Clyde Wagoner to Steinmetz, 12 Aug. 1915; H. W. Hillman to Steinmetz, 14 Aug. 1915; Steinmetz to Editor, *New York Press,* 16 Aug. 1915, CPSS; *New York Press,* 17 Aug. 1915; and F. G. R. Gordon to Editor, *Patterson* (N.J.) *Press,* 20 Aug. 1915. Similar letters by Gordon appeared in the *Providence* (R.I.) *Bulletin* and *Journal* and the *New York Press.* Steinmetz was also opposed by the *Scranton* (Pa.) *Leader,* 30 Aug. 1915.

61. *New York Press,* 13 Sept. 1915; and *DAB,* supplement 1, s.v. "Carty, John Joseph." Brownell's *Philadelphia Ledger* (14 Sept. 1915) criticized Daniels for not naming Tesla, Steinmetz, Wright, and others for the board. The AIEE board of directors made the selection from a list of names suggested by the board, past

presidents, and chairmen of sections. See BoD Minutes, 10 Aug. 1915, IEEEA; and *Proceedings of the AIEE*, 34 (Oct. 1915): 234–235.

62. Steinmetz to Henry Ford, 29 Nov. 1915; Steinmetz to Editor, *New York American*, 18 Oct. 1915, CPSS; and *New York Times*, 12 March 1916, VI, p. 8. He was also a delegate from the Schenectady German-American Alliance who was scheduled to speak at the Friends of Peace conference in Chicago in early Sept. 1915. But the major newspaper accounts do not list him as a speaker. He probably did not attend because the Socialist party did not approve of sending delegates to the conference. See *Schenectady Union-Star*, 13 Aug. 1915; *Citizen*, 3 Sept. 1915; *Chicago Daily Tribune*, 5–7 Sept. 1915; and *New York Times*, 5 Sept. 1915, p. 2.

63. Steinmetz to Editor, *New York American*, 18 Oct. 1915, CPSS. He made similar comments regarding the need to organize industry and the military efficiently, based on democratic principles, as had been done in education, in Steinmetz to Erman Ridgway, 13 Jan. 1916, pp. 8–9, CPSS.

64. J. L. R. Hayden to William H. Briggs, 4 April 1916; Steinmetz to Briggs, 10 April (quotation), 15 April, 14 July, 2 Aug. 1916; and Steinmetz to Harper and Brothers, 25 Sept. 1916, CPSS.

65. Steinmetz, *America and the New Epoch* (New York: Harper and Brothers, 1916), introduction, n.p.

66. Ibid., p. 33; and Karl Kautsky, *Das Erfurter Programm in seinem grundsätzlichen Theil*, 3rd ed. (Stuttgart, 1892), Accession No. 607850, NYSL. Despite the argument from historical materialism, *America and the New Epoch* presents a weaker form of technological determinism than given in his Camp Co-operation speech on electricity and socialism. In the book, the corporation, not technology, is the main force behind the evolution to socialism. For an argument that Marx was not a strong technological determinist, see Donald MacKenzie, "Marx and the Machine," *Technology and Culture*, 25 (1984): 473–502.

67. Steinmetz, *New Epoch*, pp. 49, 50.

68. Friedrich Engels, *Socialism: Utopian and Scientific*, trans. Edward Aveling (Chicago, 1892), pp. 138–139.

69. Steinmetz, *New Epoch*, pp. 80, 84.

70. Ralph H. Bowen, *German Theories of the Corporative State* (New York: Whittlesey House, 1947), ch. 4; and W. O. Henderson, *The Rise of German Industrial Power, 1834–1914* (Berkeley: Univ. of California Press, 1975), pp. 178–185, 229–233.

71. Steinmetz, *New Epoch*, pp. 120, 124.

72. Ibid., p. 143.

73. Ibid., pp. 154, 156, 159, 160.

74. Ibid., p. 166.

75. Ibid., pp. 167, 176.

76. Ibid., pp. 178, 185, 186.

77. Ibid., pp. 191, 192, 193. These racial stereotypes were probably also reinforced by the habit of employers hiring workers based on what they thought different races were best suited for. See Daniel Nelson, *Managers and Workers: Origins of the New Factory System in the United States* (Madison: Univ. of Wisconsin Press, 1975), p. 81.

78. Steinmetz, *New Epoch*, pp. 199–216. On the paternalism of welfare capitalism, see Stuart Brandes, *American Welfare Capitalism, 1880–1940* (Chicago: Univer-

sity of Chicago Press, 1976). Steinmetz did write an introduction for Charles M. Ripley, *Romance of a Great Factory* (Schenectady: Gazette Press, 1919).

79. Steinmetz, *New Epoch*, pp. 228–229, 222.

80. *New Republic*, 9 (1916): 103; *Appeal to Reason*, 6 Jan. 1917; *Citizen*, 12 Jan. 1917; *New York Call*, 18 March 1917; and *Bookman*, 45 (1917): 180–181. On Kelly and the Socialist papers, see Weinstein, *Decline*, pp. 57–58, 85–86; and Joseph R. Conlin, ed., *The American Radical Press*, 2 vols. (Westport, Conn.: Greenwood, 1974), 1: 50–59. The *Citizen* (9 Nov. 1917) also reprinted his chapter on the racial analysis of the American nation.

81. *Intercollegiate Socialist*, 5 (Feb.–March 1917): 22–24, quotation on p. 24; *Dial*, 62 (1917): 133–134; and Randolph Bourne, "American Use for German Ideals," in Bourne, *War and the Intellectuals*, ed. Carl Resek (New York: Harper and Row, 1964), pp. 48–52, quotation on p. 50. On Bourne, see ibid., pp. vi-xv; and Olaf Hansen, "Introduction," in Bourne, *Radical Will, Selected Writings, 1911–1918*, ed. Hansen (New York: Urizen Books, 1977), pp. 17–62.

82. *New York Times Review of Books*, 21 (1916): 574; *New York Tribune*, 7 Jan. 1917 (quotation); *New York Evening Mail*, 15 June 1917; *Independent*, 90 (7 April 1917): 85–86; and *Nation*, 104 (1917): 557. On the magazines, see Frank L. Mott, *A History of American Magazines*, 5 vols. (Cambridge, Mass.: Harvard Univ. Press, 1957–1968), 2: 378, 3: 349.

83. Steinmetz to W. H. Briggs, 19 March 1917; Briggs to Steinmetz, 21 March 1917, CPSS; and Harry W. Laidler, comp., *Study Courses in Socialism* (New York: ISS, 1919), pp. 13, 18. Laidler also cited the book in his *Socialism in Thought and Action* (New York, 1920), p. 187.

84. *Citizen*, 10 March 1911; and Alex Scott, "Socialism and the Trust," *New York Call*, 5 Aug. 1912.

85. Hillquit, *Socialism*, pp. 141–142; Edmond Kelly, *Twentieth-Century Socialism* (London, 1910), p. 318; and Spargo, *Applied Socialism*, pp. 128–129. Steinmetz's copy of a 1914 edition of Kelly's book is in NYSL, Accession No. A607420A.

86. Bowen, *German Theories*, esp. chs. 4–5; Hartmut Pogge von Strandmann, ed., *Walther Rathenau, Industrialist, Banker, Intellectual, and Politician: Notes and Diaries, 1907–1922* (Oxford: Clarendon Press, 1985), pp. 1–26, 186–194, 238–239; and Walther Rathenau, *Die neue Wirtschaft* (Berlin: S. Fischer, 1919). The latter was first published in 1918.

87. Henderson, *German Industrial Power*, pp. 179, 190–191, 195–198; and Thomas P. Hughes, *Networks of Power: Electrification in Western Society, 1880–1930* (Baltimore: Johns Hopkins Univ. Press, 1983), pp. 76–77, 126, 178–181.

88. Sklar, *Reconstruction*, pp. 154–166.

89. *Iron Age*, 98 (1916): 1118–1120, 1172–1174, quotations on p. 1173; and H. M. Gitelman, "Management's Crisis of Confidence and the Origin of the National Industrial Conference Board, 1914–1916," *Business History Review*, 58 (1984): 153–177. The similarity between Rice's speech and Steinmetz's book is noted by David Montgomery, *The Fall of the House of Labor: The Workplace, the State, and American Labor Activism, 1865–1925* (New York: Cambridge Univ. Press, 1987), p. 355.

Chapter 9: Building a New Epoch

1. David M. Kennedy, *Over Here: The First World War and American Society* (New York: Oxford Univ. Press, 1980), pp. 93–143, 253–256, 304–305; John Brooks, *Telephone: The First Hundred Years* (New York: Harper and Row, 1975), pp. 150–159; and Walter Lippmann to J. G. Phelps Stokes, 1 May 1917, quoted in James Weinstein, *The Decline of Socialism in America, 1912–1925* (New York: Monthly Review Press, 1967), p. 132. On Wilson's middle way, see Martin J. Sklar, *The Corporate Reconstruction of American Capitalism, 1890–1916: The Market, the Law, and Politics* (New York: Cambridge Univ. Press, 1988), pp. 401–412.

2. *Denver Times*, 28 Dec. 1917; and Robert Cuff, *The War Industries Board: Business-Government Relations in World War I* (Baltimore: Johns Hopkins Univ. Press, 1973), chs. 3, 5, 10.

3. H. C. Senior to J. W. Beatson, 25 Nov. 1916, CPSS; *Lynn Evening News*, 12 Jan. 1917; Harry Garfield to Woodrow Wilson, 16 Jan. [1917]; Wilson to Garfield, 16 Jan. 1917, WW; and Robert Cuff, "Harry Garfield, the Fuel Administration, and the Search for a Cooperative Order during World War I," *American Quarterly*, 30 (1978): 39–53, quotations on pp. 42, 44. On Filene, see Kim McQuaid, *A Response to Industrialism: Liberal Businessmen and the Evolving Spectrum of Capitalist Reform, 1886–1960* (New York: Garland, 1986), ch. 3.

4. George Wise, "Schenectady, Strikes, and Socialists, 1896–1923," unpublished ms., 1985, p. 27; Ida M. Tarbell, *Owen D. Young: A New Type of Industrial Leader* (New York: Macmillan, 1932), pp. 117–119; Wise, *Willis R. Whitney, General Electric, and the Origins of U.S. Industrial Research* (New York: Columbia Univ. Press, 1985), pp. 187–194; and James E. Brittain, "Creative Engineering in a Corporate Setting," *Proceedings of the IEEE*, 64 (1976): 1413–1417.

5. *Schenectady Union-Star*, 17 April (quotation) and 20 April 1917; and *Schenectady Gazette*, 12 April 1917.

6. *Citizen*, 23 March, 1 June, 6 July 1917; *Schenectady Union-Star*, 23 April, 29 May, 2, 6, and 7 June 1917; and Charles M. Ripley, *Romance of a Great Factory* (Schenectady: Gazette Press, 1919), pp. 176–204. Another colleague, J. T. H. Dempster, Steinmetz's Socialist aid in city government, was secretary of the council of home defense; see *Schenectady Union-Star*, 11 April 1917.

7. *New York World*, 1 Oct. 1917; *Knickerbocker Press*, 4 Nov. 1917; Grete Mache to Steinmetz, 30 March 1910, CPSS; Steinmetz to Clara Steinmetz, 31 March 1911, CPSF; Mache to Steinmetz, 26 March 1912, CPSS; Steinmetz to Clara Steinmetz, 2 June 1912, CPSF; Steinmetz to J. O. Meyer, 2 and 5 Dec. 1914, CPSS; Steinmetz to Transatlantic Trust Co., 4 Dec. 1916; and Steinmetz to Clara Steinmetz, 30 Jan. 1915, CPSF.

8. D. W. Brunton to George Shanklin, 16 March 1918; Steinmetz to Shanklin, 20 March 1918; Steinmetz to E. W. Rice, Jr., 17 June 1918; and Rice to Steinmetz, 10 July 1918, CPSS.

9. Steinmetz, "War or Arbitration," 24 June 1915, CPSS, quotation on p. 2; and "Dr. Steinmetz Talks on Effects of War," *Fort Wayne* (Ind.) *News*, 24 May 1917. On Socialists in the war, see Weinstein, *Decline of Socialism*, ch. 3.

10. J. L. R. Hayden to George Lunn, 9 and 25 April 1918; Lunn to Hayden 12 April 1918; D. W. Brunton to Steinmetz, 23 May 1918; F. R. Moulton to Hayden, 11

June 1918; Steinmetz to E. W. Rice, Jr., 17 June 1918; Clifton Draper to Steinmetz, 16 Oct. 1918, CPSS; Steinmetz to J. W. Joyes, 9 Oct. 1918; Hayden to Alfred H. White, 26 Oct. 1918, GEH; and *New York World*, 12 May 1918.

11. Steinmetz, "America's Energy Supply," *Trans AIEE*, 37 (1918): 985–1009, quotation on p. 987; *Philadelphia Evening Telegram*, 26 June 1918; Cuff, "Harry Garfield," p. 47; and Kennedy, *Over Here*, pp. 123–125. Examples of his later articles on energy are "Stopping Our Coal Leaks," *Scientific American*, 124 (1921): 325, 338; and "The White Revolution," *Survey*, 25 March 1922, pp. 1035–1037.

12. *New York Herald Tribune*, 21 Aug. 1932; *New York Times*, 21 Aug. 1932, p. F9; Joseph Dorfman, *Thorstein Veblen and His America* (New York: Viking, 1934), p. 510; and William E. Akin, *Technocracy and the American Dream: The Technocrat Movement, 1900–1941* (Berkeley: Univ. of California Press, 1977), chs. 2–4.

13. H. B. Brougham to Steinmetz, 12 Dec. 1916, Folder V-605, CPSC; Leon P. Alford, *Henry Lawrence Gantt* (New York: Harper and Brothers, 1934), pp. 264–269, 375, 396; Edwin Layton, "Veblen and the Engineers," *American Quarterly*, 14 (1962): 64–72; Samuel Haber, *Efficiency and Uplift: Scientific Management in the Progressive Era, 1890–1920* (Chicago: Univ. of Chicago Press, 1964), pp. 46–48; Robert Cuff, "Woodrow Wilson's Missionary to American Business, 1914–1915: A Note," *Business History Review*, 43 (1969): 545–551; and Akin, *Technocracy*, pp. 11–12, 27–32. As noted in Chapter 8, Steinmetz knew at least one member of the scientific management group, Morris Cooke. In 1916 Cooke wrote Steinmetz about his technocratic proposals in the *Collier's* article, asking if the ASME would be among his corporatist organizations (Cooke to Steinmetz, 6 April 1916, CPSS). Although both men could be labeled "technocratic progressives" (see John M. Jordan, "Technic and Ideology: The Engineering Ideal and American Political Culture, 1892–1934" [Ph.D. diss., Univ. of Michigan, 1989]), they held much different views. Cooke saw himself as a champion of the people over monopoly, while Steinmetz favored cartels and centralized technocracy.

14. Dorfman, *Veblen*, pp. 459–461; and Akin, *Technocracy*, pp. 34–35. Hoover is quoted by Donald Stabile, *Prophets of Order: The Rise of the New Class, Technocracy, and Socialism in America* (Boston: South End Press, 1984), pp. 224–225.

15. [Howard Scott], "Political Schemes in Industry," *One Big Union Monthly*, 2 (Oct. 1920): 6–10, quotation on p. 9; Haber, *Efficiency and Uplift*, pp. 154–156; and Akin, *Technocracy*, pp. 36–42. Compare [Scott], "Wastes in the Coal Industry," *Industrial Pioneer*, 1 (July 1921): 53–58, with Steinmetz, "America's Energy Supply."

16. Steinmetz to William Greene, 27 March 1913, CPSS.

17. *New York Call*, 25–29 Nov. and 2 Dec. 1913; *Citizen*, 21 and 28 Nov., 5 Dec. 1913; and Wise, "Schenectady," p. 25.

18. *New York Call*, 4–7, 19, 21–23, and 25–27 Oct. 1915; *Citizen*, 8 and 22 Oct. 1915; and Wise, "Schenectady," pp. 28–31. William Le Roy Emmet recalled that "Steinmetz declared himself an ardent socialist" during "all of these activities." See his *Autobiography of an Engineer* (New York: ASME, 1940), p. 154. Emmet, however, was probably referring to Steinmetz's joining Lunn's administration and running for office rather than to his support of the strikes.

19. *New York Call*, 30–31 Oct., 3 and 5–6 Nov. 1915; and *Citizen*, 5 Nov. 1915.

20. *New York Times*, 12 Nov. 1915; rpt. in *Citizen*, 19 Nov. 1915; and John M. Jordan, "Society Improved the Way You Can Improve a Dynamo: Charles P. Stein-

metz and the Politics of Efficiency," *Technology and Culture*, 30 (1989): 57–82, on p. 80. On scientific management at GE, see Ronald W. Schatz, *The Electrical Workers: A History of Labor at General Electric and Westinghouse, 1923–1960* (Urbana: Univ. of Illinois Press, 1983), pp. 42–46; and Ripley, *Romance of a Great Factory,* p. 79. Another difference between Steinmetz and the scientific management experts was that they thought welfare capitalism was a "joke." See Daniel Nelson, *Managers and Workers: Origins of the New Factory System in the United States* (Madison: Univ. of Wisconsin Press, 1975), p. 59.

21. Steinmetz, *America and the New Epoch* (New York: Harper and Brothers, 1916), pp. 47–48, 58–59, p. 125 (quotation); and Steinmetz, "Industrial Efficiency and Political Waste," *Harper's*, 133 (1916): 925–928.

22. Steinmetz, "Notes on Welfare Work," March 1914, pp. 6, 9, CPSS. On the wage dividend at Chicago's Commonwealth Edison company, see Forrest McDonald, *Insull* (Chicago: Univ. of Chicago Press, 1962), p. 112.

23. *Citizen*, 31 March 1916.

24. Charles M. Ripley, *Life in a Large Manufacturing Plant* (Schenectady: GE, 1919); *Schenectady Works News*, 19 Dec. 1924; Schatz, *Electrical Workers*, chs. 1–2; David Montgomery, *The Fall of the House of Labor: The Workplace, the State, and American Labor Activism, 1865–1925* (New York: Cambridge Univ. Press, 1987), pp. 438–457; Wise, "Schenectady"; and Nelson, *Managers and Workers*, pp. 115–116.

25. *American Magazine*, 87 (April 1919): 9–11, 132, 134–135; and *Schenectady Works News*, 11 April 1919.

26. Steinmetz, "Industrial Cooperation," in *Steinmetz the Philosopher*, comp. Ernest Caldecott and Philip L. Alger (Schenectady: Mohawk Development, 1965), pp. 46–62, quotations on pp. 53, 55, 58. On the talk to the machinists, see *Citizen*, 26 April 1919.

27. *New York Times*, 21 Jan. 1919, p. 4; and *Proceedings of Conference on Industrial Conditions in the State of New York . . .* (Albany, 1919), quotation on pp. 48–49.

28. Steinmetz, "Industrial Co-operation," 11 April 1919, CPSS, pp. 7, 11. This is a shorter and somewhat different version from the one cited in note 26. On proposals for industrial democracy, see James Gilbert, *Designing the Industrial State: The Intellectual Pursuit of Collectivism in America, 1880–1940* (Chicago: Quadrangle Books, 1972), pp. 98–113.

29. Max Horn, *The Intercollegiate Socialist Society, 1905–1921: Origins of the Modern American Student Movement* (Boulder, Colo.: Westview Press, 1979), chs. 6–7; Louis Lazarus, "A List of References on Steinmetz in the Socialist Literature," *Tamiment Institute Library Bulletin*, No. 44 (April 1965): 4–7; Harry Laidler, Organizing Secretary's Report, Dec. 1914, p. 3, ISSP; Steinmetz, "Socialism and Invention," *Socialist Review*, 8 (Dec. 1919): 3–7; Laidler, comp., *Study Courses in Socialism* (New York: ISS, 1919), pp. 13, 18; and Robert B. Westbrook, "Tribune of the Technostructure: The Popular Economics of Stuart Chase," *American Quarterly*, 32 (1980): 387–408. The League for Industrial Democracy probably owed a debt to the Guild Socialist movement in Britain, since Laidler suggested changing the name and the scope of the ISS at a meeting where he discussed his visit with prominent Guild Socialists in the summer of 1921. See *New York Call*, 19 and 20 Nov. 1921.

30. *Schenectady Works News*, 6 Jan. (quotation), 20 Jan., 3 and 17 Feb., 3 March, 16 June, 8 Dec. 1922, 21 March, 18 April, 6 and 20 June, 18 July, 15 Aug., 19 Sept.,

17 Oct., 21 Nov. 1924; John T. Broderick, *Pulling Together* (Schenectady, 1922), n.p. (introduction); Montgomery, *Fall of the House of Labor*, pp. 411, 419–426, 456–457; Larry G. Gerber, "Corporatism in Comparative Perspective: The Impact of the First World War on American and British Labor Relations," *Business History Review*, 62 (1988): 93–127; Nelson, *Managers and Workers*, pp. 156–167; and Stuart Brandes, *American Welfare Capitalism, 1880–1940* (Chicago: Univ. of Chicago Press, 1976), pp. 119–134.

31. *Citizen*, 23 March, 27 April, 16 Nov. (headline), 30 Nov., 14 Dec. 1917; and *Schenectady Union-Star*, 21 March 1917.

32. Quotations cited in Jonathan Coopersmith, "The Electrification of Russia, 1880 to 1925" (Ph.D. diss., Oxford Univ., 1985), pp. 144, 168–169.

33. Ibid., chs. 6–7; Thomas P. Hughes, *Networks of Power: Electrification in Western Society, 1880–1930* (Baltimore: Johns Hopkins Univ. Press, 1983), chs. 12–14; and Hughes, "The Industrial Revolution That Never Came," *American Heritage of Invention and Technology*, 3 (Winter 1989): 58–64.

34. *Citizen*, 1 Jan. 1921.

35. *Citizen*, 29 April 1921; and [Steinmetz], untitled three-page mss., Sept. 1922, quotations on p. 1, CPSS.

36. Steinmetz to Lenin, 16 Feb. 1922; and Lenin to Steinmetz, 12 April 1922, Special Collections, NYSL. The complete correspondence was widely published at the time. See, for example, *Soviet Russia*, 6 (1922): 426 (from the *Worker*); *Nation*, 115 (1922): 78; and *Journal of Commerce* (Chicago), 13 Dec. 1922.

37. *Citizen*, 17 and 24 Oct. 1913, 2 Jan. 1914; Harold Ware, "Tovarish Lenin to Dr. Steinmetz," *Soviet Russia Pictorial*, 8 (1923): 147, 154–155; Ware to Steinmetz, 4 Aug. 1923, CPSS; Ella Reeve Bloor, *We Are Many: An Autobiography* (New York: International Publishers, 1940), pp. 118–119, 266–279, quotation on p. 276; Jessica Smith, "Some Memoirs of Russia in Lenin's Time," in Daniel Mason et al., *Lenin's Impact on the U.S.* (New York: NWR Press, 1970), pp. 95–104; and Sender Garlin, *Charles P. Steinmetz: Scientist and Socialist . . .* (New York: American Institute for Marxist Studies, 1977, occasional paper 22), pp. 16–17, 32–33, 36–37. Ware was the son of Bloor, a prominent Communist who had known Steinmetz when she was an organizer for the national Socialist party in Schenectady before the war.

38. Telegram, Peter Bogdanoff to Steinmetz, 11 June 1922, copy in GEH.

39. Steinmetz, "The Soviet Plan to Electrify Russia," *Electrical World*, 80 (1922): 715–719, quotation on p. 719; and Steinmetz, "Russia's First Regional Power Station," ibid., pp. 1155–1158.

40. Ibid., pp. 701, 1141, and 412.

41. P. L. Guercken to General Electric, 15 Jan. 1919; Thomas N. Perkins to Gerard Swope, 15 April 1920; Charles Coffin to E. W. Rice, Jr., A. W. Burchard, O. D. Young, and M. A. Oudin, 31 Jan. 1921, ODY; and *Schenectady Works News* (25 Feb. 1921).

42. Foreign Letter, Whaley-Eaton Service, 15 Dec. 1921, ODY; *New York Times*, 22 April 1922, p. 1; Hartmut Pogge von Strandmann, ed., *Walter Rathenau, Industrialist, Banker, Intellectual, and Politician: Notes and Diaries, 1907–1922* (Oxford: Clarendon Press, 1985), pp. 22, 265; Anthony Sutton, *Western Technology and Soviet Economic Development, 1917 to 1930* (Stanford: Hoover Institution, 1968), pp. 186–187, 198; and GE Press Release, 18 Oct. 1928, L3597–98, JWH.

43. George Galvin to Steinmetz, 7 July 1915, CPSS; Steinmetz to Lenin, 16 Feb. 1922, NYSL; Smith, "Memoirs of Russia," p. 101; A. A. Heller to Steinmetz, 22 April and 8 June 1922, copies in GEH; Heller, *The Industrial Revival in Soviet Russia* (New York, 1922), pp. xi–xv; Theodore Draper, *American Communism and Soviet Russia: The Formative Period* (1960; rpt. New York: Random House, 1986), p. 204; and Garlin, *Steinmetz*, pp. 18–19, 38. Heller resigned as president of the American Socialist Society in May 1912; Steinmetz was listed as a member in April 1920. See the society's BoD minutes, 19 May 1921, and membership list, April 1920, reel 30, RSR.

44. Four Russian engineers also conferred with Steinmetz at Camp Mohawk in the spring of 1919, but the purpose of their visit is unknown. See Emil J. Remscheid, *Recollections of Steinmetz* (Schenectady: GE, 1977), p. 32.

45. *New York Call*, 3 July (quotation) and 11 July 1922. On Karapetoff's running as state engineer, see, for example, *Citizen*, 30 Oct. 1914 and 9 July 1920. The Socialist party named Steinmetz as its candidate for the position in July. The Farmer-Labor party (1919–1923), which was considering joining the Socialist party for the campaign, said that it would probably accept Steinmetz. In October the New York American Labor party was formed as a "political partnership of the Socialist Party, the Farmer-Labor Party, and a large number of progressive organizations" for the 1922 campaign. See *New York Times*, 3 July 1922, p. 5; 16 Oct. 1922, p. 2. On the history of the Farmer-Labor party, see Weinstein, *Decline of Socialism*, pp. 222–272.

46. *New York Times*, 8 July 1922, p. 2; 13 Oct. 1922; and *New York Call*, 13 Oct. (quotation), 16, 19, and 20 Oct. 1922. On Edison's visit, see Chapter 10 below.

47. William Feigenbaum, "Steinmetz Would Chain Roaring Niagara Currents," *New York Call*, 16 July 1922; and Steinmetz, "Mobilizing Niagara to Aid Civilization," *Electrical World*, 71 (1918): 399.

48. Steinmetz, "America's Energy Supply"; Steinmetz, "Hydro-electric Power Collection," *General Electric Review*, 22 (1919): 960–963; and *New York Times*, 29 Oct. 1922, VIII, p. 2.

49. *Electrical World*, 80 (1922): 919–920.

50. *New York Call*, 9 Nov. 1922; *New York Journal*, 9 Nov. 1922; *Brooklyn Eagle*, 9 Nov. 1922; and *New York Times*, 8 Dec. 1922, p. 10.

51. See, for example, "Who Licked Steinmetz?" *Sandusky* (Ohio) *Register*, 27 Dec. 1922; and many similar clippings in CPSC.

52. *New York Times*, 9 Nov. 1922, p. 18.

53. *New York Times*, 9 Dec. 1922, p. 14. On identifying industrial cooperation with socialism in 1919, see Steinmetz, "Socialism and Invention."

54. Excerpts from the speech were reprinted in many newspapers. See, for example, *New Britain* (Conn.) *Herald*, 3 Feb. 1923; and *Spartansburg* (S.C.) *Herald*, 4 Feb. 1923.

55. Gerber, "Corporatism in Comparative Perspective," p. 124; Charles S. Maier, "Between Taylorism and Technocracy: European Ideologies and the Vision of Industrial Productivity in the 1920s," *Journal of Contemporary History*, 5, No. 2 (1970): 27–61; Herbert Hoover, *American Individualism* (New York, 1922), quotation on p. 45; Ellis Hawley, "Herbert Hoover, the Commerce Secretariat, and the Vision of an 'Associative State,' 1921–1928," *Journal of American History*, 61 (1974): 116–140; and Hawley, *The Great War and the Search for a Modern Order* (New York: St. Martin's

Press, 1979). Steinmetz took note of the corporate state developing in Italy and called it a "dictatorship of the middle class." See Steinmetz, "Industrial Cooperation," March 1923, p. 13, CPSU.

56. *New York Times*, 11 Dec. 1922, p. 16; *Baltimore American*, 12 Dec. 1922; *Butte* (Mont.) *Post*, 21 Dec. 1922; *New York World* editorial rpt. in *Birmingham* (Ala.) *Age-Herald*, 18 Dec. 1922; and H.K. [Henry Kuhn], "Revamping Capitalism a la Steinmetz," *Weekly People*, 30 Dec. 1922.

57. Josephine Y. Case and Everett N. Case, *Owen D. Young and American Enterprise: A Biography* (Boston: David R. Godine, 1982), chs. 15–16; Loth, *Swope*, chs. 14–15; Kim McQuaid, "Young, Swope, and General Electric's 'New Capitalism': A Study in Corporate Liberalism," *American Journal of Economics and Sociology*, 36 (1977): 323–334; and McQuaid, *Response to Industrialism*, pp. 109ff.

58. [Atherton Brownell], "Report of an Investigation into Industrial Conditions in the Several Plants of the General Electric Company, together with the Recommendations of a Plan to Improve Them" (preliminary draft), n.d. [1919], p. 4, ODY. The report is discussed in Schatz, *Electrical Workers*, pp. 22, 37; and Montgomery, *Fall of the House of Labor*, pp. 442–455.

59. Camp Diary Entry, 27 Sept. 1922, rpt. in Remscheid, *Recollections of Steinmetz*, p. 22; and Loth, *Swope*, pp. 137–139. Swope later praised Steinmetz's concern for social issues when he gave the Tenth Steinmetz Memorial Lecture in 1936; see Swope, "An Engineering View of and from Steinmetz," *Electrical Engineering*, (New York) 55 (1936): 572–574.

60. Steinmetz, "Industrial Cooperation," quotations on pp. 15, 16, 17. The letter to Swope is cited in Wise, "Schenectady," p. 37.

61. Gerard Swope, "Plan for Stabilization of Industry . . . ," U.S. Bureau of Labor Statistics, *Monthly Labor Review*, Nov. 1931, pp. 45–53, quotation on p. 46. Young encouraged Swope to make this speech and introduced it. See Case and Case, *Young*, pp. 543–546.

62. Ellis W. Hawley, *The New Deal and the Problem of Monopoly* (Princeton: Princeton Univ. Press, 1966), quotation on p. 42; Hawley, "Herbert Hoover"; "A Personal Sketch of Gerard Swope," *Schenectady Works News*, 7 July 1922; Loth, *Swope*, pp. 84–85, 100–101, ch. 14; Kim McQuaid, "Corporate Liberalism in the American Business Community, 1920–1940," *Business History Review*, 52 (1978): 342–368; Cuff, *War Industries Board*, p. 165; Pogge von Strandmann, *Rathenau*, pp. 10–11, 256; and Rathenau, *Hauptwerke und Gespräche*, 6 vols., ed. Ernst Schulin (Munich: Gotthold Müller, 1977), 2: 897–898. Rathenau incorporated his "new economy" into a broader work, *The New Society*, trans. unknown (London, 1921). Jordan, "Technic and Ideology," pp. 340–341, also notes the similarity between the corporatism of Steinmetz and Rathenau.

63. Loth, *Swope*, ch. 15; McQuaid, "Corporate Liberalism," pp. 354–363; Hawley, *New Deal*, chs. 1–6; and Gerber, "Corporatism in Comparative Perspective," p. 126. Swope later distanced himself from the NRA, because it gave too large a role to the government, and in Nov. 1933 he proposed a National Chamber of Commerce and Industry, an association of trade associations, to replace the NRA. The new body would have fostered the industrial self-government favored by Swope—and Steinmetz. See Donald R. Brand, *Corporatism and the Rule of Law: A*

Study of the National Recovery Administration (Ithaca: Cornell Univ. Press, 1988), pp. 127–132.

Chapter 10: Modern Jove

1. Wyn Wachhorst, *Thomas Alva Edison: An American Myth* (Cambridge, Mass.: MIT Press, 1981); David Hounshell, "Edison and the Pure Science Ideal in America," *Science,* 207 (1980): 612–617; and Marcel C. LaFollette, *Making Science Our Own: Public Images of Science, 1910–1955* (Chicago: Univ. of Chicago Press, 1990), pp. 51, 135, 159–162. The distinction between inventors and scientists was not made at the time in science fiction either; see Thomas D. Clayton, "The Scientist as Hero in American Science Fiction, 1880–1920," *Extrapolation,* 7 (Dec. 1965): 18–28.

2. *Cincinnati Times-Star,* 25 Aug. 1915; and *New York Times,* 11 Feb. 1916.

3. *Baltimore News,* 23 Aug. 1916; and Clyde D. Wagoner, "Steinmetz Revisited: The Myth and the Man," *General Electric Review,* 60 (1957): 16–21. Even his private shorthand system made good copy; see *New York Journal,* 18 June 1920.

4. John Dos Passos, "Edison and Steinmetz: Medicine Men," *New Republic,* 61 (1929): 103–105, quotations on p. 104.

5. Leo Marx, *The Machine in the Garden: Technology and the Pastoral Ideal in America* (New York: Oxford Univ. Press, 1964), esp. pp. 220–222; Wachhorst, *Edison,* pp. 39, 118; Cecelia Tichi, *Shifting Gears: Technology, Literature, Culture in Modernist America,* (Chapel Hill: Univ. of North Carolina Press, 1987), ch. 3; and Elizabeth Ammons, "The Engineer As Cultural Hero and Willa Cather's First Novel, *Alexander's Bridge,*" *American Quarterly,* 38 (1986): 746–760.

6. *New York Times,* 20 Aug. 1923, p. 1; and *New York Herald,* 21 Aug. 1923.

7. *New York Times,* 1 Nov. 1908, V, p. 9; 12 Nov. 1911, V, p. 4; 2 Nov. 1913; 12 July 1914, p. 3; *Knickerbocker Press,* 12 July 1914; *Cincinnati Times-Star,* 25 Aug. 1915; Donald Wilhelm, "The Oracle of Schenectady," *Illustrated World,* 25 (1916): 158–162; Mary B. Mullett, "The Little Giant of Schenectady," *American Magazine,* 94 (22 Oct. 1922): 19, 123–128, 130; *Schenectady Works News,* 22 Oct. 1937 (clipping in Steinmetz Faculty File, SL); and Larry Hart, *Schenectady's Golden Era, 1880–1930,* 3rd ed. (Schenectady: Old Dorp Books, 1974), p. 18.

8. Proteus [Steinmetz] to Hinz and Mrs. Hinz [Heinrich Lux], 25 Oct. 1893, in Justin G. Turner, "Steinmetz," *Manuscripts,* 15 (Fall 1963): 27–34, quotation on p. 32; and *Schenectady Works News,* 22 Oct. 1937.

9. *New York Times,* 3 March 1922, p. 1; *New York Tribune,* 3 March 1922; *Washington Post,* 3 March 1922; *Chicago Tribune,* 3 March 1922; *San Francisco Chronicle,* 3 March 1922; and *Literary Digest,* 73 (15 April 1922): 78.

10. "The Romance of an Engineering Genius," *Current Opinion,* 73 (1923): 339–341; and John W. Hammond, *Charles Proteus Steinmetz: A Biography* (New York: Century, 1924), pp. 331–344. A detailed description of the effects of the lightning strike is given by former lab assistant Emil J. Remscheid in *Recollections of Steinmetz* (Schenectady: GE, 1977), pp. 23–27.

11. *New York Times,* 19 Oct. 1922, pp. 1, 3; *Current Opinion,* 73 (1922): 777–778; and *Philadelphia* _____, 24 Oct. 1922 [name illegible on clipping]. Examples of the clippings are *Cincinnati Times-Star,* 19 Oct. 1922; *New Haven* (Conn.) *Register,* 23 Oct. 1922; *Charlotte* (N.C.) *Observer,* 25 Oct. 1922; *Elgin* (Ill.) *Courier,* 26 Oct. 1922;

New York Times, 29 Oct. 1922, VI, p. 2; *New York Call,* 29 Oct. 1922; *Portland Oregonian,* 29 Oct. 1922; *Redondo Beach* (Calif.) *Reflex,* 1 Dec. 1922; and *Boston Post,* 4 March 1923. On his subscription to a clipping service, see H. C. Senior to Henry Romeike, Inc., 3 Feb. 1915, CPSS.

12. Matthew Josephson, *Edison* (New York: McGraw-Hill, 1959), p. 466; Steinmetz to Edison, 11 Oct. 1910, TAE; W. S. Andrews to H. F. Miller, 12 and 19 Jan. 1911, TAE; and *New York Evening Mail,* 12 Aug. 1915.

13. *New York Times,* 11 Oct. 1914, IV, p. 1; 11 Feb. 1916, p. 20; *Knickerbocker Press,* 14 July 1915; *Cincinnati Times-Star,* 25 Aug. 1915; and Steinmetz, "Edison: The Genius," *Transactions of the Illuminating Engineering Society,* 11 (1916): 623–624. The "gnarled genius" quotation comes from "GE's Third Generation: Wilson and Reed," *Fortune,* 21 (Jan. 1940): 101–104.

14. *New York Times,* 30 Jan. 1922, p. 18; 20 April 1922, p. 19; 24 April 1922, p. 5; *New York Sun,* 22 Nov. 1922; and Wachhorst, *Edison,* pp. 137–138. The public did have an ambivalent attitude toward chemists because of chemical warfare; see Hugh R. Slotten, "Humane Chemistry or Scientific Barbarism? American Responses to World War I Poison Gas, 1915–1930," *Journal of American History,* 77 (1990): 476–498.

15. Arthur J. Palmer, *Edison: Inspiration to Youth* (Milan, Ohio: Edison Birthplace Museum, 1928; rpt. 1962), p. 67.

16. See, for example, Wachhorst, *Edison,* pp. 50–51; and James O. Robertson, *American Myth, American Reality* (New York: Hill and Wang, 1980), esp. ch. 5. The contradictions in the Edison myth are also evident in Frank L. Dyer and Thomas C. Martin, *Edison: His Life and Inventions,* 2 vols. (New York: Harper and Brothers, 1910), which served as the basis for the 1927 booklet and also for William H. Meadowcroft, *The Boy's Life of Edison* (New York: Harper and Brothers, 1911).

17. Steinmetz, "The Oxide Film Lightning Arrester," *Trans AIEE,* 37 (1918): 871–880; and Crosby Field, "The Oxide Film Lightning Arrester," ibid., pp. 881–890. Field patented the oxide-film arrester.

18. E. E. F. Creighton, "Protective Apparatus Engineering," *Trans AIEE,* 26 (1907): 1049–1095; N. A. Lougee, "Performance and Life Tests on the Oxide Film Lightning Arrester," *Trans AIEE,* 39 (1920): 1981–1994; and F. W. Peek, "The Effect of Transient Voltages on Dielectrics," *Trans AIEE,* 34 (1915): 1857–1909.

19. F. L. Hunt, "Lightning Arresters," *Trans AIEE,* 40 (1921): 837–842, quotation on p. 841.

20. J. L. R. Hayden and N. A. Lougee, "A Generator for Making Lightning," *General Electric Review,* 24 (1921): 946–948; *New York Times,* 13 Sept. 1921, p. 1; and *Schenectady Works News,* 7 Oct. 1921.

21. Saul Dushman, "A New Device for Rectifying High Tension Alternating Currents: The Kenotron," *General Electric Review,* 18 (1915): 156–167; and Irving Langmuir, "The Pure Electron Discharge and Its Applications in Radio Telegraphy and Telephony," ibid., 327–339.

22. J. L. R. Hayden and N. A. Lougee, "Types of Lightning Arresters," *General Electric Review,* 24 (1921): 1018–1022; E. E. F. Creighton, "On Deviations from Standard Practice in Lightning Arresters," *Trans AIEE,* 41 (1922): 52–59; and Lougee, "Discussion," ibid., 62.

23. Joseph Slepian, "Discussion," *Trans AIEE,* 41 (1922): 106–107; and Sle-

pian, U.S. Patent No. 1,509,493 (filed 11 Feb. 1922, issued 23 Sept. 1924).

24. See, for example, *Electrical World*, 79 (1922): 470; *Scientific American*, 127 (1922): 26; and Mullett, "Little Giant."

25. F. L. Hunt, "Subcommittee on Lightning Arresters," *Journal of the AIEE*, 41 (1922): 896–897; *Electrical World*, 81 (1923): 408; *Schenectady Works News*, 2 March 1923; A. L. Atherton, "1922 Developments in Autovalve Lightning Aresters," *Trans AIEE*, 42 (1923): 179–184; and W. F. Young, "Tests on 22-kV and 4-kV Lightning Arresters," *Trans AIEE*, 43 (1924): 573–578.

26. *Milwaukee Journal*, n.d. [1923]; *Current Opinion*, 75 (1923): 216–218; and *New York Times*, 18 Sept. 1921, VII, pp. 3, 7. Examples of the AP story are in *Pocatello* (Idaho) *Tribune*, 6 June 1923; and *Boston Transcript*, 6 June 1923, both of which also mention that Faccioli was a "cripple." A measure of Steinmetz's renown, as compared with Peek's, was given by *Outlook*, 134 (27 June 1923), which cited Steinmetz on the work at Pittsfield but did not mention Peek.

27. David E. Nye, *Image Worlds: Corporate Identities at General Electric, 1890–1930* (Cambridge, Mass.: MIT Press, 1985), chs. 7–8; Pamela Lurito, "The Message Was Electric," *IEEE Spectrum*, 21 (Sept. 1984): 84–95; Alan R. Raucher, *Public Relations and Business, 1900–1929* (Baltimore: Johns Hopkins Press, 1968), pp. 80–91; David Loth, *Swope of GE* (New York: Simon and Schuster, 1958), pp. 130, 131, 144; and "GE's Third Generation," pp. 101–102. A trade association of mostly utility companies founded in 1885 and based in New York City, the NELA should not be confused with a related organization with the same acronym, the National Electric Lamp Association—the holding company of lamp manufacturers established in Cleveland in 1901, which was the subject of the 1911 antitrust suit against GE. On this distinction, which Nye conflates in his otherwise fine study, see Paul W. Keating, *Lamps for a Brighter America: A History of the General Electric Lamp Business* (New York: McGraw-Hill, 1954), p. 55.

28. *New York Times*, 16 Jan. (quotation), 18, 20, and 21 Jan. 1922, 20 and 26 May 1922.

29. *New York Times*, 21 March 1924; Harry M. Daugherty, "True Public Ownership," *NELA Proceedings*, 1922, pp. 188–190; Owen Young to E. W. Rice, Jr., 13 Jan. 1923, ODY; Arthur A. Bright, Jr., *The Electric Lamp Industry: Technological Change and Economic Development from 1800 to 1947* (New York: Macmillan, 1949), pp. 253–255; and Keating, *Lamps*, pp. 135–137. Daugherty's assistant read his paper before the NELA since he could not be at the meeting. The NELA used the address in its publicity campaign against public ownership; see Ernest Gruening, *The Public Pays: A Study of Power Propaganda* (New York: Vanguard Press, 1931), p. 214.

30. *Schenectady Works News*, 3 Nov. 1922; *General Electric Review*, 25 (1922): 714–717; and *New York Times*, 19 Oct. 1922, p. 1; 1 May 1923, p. 23; 28 Jan. 1924, p. 28; 21 March 1924, p. 27.

31. Clyde Wagoner, "The General Electric News Bureau," [c. 1955], quoted in John M. Jordan, "Society Improved the Way You Can Improve a Dynamo: Charles P. Steinmetz and the Politics of Efficiency," *Technology and Culture*, 30 (1989): 57–82, on pp. 63–64; Wagoner to Steinmetz, 29 Jan. 1923, CPSS; *New Britain* (Conn.) *Herald*, 3 Feb. 1923; *Perth Amboy* (N.J.) *News*, 3 Jan. 1923; *Spartansburg* (S.C.) *Herald*, 4 Feb. 1923; and *Fort Wayne* (Ind.) *News Service*, 7 Feb. 1923.

32. Edward McKernon to C. D. Wagoner, 2 March 1922, GEH; *New York Times*,

9 Sept. 1923, VI, p. 5; and *Schenectady Works News,* 3 March 1922 and 21 Sept. 1923. A camp photo also appeared in *New York Tribune,* 21 Oct. 1923. In "Steinmetz Revisited" Wagoner said that on his last visit to the camp he found Steinmetz reading a novel called *Lunatics at Large,* the same anecdote that appears in the *Schenectady Works News* in 1923.

33. J. E. Watson to Steinmetz, 23 Nov. and 12 Dec. 1922, CPSS.

34. Steinmetz, *America and the New Epoch* (New York: Harper and Brothers, 1916), pp. 211–216.

35. [Atherton Brownell], "Report of an Investigation into Industrial Conditions in the Several Plants of the General Electric Company, together with the Recommendations of a Plan to Improve Them" (preliminary draft), n.d. [1919], pp. 3, 10–11, 13–17, ODY; Steinmetz, "Electricity and Civilization," *Harper's,* 144 (1922): 227–233; Steinmetz, "The White Revolution," *Survey,* 25 March 1922, pp. 1035–1037; Steinmetz, "Back of the Electric Button," *Good Housekeeping,* 76 (May 1923): 48, 218–220; and Nye, *Image Worlds,* pp. 125–126. I would like to thank Ross Bassett for pointing out the connection between Steinmetz's articles and the ad campaigns.

36. Steinmetz to Atherton Brownell, 8 Feb. 1916, CPSS.

37. Steinmetz, "Commission Control," *Collier's,* 57 (8 April 1916): 17, 27 (quotation); Steinmetz, *New Epoch,* p. 225; and *New York Times,* 27 Nov. 1921, II, p. 3. The Federal Trade Commission did regulate interstate transmission of electricity in the 1930s, an idea Morris Cooke advocated and probably picked up from Steinmetz. See Jean Christie, *Morris Llewellyn Cooke: Progressive Engineer* (New York: Garland, 1983), pp. 80, 98, 128; and Cooke to Steinmetz, 6 April 1916, CPSS.

38. *New York Times,* 13 Oct. 1922, p. 6; *New York Call,* 13, 16, and 20 Oct. 1922.

39. Carl D. Thompson, *Confessions of the Power Trust* (New York: E. P. Dutton, 1932), pp. 513–528, 542–553; Thompson, *Public Ownership* (New York: Thomas Y. Crowell, 1925), p. 269; Preston J. Hubbard, *Origins of the TVA: The Muscle Shoals Controversy, 1920–1932* (Nashville: Vanderbilt Univ. Press, 1961), chs. 1–3; and Paula Eldot, *Governor Alfred E. Smith: The Politician as Reformer* (New York: Garland, 1983), pp. 239–241.

40. *New York Times,* 4 Jan. 1923, p. 11; 16 Jan. 1923, p. 23; *Albany News,* 23 Jan. 1923; *Buffalo Commercial,* 26 Jan. 1923; *Buffalo Express,* 26 Jan. 1923; and Eldot, *Smith,* ch. 7.

41. *New Britain* (Conn.) *Herald,* 3 Feb. 1923.

42. See, for example, New Bedford (Mass.) *Times,* 12 Feb. 1923; *Rio Vista* (Calif.) *River News,* 14 July 1923; and *Douglas* (Wyo.) *Enterprise,* 14 Aug. 1923.

43. *Pueblo* (Colo.) *Star-Journal,* 31 Aug. 1923; and *Denver Post,* 4 and 5 Sept. 1923. On the Rocky Mountain committee, see Gruening, *The Public Pays,* pp. 38, 42–50, 106, 117–118, 135–136, 158–169, 186. Like many notable electrical engineers, Steinmetz was an honorary member of the NELA.

44. *Literary Digest,* 69 (16 April 1921): 32, 34; *New York Times,* 4 April 1921, p. 5; 18 May 1921, p. 18; Marshall Missner, "Why Einstein Became Famous in America," *Social Studies of Science,* 15 (1985): 267–291; and Paul A. Carter, *Another Part of the Twenties* (New York: Columbia Univ. Press, 1977), pp. 63–71.

45. *New York Times,* 20 Oct. 1929, X, p. 16; 22 Oct. 1929, p. 3; 23 Oct. 1929, p. 3; Ronald Tobey, *The American Ideology of National Science, 1919–1930* (Pittsburgh:

Univ. of Pittsburgh Press, 1971), ch. 4; and LaFollette, *Making Science Our Own*, pp. 84–86, 101–102.

46. *Schenectady Works News*, 20 May 1921; and *Worldwide Wireless*, 2 (June 1921): 6–8.

47. Steinmetz, Camp Diary, 25 April, 8 and 24 June 1921; *Citizen*, 29 April 1921; Steinmetz, "Einstein's Theory of Relativity," *General Electric Review*, 24 (1921): 1014–1015; *New York Times*, 30 Jan. 1922, p. 10; Steinmetz, *Four Lectures on Relativity and Space* (New York, 1923), p. vi; and Ernst Caldecott and Philip L. Alger, comps., *Steinmetz the Philosopher* (Schenectady: Mohawk Development, 1965), p. 171.

48. *New York Times*, 20 April 1922, p. 19 (quotation); 30 April 1922, VIII, p. 8; *Schenectady Works News*, 5 May 1922; *Cleveland Press*, 3 Dec. 1922; and *Los Angeles Examiner*, 21 Dec. 1922. See also Steinmetz, "There Are No Ether Waves," *Popular Radio*, July 1922, pp. 161–166.

49. Steinmetz, *Relativity and Space*; *Book Review Digest*, 1923–1925; and *Electrical World*, 81 (1923): 877.

50. "Science and Religion," *Science*, 57 (1923): 630–631; and Carter, *Another Part of the Twenties*, pp. 42–45. Erwin N. Hiebert, "Modern Physics and Christian Faith," in *God and Nature: Historical Essays on the Encounter between Christianity and Science*, ed. David C. Lindberg and Ronald L. Numbers (Berkeley: Univ. of California Press, 1986), pp. 424–447, points out that the theory of relativity was used on both sides of the debate.

51. Ernest Caldecott, "The All-Round Thinker," in *Steinmetz the Philosopher*, pp. 164–168; *Literary Digest*, 75 (4 Nov. 1922): 46; *New York Evening World*, 5 Dec. 1922; Steinmetz, "Science and Religion," *Harper's*, 144 (1922): 296–302; and "The Place of Religion in Modern Scientific Civilization," in *Steinmetz the Philosopher*, pp. 91–95, quotations on p. 95.

52. "Science No Refutation of Religion," *Literary Digest*, 75 (25 Nov. 1922): 32–33, quotation on p. 32; *New York Times*, 8 Nov. 1922, p. 14; *Philadelphia Public Ledger*, 10 April 1923; and *Osh Kosh* (Wis.) *Northwestern*, 2 June 1923 (quoting *Philadelphia Bulletin*). On criticism, see, for example, *Knickerbocker Press*, 13 Nov. 1922; *New York Times*, 19 Nov. 1922, VIII, p. 8 (letter to the editor); and *Birmingham* (Ala.) *News*, 22 Nov. 1922.

53. Steinmetz to B. G. Lamme, 28 April 1914, CPSS; Lamme, *An Autobiography* (New York: G. P. Putnam's Sons, 1926), pp. 229–232; H. C. Senior to G. S. Andrews, 2 Oct. 1915, CPSS; and Steinmetz book collection, NYSL.

54. E. C. Wolf to Steinmetz, 28 May 1915, CPSS; and Steinmetz, "You Will Think This a Dream," *Ladies' Home Journal*, 32 (Sept. 1915): 12. On the progressive virtues of electricity, see Howard P. Segal, *Technological Utopianism in American Culture* (Chicago: Univ. of Chicago Press, 1985), pp. 23–26; and James W. Carey and John J. Quirk, "The Mythos of the Electronic Revolution," *American Quarterly*, 39 (1970): 219–241, 395–424.

55. H. W. Hillman, "Some Reasons for the Success of the New Electric-Heating Devices in the Home," *General Electric Review*, 7 (1906): 93–95; Errett L. Callahan, "Commercial Importance of Heating Devices to the Central Station," *General Electric Review*, 8 (1907): 56–61; Siegfried Giedion, *Mechanization Takes Command* (New York, 1948), pp. 542–547; U.S. Bureau of the Census, *The Statistical History of the*

United States from Colonial Times to the Present, 1-vol. rev. ed. (Stamford, Conn.: Fairfield, 1965), pp. 420, 510; Ruth Schwartz Cowan, *More Work for Mother: The Ironies of Household Technology from the Open Hearth to the Microwave* (New York: Basic Books, 1983); Mark Rose, "Urban Environments and Technological Innovation: Energy Choices in Denver and Kansas City, 1900–1940," *Technology and Culture*, 25 (1984): 503–539; Fred E. H. Schroeder, "More 'Small Things Forgotten': Domestic Electrical Plugs and Receptacles, 1881–1931," *Technology and Culture*, 27 (1986): 525–543; and David E. Nye, *Electrifying America: Social Meanings of a New Technology, 1880–1940* (Cambridge, Mass.: MIT Press, 1990), ch. 6. Hillman was later involved with Steinmetz in an electric car company. See note 69 below.

56. Steinmetz, *New Epoch*, pp. 58–59; *New York Call*, 24 May 1918; *Philadelphia Public Ledger*, 9 Nov. 1917; Steinmetz, "Electricity and Civilization"; and Steinmetz, "White Revolution."

57. John D. Sherman, "Steinmetz, Electrical Wizard, in the Role of Prophet," *Solano* (Calif.) *Courier*, 11 April 1923.

58. *New York Globe*, 7 April 1923. See also *Knickerbocker Press*, 25 Feb. 1923; and *Pittsburgh Gazette Times*, 1 April 1923. For a more accurate statement, see Steinmetz, "Radio Power Transmission's Improbability," *Wireless Age*, 10 (Oct. 1922): 63.

59. *New Haven* (Conn.) *Register*, 17 Feb. 1923. See also *New York Times*, 15 Feb. 1923; St. Paul (Minn.) *Dispatch*, 24 Feb. 1923; and *New York Herald*, 29 July 1923.

60. Steinmetz, "A Hundred Years from Now," *American Legion Magazine*, 17 Aug. 1923, pp. 5–6, 20, 21.

61. *New York Times*, 20 Aug. 1923, p. 1; and *Wall Street Journal*, 21 Aug. 1923. For examples of the different forms of the story, see New York *Tribune*, 20 Aug. 1923; Boston *Herald*, 20 Aug. 1923; *Keokuk* (Iowa) *Gate*, 20 Aug. 1923; and *Ellensburg* (Wash.) *Record*, 20 Aug. 1923.

62. *Chicago News*, 21 Aug. 1923; *New York Call*, 21 Aug. 1923; and *Pittsburgh Chronicle Telegram*, 21 Aug. 1923 (quotation).

63. Herbert Kaufman, "Steinmetz Pictures Conditions One Hundred Years Hence," *Syracuse Telegram*, 21 Aug. 1923.

64. *New York Times*, 21 Aug. 1923, p. 16; *New York Tribune*, 21 Aug. 1923; *New York Herald*, 21 Aug. 1923; and Heywood Broun, "It Seems to Me," *Atlantic City Union*, 22 Aug. 1923.

65. *Pensacola* (Fla.) *News*, 23 Sept. 1923.

66. George Wise, "Predictions of the Future of Technology, 1890–1940" (Ph.D. diss., Boston Univ., 1976), pp. 167–169, 318–320. See also Joseph Martinso, "On Charles P. Steinmetz As Prophet," *Futurist*, Oct. 1974, p. 227.

67. Richard H. Schallenberg, "Prospects for the Electric Vehicle: A Historical Perspective," *IEEE Transactions on Education*, E-23 (1980): 137–143; and W. Bernard Carlson, "Thomas Edison As a Manager of R&D: The Case of the Alkaline Storage Battery, 1898–1915," *IEEE Technology and Society Magazine*, 7 (Dec. 1988): 4–12.

68. *New York Times*, 29 Jan. 1914, p. 7; and "Address of Dr. Charles P. Steinmetz," *NELA Proceedings, Technical Sessions*, 1914, pp. 162–172, quotation on p. 164. Steinmetz also gave the Detroit Electric company permission to use a photograph of him in the car in its ads. See H. C. Senior to F. E. Price, 14 Sept. 1914, CPSS.

69. Steinmetz to H. W. Hillman, 6 June 1914, CPSS; *New York Times*, 28 March

1915, V, pp. 14–15; *Cleveland Plain Dealer,* 18 April 1915; *Current Opinion,* 1 May 1915, pp. 338–339.

70. *New York Times,* 30 Jan. 1920, p. 18; 31 March 1920, p. 22; and "Steinmetz Electric Motor Car Corporation," n.d., GEH. Steinmetz was also sued for breach of contract in connection with forming the syndicate. See *Schenectady Union-Star,* 10 Aug. 1920; and *New York Times,* 10 Aug. 1920. The suit was apparently dropped.

71. *New York Times,* 19 Feb. 1922, VII, p. 12; and *New York Call,* 19 Feb. and 23 April 1922. On his positions in the company, see Steinmetz to Gerard Swope, 23 Nov. 1922, EWR.

72. *New York Times,* 24 Oct. 1922, p. 7; *Dayton* (Ohio) *News,* 12 Nov. 1922; *Jackson* (Mich.) *News,* 12 Nov. 1922; and *Roanoke* (Va.) *Times,* 12 Nov. 1922. Family papers in the care of Robert Koolakian, Syracuse, N.Y., indicate that Steinmetz may have been involved with H. B. Azadian of Syracuse in an electric-car project in 1913.

73. *Printer's Ink,* 1 March 1923; and *Journal of Commerce* (New York), 27 March and 2 June 1923.

74. *New York Tribune,* 4 June 1923; and *New York Times,* 15 and 17 Dec. 1923.

75. See, for example, *Minneapolis News,* 24 Feb. 1923; *Worcester* (Mass.) *Post,* 3 March 1923; and *Des Moines* (Iowa) *News,* 12 April 1923.

76. J. L. R. Hayden to J. S. Murphy, 26 July 1923; *Denver Post,* 30 Aug. 1923; *Los Angeles Daily Times,* 18 Sept. 1923; *San Francisco Chronicle,* 23–25 Sept. 1923; Hayden to Mary Hun, 21 Nov. 1923; Hammond, *Steinmetz,* pp. 463–473; and Hayden to F. C. Pratt, 12 Oct. 1923 (quotations), CPSS.

77. *New York Times,* 22 Oct. 1923, p. 19; and Hammond, *Steinmetz,* pp. 474–475. Hammond says that J. L. R. Hayden spoke to him just after he awoke, while newspaper accounts said it was a nurse. See *Schenectady Union-Star,* 26 Oct. 1923; and *New York Times,* 27 Oct. 1923, p. 1.

78. *Literary Digest,* 79 (17 Nov. 1923): 38 (quotation), 40, 52.

79. *New York Times,* 27 Oct. 1923, pp. 1, 11; *San Francisco Chronicle,* 27 Oct. 1923; *New York Tribune,* 27 Oct. 1923; and *Washington Post,* 27 Oct. 1923. The subsequent stories appeared from 28 Oct. to 1 Nov. 1923.

80. *New York Times,* 28 Oct. 1923, p. 23; 30 Oct. 1923, p. 19; *Schenectady Gazette,* 1 Nov. 1923; and *Schenectady Works News,* 2 Nov. 1923.

81. *New York Tribune,* 28 Oct. 1923 (quotation); *New York Times,* 1 Nov. 1923, p. 1; *New York Leader,* 1 Nov. 1923; and *Literary Digest,* 79 (15 Dec. 1923): 31.

82. Steinmetz Estate Tax Appraisal, 5 Nov. 1924, File 1060-18, SC. He was also part owner of the Schenectady Hygienic Laboratory, which sold liquid soap to GE; see Steinmetz to Gerard Swope, 23 Nov. 1922, EWR. Clara died in 1940 and was buried in Steinmetz's lot in Vale Cemetery; see *Schenectady Gazette,* 19 Feb. 1940.

83. *Cornell Daily Sun,* 29 Oct. 1923; and *New York Times,* 1 Nov. 1923, p. 20. On Karapetoff's relationship with GE and his attitude toward Steinmetz, see "Vladimir Karapetoff, Personal Data," 1942; and Karapetoff to his wife, 15 Feb. 1922 and n.d. [c. Nov. 1923], VK. Faculty at the Cornell School of Electrical Engineering tell the story that shortly after Steinmetz's death, Karapetoff wore a crimson cape to class, symbolizing his ascent to the throne of E.E. (conversation with Terrence Fine, Fall 1987).

84. *New York Times,* 27 Oct. 1923, p. 1; 22 Nov. 1923, p. 18 (quotations).

85. *New York Globe*, 3 Nov. 1923; Michael Pupin, *From Immigrant to Inventor* (New York, 1923); and *Literary Digest*, 79 (17 Nov. 1923): 52.

86. The point is made in George Wise, *Willis R. Whitney, General Electric, and the Origins of U.S. Industrial Research* (New York: Columbia Univ. Press, 1985), p. 214.

87. *New York Times*, 21 March 1924, p. 27; 31 Dec. 1924, pp. 1, 4; Thompson, *Power Trust*, pp. xvii-xix, 3–7; Philip J. Funigiello, *Toward a National Power Policy: The New Deal and the Electric Utility Industry, 1933–1941* (Pittsburgh: Univ. of Pittsburgh Press, 1973), chs. 1–3; Ellis W. Hawley, *The New Deal and the Problem of Monopoly* (Princeton: Princeton Univ. Press, 1966), ch. 17; Forrest McDonald, *Insull* (Chicago: Univ. of Chicago Press, 1962); Thomas K. McGraw, *TVA and the Power Fight, 1933–1939* (Philadelphia: J. B. Lippincott, 1971), pp. 11–12, 21–23; and Hugh G. J. Aitken, *The Continuous Wave: Technology and American Radio, 1900–1932* (Princeton: Princeton Univ. Press, 1985), ch. 10.

88. *Schenectady Works News*, 16 Nov. 1923, 20 June, 18 July, 17 Oct. 1924; and "Contribution of Charles P. Steinmetz to the Development of High-Tension Transmission," 15 Jan. 1924, L2015, JWH. The company also published articles on Steinmetz in the December 1923 issues of the *General Electric Review* and the *GE Monogram*.

89. Hammond, news releases, n.d. [1922], L1005–09, JWH; *Who Was Who in America*, vol. 1, s.v. "Hammond, John W."; Hammond, *Steinmetz*, p. v; and Hammond, *Men and Volts: The Story of the General Electric Company* (New York: Lippincott, 1941).

90. Hammond, "History of General Electric Company, Suggestive Thoughts," n.d. [ca. 1925], L1983–90, quotations on L1983, L1987, JWH.

91. Hammond, *Steinmetz*, photographs facing pp. 324, 360, 400. Although Hammond was careful to say simply that the photograph was taken during Einstein's American visit in 1921, the caption of the brushed-out photograph sent to publishers by GE changed over the years to indicate that Einstein had visited Steinmetz in Schenectady. See, for example, *Life*, 23 April 1965, p. 53; and *Cornell Engineering Quarterly*, Autumn 1989, p. 50, and Winter 1990, pp. 59–60. Showing a lapse of institutional memory, the *Schenectady Works News* published the unretouched, original photograph on 20 May 1921 and the doctored version on 18 July 1924.

92. *Schenectady Works News*, 21 Nov. 1924; Hammond, "Charles Proteus Steinmetz," *Mentor*, 13 (May 1925): 3–22; and Hammond, *A Magician of Science: The Boy's Life of Steinmetz* (New York: Century, 1926).

93. *GE Monogram*, Aug. 1930; and *Schenectady Gazette*, 17 Dec. 1936, clippings in Steinmetz's Faculty File, SL.

94. Jordan, "Politics of Efficiency," p. 63; *Schenectady Gazette*, 23 March 1934, clipping in Steinmetz's Faculty File, SL; "Fiftieth Anniversary Dinner," 9 April 1934, DCJ; D. C. Jackson to Philip Alger, 14 April 1934; Alger to Jackson, 18 April 1934, DCJ; and *Scientific American*, 150 (1934): 316. Wagoner wrote about Steinmetz as late as 1963; see *Schenectady Gazette*, 9 April 1963. His place has been taken by local historian Larry Hart.

95. *Official Guidebook of the World's Fair of 1934* (Chicago, 1934), pp. 98–99, 174; *Official Guidebook of the New York World's Fair of 1939* (New York, 1939), p. 184; *New York Times*, 31 March 1938, 17 Jan., 29 March, 28 April, 21 July (quotations), 8 Oct., 1 Nov. 1939; "10,000,000-Volt Sparks . . . ," *General Electric Review*, 42 (1939): 420–426; and *Electrical World*, 112 (1939): 1094. A bust of Steinmetz was also placed in the

Hall of Fame at the 1939 fair; see *Schenectady Gazette*, 5 Oct. 1938.

96. *New York Times*, 19 March 1934, 5 and 7 June 1936, 19 June 1938, 24 April 1941; *Laws of the State of New York, 1936* (Albany, 1936), Chapter 853; *Laws of the State of New York, 1941* (Albany, 1941), Chapter 617; "The Steinmetz Memorial Collection," New York State Library, *125th Annual Report* (New York, 1942), pp. 37–38; and Larry Hart, *Steinmetz in Schenectady: A Picture Story of Three Memorable Decades* (Schenectady: Old Dorp Books, 1978), pp. 305–311.

97. GE, "The Story of Steinmetz," 1940, copy in Steinmetz Folder, IEEEA; "Steinmetz: Latter-Day Vulcan," copy in GER; Albert Rosenfeld, "Centennial of the Wizard Steinmetz: The Thunderer's Legacy," *Life*, 58 (23 April 1965): 53–54, 57–58; and "A Stamp in Commemoration of Charles Proteus Steinmetz," *Congressional Record*, 6 April 1965, p. 6934, copy in CPSU. A file of the centennial committee's activities is in GER.

98. W. W. Trench, "Steinmetz: His Life and Philosophy," 9 April 1940, Steinmetz Folder, IEEEA.

99. Quoted in Hughes, *Networks of Power*, p. 312.

100. *New York Leader*, 26, 27, and 28 Oct. (quotation) 1923; *New York Times*, 29 Oct. 1923, p. 23; 26 Nov. 1923, p. 9; *Soviet Russia Pictorial*, 8 (1923): 279; and *Albany News*, 14 Dec. 1925, clipping in Socialist Party File, SCHS. A group of German-American Socialist engineers were also favorably impressed with Steinmetz's socialism in 1924; see Hans-Joachim Braun, "A Technological Community in the United States: The National Association of German-American Technologists, 1884–1941," *Amerikastudien*, 30 (1985): 447–463, on p. 462.

101. McAlister Coleman, *Pioneers of Freedom* (New York: Vanguard Press, 1929), quotations on pp. v, 193; "Men and Women of Distinction Who Are Socialists or Were While Living," 16 Sept. 1931, reel 78, SPA; Ella Reeve Bloor, *We Are Many: An Autobiography* (New York: International Publishers, 1940), pp. 119–120, 276; and August Claessens, draft ms. of "I Visit a Genius" chapter of *Didn't We Have Fun! . . .*, n.d. [c. 1950], AC. Norman Thomas recalled, "I knew Steinmetz slightly and greatly rejoiced that we had him in the Party [but] one feels that he somehow had a confidence in big corporations—General Electric treated him very well—a little beyond what the record would assure." See Thomas to Ernest Caldecott, 3 Nov. 1964, CPSU.

102. Quoted in LaFollette, *Making Science Our Own*, p. 55.

103. John Dos Passos, *USA: The 42nd Parallel* (New York: Harcourt and Brace, 1937), pp. 379–382.

104. Geoffrey C. Upward, *A Home for Our Heritage: The Building of Greenfield Village and Henry Ford Museum, 1929–1979* (Dearborn, Mich.: Ford Museum, 1979), p. 81; Jonathan Leonard, *Loki: The Life of Charles Proteus Steinmetz* (Garden City, N.Y.: Doubleday, 1929); Lewis W. Moyer, "Lightning in the Laboratory: A Radio Drama Based on the Life of Charles Proteus Steinmetz, *Scholastic*, 33 (10 Dec. 1938): 17E-19E, 24E; John A. Miller, *Modern Jupiter: The Story of Charles Proteus Steinmetz* (New York: ASME, 1958); Floyd Miller, *The Man Who Tamed Lightning: Charles Proteus Steinmetz* (New York: Scholastic Book Services, 1965); and Rosenfeld, "Centennial of the Wizard Steinmetz."

105. "Centennial Hall of Fame," *IEEE Spectrum*, April 1984, pp. 64–66; Dorothy Markey, *The Little Giant of Schenectady* (New York: Aladdin Books, 1956); Sig-

mund A. Lavine, *Steinmetz: Maker of Lightning* (New York: Dodd, Mead, 1959); Floyd Miller, *The Electrical Genius of Liberty Hall: Charles Proteus Steinmetz* (New York: McGraw-Hill, 1962); Anne Welsh Guy, *Steinmetz: Wizard of Light* (New York: Alfred A. Knopf, 1965); and Erick Berry, *Charles Proteus Steinmetz: Wizard of Electricity* (New York: Macmillan, 1966). One woman who grew up in the 1920s recently recalled that Steinmetz, "like Gertrude Stein's Darwin, was 'all over my childhood.'" See Rita Guerlac to the Editor, *Cornell Engineering Quarterly*, 24 (Winter 1990), p. 59. I have not found any archival material to corroborate the relationship between Intrepid and Steinmetz, or to place Steinmetz in Britain at the time stated by Stevenson.

106. Wachhorst, *Edison*, pp. 7, 175; Warren I. Susman, "Culture Heroes: Ford, Barton, Ruth," in *Culture As History: The Transformation of American Society in the Twentieth Century* (New York: Pantheon, 1984), pp. 122–149; David L. Lewis, *The Public Image of Henry Ford* (Detroit: Wayne State Univ. Press, 1976); and Reynold M. Wik, *Henry Ford and Grass-Roots America* (Ann Arbor: Univ. of Michigan Press, 1972).

107. "Honoring EE Pioneers," *IEEE Spectrum*, Sept. 1983, p. 61.

Bibliographic Note

Archives

All archives cited in this book are listed at the beginning of the notes. The main sources were the several collections of Steinmetz papers in Schenectady, New York. The Schenectady County Historical Society has the largest and most comprehensive collection, ranging from family documents of the 1850s to Steinmetz's speeches of the 1920s. Particularly noteworthy are Breslau student notebooks, family letters from the 1890s, engineering notebooks from his work at Eickemeyer's and Lynn, and correspondence from about 1908 to 1920. The gaps in this record are filled to some degree by correspondence and reports in the valuable John W. Hammond Collection on the history of General Electric at the Hall of History Foundation; a smaller amount of similar material from the early 1890s in the Edwin W. Rice, Jr., Papers at Schaffer Library, Union College; and various records gathered by George Wise at the GE Research and Development Center. When this book was in press, the Hall of History Foundation informed me of about fifty Steinmetz letters it had obtained from the estate of his late grandson, Joseph Steinmetz Hayden, which I was able to incorporate into the book. Schaffer Library, which has been collecting in the areas of science, technology, and business for some time, has two boxes of Steinmetz Papers, dealing mostly with his connection with the college, and the voluminous papers of Ernst F. W. Alexanderson and Ernst J. Berg. The Berg Papers, which deserve more exposure in the history of technology, are especially strong in the day-to-day life at Liberty Street, business affairs after 1900, and Berg's teaching at the University of Illinois and Union College. The City Hall History Center houses the Van Vechten Collection on Schenectady socialism, a few Steinmetz letters, and about twelve letter boxes full of Steinmetz newspaper clippings—an invaluable source for studying his public image.

Archives of interest outside Schenectady include Steinmetz's personal books at the New York State Library in Albany and the Steinmetz Papers at the Henry Ford Museum, which has a cache of letters and postcards written to his sister Clara. Scattered material may also be found in the Elihu Thomson Papers at the American Philosophical Society; the Thomas S. Fiske Papers at Columbia University; and the Owen D. Young Papers at St. Lawrence University. Particularly important for the professionalization of electrical engineering in this period are the archives of the Institute of Electrical and Electronics Engineers and the Dugald C. Jackson Papers at MIT. An important group of archives on the history of American socialism is located in the Tamiment Library at New York University.

Periodicals

The international electrical journals were a major source for the first half of this book. Especially important were the *Transactions of the American Institute of Electrical Engineers, Electrical World, Electrical Engineer,* and *Electrical Review* in the United States; the *Electrician* in Britain; the *Elektrotechnische Zeitschrift* in Germany; and *La lumière électrique* in France. Information about technical, business, and labor activities at GE and Westinghouse was provided by *General Electric Review, Electric Journal,* and the *Schenectady Works News.*

In addition to using Steinmetz's newspaper clipping collection, I consulted the following newspapers in microfilm: *New York Times, New York Call, New York Tribune, New York World, New York Herald, Washington Post, Chicago Tribune, San Francisco Chronicle, Los Angeles Times, Schenectady Gazette, Schenectady Union-Star,* and *Schenectady Citizen.*

Steinmetz Studies

The first full-length biography was John W. Hammond's *Charles P. Steinmetz: A Biography* (New York: Century, 1924), whose strengths and weaknesses have been discussed in this book. Chief among the journalistic studies that followed was Jonathan Leonard's *Loki: The Life of Charles Proteus Steinmetz* (Garden City, N.Y.: Doubleday, 1929), which gives a lively account of Steinmetz's public image but also makes up conversations between characters in the book. Several juvenile biographies are listed in the notes to Chapter 10. Much more credible are two books by John A. Miller: *Workshop of Engineers: The Story of the General Engineering Laboratory of the General Electric Company* (Schenectady: GE, 1953); and *Modern Jupiter: The Story of Charles Proteus Steinmetz* (New York: American Society of Mechanical Engineers, 1958).

More scholarly studies began to appear as the centennial of Steinmetz's birth approached in 1965. The first of these were Justin G. Turner, "Steinmetz," *Manuscripts,* 15 (Fall 1963): 27–34, which reprints an important 1893 letter from Steinmetz to Heinrich Lux; Louis Lazarus, "A List of References on Steinmetz in the Socialist Literature," *Tamiment Institute Library Bulletin*, no. 44 (April 1965): 4–7; and Ernest Caldecott and Philip L. Alger, comp., *Steinmetz the Philosopher* (Schenectady: Mohawk Development, 1965), an annotated collection of nontechnical articles and speeches.

James E. Brittain wrote good chapters on Steinmetz's work in magnetic hysteresis and complex quantities in "B. A. Behrend and the Beginnings of Electrical Engineering, 1870–1920" (Ph.D. diss., Case Western Reserve Univ., 1970) and explored his organizational style in "C. P. Steinmetz and E. F. W. Alexanderson: Creative Engineering in a Corporate Setting," *Proceedings of the IEEE*, 64 (1976): 1413–1417. Other historians turned to his political writings: James Gilbert, *Designing the Industrial State: The Intellectual Pursuit of Collectivism in America, 1880–1940* (Chicago: Quadrangle Books, 1972), which has a chapter on Steinmetz; Gilbert, "Collectivism and Charles Steinmetz," *Business History Review*, 47 (1974): 520–540; Sender Garlin, *Charles P. Steinmetz: Scientist and Socialist* (New York: American Institute for Marxist Studies, 1977, occasional paper 22); and, most recently, John M. Jordan, "Society Improved the Way You Can Improve a Dynamo: Charles P. Steinmetz and the Politics of Efficiency," *Technology and Culture*, 30 (1989): 57–82. I wrote a Ph.D. dissertation and three articles on Steinmetz in the 1980s.

Recent popular accounts are by former lab assistant Emil J. Remscheid, *Recollections of Steinmetz* (Schenectady: GE, 1977), which contains entries from Remscheid's camp diary; and local historian Larry Hart's *Steinmetz in Schenectady: A Picture Story of Three Memorable Decades* (Schenectady: Old Dorp Books, 1978), which reprints over two hundred glass-plate-negative photographs taken by Steinmetz.

Secondary Sources

Although all sources are cited fully in the notes, I would like to acknowledge my debt to several key books and articles.

In the area of electrical history these include Harold C. Passer, *The Electrical Manufacturers, 1875–1900* (Cambridge, Mass.: Harvard Univ. Press, 1953); James E. Brittain, "B. A. Behrend and the Beginnings of Electrical Engineering, 1870–1920" (Ph.D. diss., Case Western Reserve Univ., 1970); Thomas P. Hughes, *Networks of Power: Electrification in Western Society, 1880–1930* (Baltimore: Johns Hopkins Univ. Press, 1983); Ronald W. Schatz, *The Electrical Workers: A History of Labor at General Electric and Westinghouse, 1923–1960* (Urbana: Univ. of Illinois Press, 1983); W. Bernard Carlson, "Invention, Science, and Business: The Professional Career of Elihu Thomson, 1870–1900" (Ph.D. diss., Univ. of Pennsylvania, 1984);

David E. Nye, *Image Worlds: Corporate Identities at General Electric, 1890–1930* (Cambridge, Mass.: MIT Press, 1985); Leonard S. Reich, *The Making of American Industrial Research: Science and Business at GE and Bell, 1876–1926* (New York: Cambridge Univ. Press, 1985); George Wise, *Willis R. Whitney, General Electric, and the Origins of U.S. Industrial Research* (New York: Columbia Univ. Press, 1985); and Nye's path-breaking and insightful *Electrifying America: Social Meanings of a New Technology, 1880–1940* (Cambridge, Mass.: MIT Press, 1990). The rich lode of Edison archives at West Orange, New Jersey, has yielded two fine studies thus far: Robert Friedel, Paul Israel and Bernard Finn, *Edison's Electric Light: Biography of an Invention* (New Brunswick: Rutgers Univ. Press, 1986); and Andre Millard, *Edison and the Business of Innovation* (Baltimore: Johns Hopkins Univ. Press, 1990).

The complex relationships among science, technology, the corporation, politics, and labor in this period are explored in Monte A. Calvert, *The Mechanical Engineer in America, 1830–1910: Professional Cultures in Conflict* (Baltimore: Johns Hopkins Press, 1967); Edwin T. Layton, Jr., "Mirror-Image Twins: The Communities of Science and Technology in Nineteenth-Century America," *Technology and Culture*, 12 (1971): 562–580; Daniel Nelson, *Managers and Workers: Origins of the New Factory System in the United States, 1880–1920* (Madison: Univ. of Wisconsin Press, 1975); Alfred D. Chandler, Jr., *The Visible Hand: The Managerial Revolution in American Business* (Cambridge, Mass.: Harvard Univ. Press, 1977); David F. Noble, *America by Design: Science, Technology, and the Rise of Corporate Capitalism* (New York: Knopf, 1977); Ellis W. Hawley, "The Discovery and Study of a 'Corporate Liberalism,'" *Business History Review*, 52 (1978): 309–320; Louis Galambos, "Technology, Political Economy, and Professionalization: Central Themes of the Organizational Synthesis," *Business History Review*, 57 (1983): 471–493; David Montgomery, *The Fall of the House of Labor: The Workplace, the State, and American Labor Activism, 1865–1925* (New York: Cambridge Univ. Press, 1987); and Martin J. Sklar, *The Corporate Reconstruction of American Capitalism, 1890–1916: The Market, the Law and Politics* (New York: Cambridge Univ. Press, 1988).

In writing the chapters on "engineering society" I was indebted to Samuel Haber, *Efficiency and Uplift: Scientific Management in the Progressive Era, 1890–1920* (Chicago: Univ. of Chicago Press, 1964); Robert H. Wiebe, *The Search for Order, 1877–1920* (New York: Hill and Wang, 1967); Edwin T. Layton, Jr., *The Revolt of the Engineers: Social Responsibility and the American Engineering Profession* (1971; rpt. Baltimore: Johns Hopkins Univ. Press, 1986); Daniel T. Rodgers, "In Search of Progressivism," *Reviews in American History*, 10 (Dec. 1982): 113–132; A. Michal McMahon, *The Making of a Profession: A Century of Electrical Engineering in America* (New York: IEEE

Press, 1984); and Howard P. Segal, *Technological Utopianism in American Culture* (Chicago: Univ. of Chicago Press, 1985).

Of the growing literature on the history of American socialism, I found the following particularly useful for this study: James Weinstein, *The Decline of Socialism in America, 1912–1925* (New York: Monthly Review Press, 1967); Sally M. Miller, *Victor Berger and the Promise of Constructive Socialism, 1910–1920* (Westport, Conn.: Greenwood Press, 1973); Kenneth E. Hendrickson, "Tribune of the People: George R. Lunn and the Rise and Fall of Christian Socialism in Schenectady," in *Socialism and the Cities*, ed. Bruce M. Stave (Port Washington, N.Y.: Kennikat Press, 1975), pp. 72–98; Arnold E. Kaltinick, "Socialist Administration in Four American Cities (Milwaukee, Schenectady, New Castle, Pennsylvania, and Conneaut, Ohio), 1910–1916" (Ph.D. diss., New York Univ., 1982); Nick Salvatore, *Eugene V. Debs: Citizen and Socialist* (Urbana: Illinois Univ. Press, 1982); and Donald Stabile, *Prophets of Order: The Rise of the New Class, Technocracy, and Socialism in America* (Boston: South End Press, 1984).

In examining the Steinmetz myth I relied heavily on Leo Marx, *The Machine in the Garden: Technology and the Pastoral Ideal in America* (New York: Oxford Univ. Press, 1964); James O. Robertson, *American Myth, American Reality* (New York: Hill and Wang, 1980); Wyn Wachhorst, *Thomas Alva Edison: An American Myth* (Cambridge, Mass.: MIT Press, 1981); Warren I. Susman, "Culture Heroes: Ford, Barton, Ruth," in his *Culture As History: The Transformation of American Society in the Twentieth Century* (New York: Pantheon, 1984), pp. 122–149; and Marcel C. LaFollette, *Making Science Our Own: Public Images of Science, 1910–1955* (Chicago: Univ. of Chicago Press, 1990).

Index

Books in the Series

The Mechanical Engineers in America, 1839–1910: Professional Cultures in Conflict, by Monte Calvert

American Locomotives: An Engineering History, 1830–1880, by John H. White, Jr.

Elmer Sperry: Inventor and Engineer, by Thomas Parke Hughes (Dexter Prize, 1972)

Philadelphia's Philosopher Mechanics: A History of the Franklin Institute, 1824–1865, by Bruce Sinclair (Dexter Prize, 1978)

Images and Enterprise: Technology and the American Photographic Industry, 1839–1925, by Reese V. Jenkins

The Various and Ingenious Machines of Agostino Ramelli, edited by Eugene S. Ferguson, translated by Martha Teach Gnudi

The American Railroad Passenger Car, New Series, no. 1, by John H. White, Jr.

Neptune's Gift: A History of Common Salt, New Series, no. 2, by Robert P. Multhauf

Electricity before Nationalisation: A Study of the Development of the Electricity Supply Industry in Britain to 1948, New Series, no. 3, by Leslie Hannah

Alexander Holley and the Makers of Steel, New Series, no. 4, by Jeanne McHugh

The Origins of the Turbojet Revolution, New Series, no. 5, by Edward W. Constant II (Dexter Prize, 1982)

Engineers, Managers, and Politicians: The First Fifteen Years of Nationalised Electricity Supply in Britain, New Series, no. 6, by Leslie Hannah

Stronger Than a Hundred Men: A History of the Vertical Water Wheel, New Series, no. 7, by Terry S. Reynolds

Authority, Liberty, and Automatic Machinery in Early Modern Europe, New Series, no. 8, by Otto Mayr

Inventing American Broadcasting, 1899–1922, New Series, no. 9, by Susan J. Douglas

Edison and the Business of Innovation, New Series, no. 10, by Andre Millard

What Engineers Know and How They Know It: Analytical Studies from Aeronautical History, New Series, no. 11, by Walter G. Vincenti

Alexanderson: Pioneer in American Electrical Engineering, New Series, no. 12, by James E. Brittain

Steinmetz: Engineer and Socialist, New Series, no. 13, by Ronald R. Kline